T0314312

CALCULATED VALUES

CALCULATED VALUES

Finance, Politics, and the Quantitative Age

WILLIAM DERINGER

Harvard University Press

Cambridge, Massachusetts

London, England

2018

Library of Congress Cataloging-in-Publication Data

Names: Deringer, William, 1984– author.
Title: Calculated values : finance, politics, and the quantitative age / William Deringer.
Description: Cambridge, Massachusetts : Harvard University Press, 2018. |
Includes bibliographical references and index.
Identifiers: LCCN 2017034530 | ISBN 9780674971875 (alk. paper)
Subjects: LCSH: Great Britain—Politics and government—1603–1714. |
Great Britain—Politics and government—18th century. | Quantitative
research—Great Britain—History. | Numerical calculations—History. |
Numerical calculations—Political aspects. | Persuasion (Rhetoric)—Political aspects.
Classification: LCC DA18 .D425 2018 | DDC 941.06—dc23
LC record available at https://lccn.loc.gov/2017034530

For Susanna
(and G!)

Contents

Preface:
Quantification and Its Discontents

Facts and *Figures* are the most stubborn Evidences, they neither yield to the most persuasive *Eloquence,* nor bend to the most imperious *Authority.*

> WILLIAM PULTENEY, *A State of the National Debt, As It Stood December the 24th, 1716* (1727)

Truth and Numbers are always the same; tho' the first may be obscured, yet it can never entirely lose its Lustre; and tho' the latter may be transpos'd, they can never lose their Denomination and true Value, when ranged in their proper place.

> JOHN CROOKSHANKS, *Some Seasonable Remarks on a Book Publish'd in the Month of July, 1718* (1718)

Every man, who has ever reason'd on this subject, has always prov'd his theory, whatever it was, by facts and calculations.

> DAVID HUME, "Of the Balance of Trade," *Political Discourses* (1752)

WE LIVE IN A QUANTITATIVE AGE. Across many domains, our modern political and intellectual culture features a marked confidence in the power of numerical calculations: to explain how the social world works, to measure our collective well-being, to evaluate our institutions and hold accountable the individuals who govern us. Understanding, judging, and improving society means counting, measuring, and calculating it. Numbers do not just depict but define who we collectively are, what we think, how we are doing. The economic strength of our communities is represented by a series of familiar numbers: the GDP and the CPI, the Dow Jones Industrial Average and the unemployment rate. Our political sentiments are quantified in public opinion polls and aggregated into ever more sophisticated models, whose numerical outputs become defining indicators of the political climate and reshape how politicians engage with their constituents. There are few problems that do not seem solvable, or at least more solvable, with the application of quantitative analysis. In criminal justice, law enforcement institutions turn to quantitative data to target policing efforts more efficiently and to test their efficacy, while judges and corrections departments increasingly rely on probabilistic calculations about recidivism to guide decisions about sentencing and parole. In education, researchers and administrators are developing new calculations to measure how much students are learning and determine which schools and teachers are adding the most value to students' lives. Indicators and algorithms are transforming how we evaluate the success of universities, hospitals, and NGOs, how we determine the best athletes and the best journalists, even how we pick the best potential employees and the best potential mates. Better calculations, it is imagined, will make a better society. With new monitoring technologies to calculate how we work and sleep, eat and exercise, better numbers may even make better selves.

Empowering all of these projects is a prevailing sense that *quantitative* evidence and analysis—"facts and figures"—are an especially powerful way of knowing about the world, over and above their *qualitative* counterparts. To really know something, it seems, you have to quantify it. Numerical calculations are revered for many virtues in the modern quantitative age. They are seen to be objective—they represent authentic, impersonal, and impartial facts about the world, imagined to transcend personal and political differences. They are tenacious—they resist easy manipulation and are stubborn in the face of authoritative proclamations or artful speech. They are incisive—they cut through traditional intuitions and unknown biases. (As economist Steven Levitt and co-author Stephen Dubner, whose 2005 *Freakonomics* constituted a kind of manifesto for this view, put it: "There is nothing like the sheer power of numbers to scrub away layers of confusion and contradiction.")[1] And they are conclusive—they provide the kinds of clear answers that can often settle otherwise intractable arguments. Because of these special qualities— objectivity, tenacity, incisiveness, and so on—we can say that numbers and calculations bear a certain *authority* in modern thinking.

Calculated Values is about this special authority that quantitative thinking holds in modern culture, especially in the political domain. In particular, it seeks to understand its history. The early twenty-first century has witnessed the emergence of a particularly robust version of this confidence in numbers, marked by the rising prestige of quantitative (rather than qualitative) research methods across the academic social sciences, the growing power of quantitative indicators as tools of governance and management, and the eager pursuit of new computational techniques for unlocking knowledge from "big data." But the authority of the quantitative is not new. In fact, it has a deep historical genealogy.

In the English-speaking world, the notion that numerical calculation was deserving of special esteem as a way of thinking and knowing, particularly in political contexts, first began to take hold in Britain around the turn of the eighteenth century. Prior to that point, numerical thinking had held a rather marginal place in political affairs. Though a certain amount of numerical labor had long been a practical necessity in some areas of governmental practice, like the management of royal money at the "Exchequer," numbers held no particular cachet as a medium of political knowledge. In fact, well into the seventeenth century, quantitative thinking had been held in rather low esteem by many, particularly among the political and social elite. In the decades that followed Britain's "Glorious Revolution" of 1688, though, numerical calculation became an expected, even exalted, part of politics and public discourse. Britons increasingly came to believe that numerical evidence and analysis had special virtues and thus had a critical role to play in guiding political decisions and solving public problems. One testament to this emerging reverence for the quantitative can be found in the first epigraph to this book: "*Facts* and *Figures* are the most stubborn Evidences," wrote William Pulteney, a Member of Parliament and frequent commentator on political economy, in 1727. A few years earlier, a bookkeeper, economic propagandist, and accused pirate named John Crookshanks had offered an even more lyrical expression of this numerical enthusiasm: "Truth and Numbers are always the same."

How and why did calculation attain this new prominence when it did? One possible explanation for how calculations gained this authority in the past is that, in short, they deserved it. That is, people came to believe that numerical calculations were an especially powerful and effective way of knowing because calculations consistently proved themselves to be so. Numbers were gradually recognized to have special virtues—to be objective, tenacious, incisive, and so on—because they *are* special. There is certainly no doubt that numerical calculations have frequently proved to be tremendously useful tools for thinking through political, economic, and social questions. Yet efficacy alone cannot explain the distinctive authority numbers and calculations have come to hold. One of the great achievements of decades of scholarship by historians of science has been to show that the historical success (or failure) of different techniques of knowing—laboratory experiments, mathematical models, scientific imaging technologies, and so on—are the result of complex social processes, shaped by human values and riven by human conflict. This is certainly the case when it comes to the use of numerical calculation as a tool for knowing about politics, economy, and society. As will be shown in the pages that follow, the dawn of the quantitative age in British public life was no neat story about the steady triumph of reason and science. It was a messy and contentious story, more about party politics, government secrets, and financial crisis than enlightened rationality.

Then again, the quantitative age of the twenty-first century is messy and contentious as well. However confident some people are that numerical data and analysis will cut through biases and settle hard political questions, numbers often prove themselves inadequate to that task. *Pace* William Pulteney, facts and figures often seem to bend quite readily; *pace* John Crookshanks, truth and numbers are frequently misaligned. Those questions that seem to need good numbers the most—about what drives prosperity, how to measure the public good, how to determine the most effective policies for taxation, education, healthcare, or environmental defense—these kinds of questions are stubbornly resistant to definitive numerical answers. For one thing, calculations are often extremely brittle. It can seem profoundly difficult to produce a perfect calculation, one which does not make some dubious assumption, take some data out of context, or include some computational mistake. Avoiding all "bugs" is a serious challenge, and even small errors can produce misleading results or significantly compromise the calculation's usefulness as a means to advance political discussion. The more ambitious the calculation, the more room for error; the more politically contentious, the more likely those errors, real or just apparent, will be found.

In recent years, this is a lesson that has been very publicly learned by some of the most prominent economic calculators. Consider Carmen Reinhart and Kenneth Rogoff, who gained widespread attention for a 2010 article demonstrating that nations with public debt above about 90 percent of GDP were at severe risk of economic recession. The article became a favorite reference for supporters of economic austerity policies and was prominently cited by leading conservative politicians including U.S. Representative (and later Speaker of the House) Paul Ryan and U.K. Chancellor of the Exchequer George Osborne. Yet it also became the subject of highly public criticism after Thomas Herndon, a graduate student at the University of Massachusetts, Amherst, tried to reproduce their empirical results as part of a seminar assignment and uncovered a crippling error in their underlying spreadsheet. Even when calculators do not make any such unwitting mistakes, the specter of error still looms. The most politically significant calculations invariably contain myriad points of legitimate complexity and contention, many of which might be seized upon by a hostile critic as damningly erroneous. Consider, from the other side of the political spectrum, Thomas Piketty's *Capital in the Twenty-First Century* (2014), celebrated by the political left for its painstaking quantitative analysis of long-term trends in economic inequality. Piketty's argument came under very similar kinds of attacks as Reinhart and Rogoff's, in even more public venues like *Forbes* and the *Financial Times*. Though Piketty had decisive defenses for all of the apparent mistakes identified by critics, the insinuation that his underlying data were flawed spread widely among those skeptical of *Capital*'s political implications.[2]

The potential for errors in contentious calculations is not even the most distressing thing about quantification in modern political life. Arguably even more unsettling is the fact that many politically charged numerical disputes come down to subtle differences of quantitative interpretation, where both sides could make reasonable claims to being correct. Computational disputes seem, all too often and all too quickly, to reduce to intractable disputes about preferred models, proper methods—and, indeed, political values. Political calculators so often want the tough questions to have clear, empirical answers. But questions of political calculation are very often underdetermined: available economic data regularly permit multiple, plausible interpretations, often resulting from only small changes in underlying assumptions. For example, major debates among environmental economists about whether to invest heavily in immediate efforts to mitigate climate change have turned largely upon one variable, the so-called "discount rate." Tweaking that single figure by a percentage point or two can produce an entirely different conclusion about what ought to be done about climate change today.[3]

Wherever there is flexibility in accounting procedures or modeling techniques or choice of datasets, such analytical opportunities might be exploited to produce calculations justifying pre-determined political positions. At times, such numerical gamesmanship can seem rather absurd. During the 2012 race for the U.S. Presidency, Republican challengers Mitt Romney and Paul Ryan contended—with some technical justification—that the Affordable Care Act passed in President Barack Obama's first term would "cut" $716 billion from Medicare to fund expanded healthcare coverage for uninsured Americans. The President and his supporters tried to refute the claim by pointing out that the missing billions were simply an accounting technicality—involving, in essence, money moving from one part of the proverbial ledger to another—that had no material impact on Medicare and its users. Both sides had plausible cases to make about the mysterious $716 billion, and no clear "answer" was forthcoming—not that such an answer would have done much to settle the deeper dispute about the role of the federal government in healthcare.[4]

Quantitative controversies like these cast public light on something that is well known to the statisticians, accountants, and economists who work closely with such numbers and calculations: namely, that all numerical evidence, no matter how empirically secure it may be, is the product of human efforts, assumptions, and choices. And where there is discretion, there is room for dispute. Recent political events have shown that even the most fundamental political numbers—the most stubborn facts and figures—are not as straightforward as they seem, and thus not safe from reinterpretation and controversy. What might seem like clear accounting figures, like the size of a nation's public debt, have been revealed to be highly complicated feats of calculation, allowing a range of plausible answers. A telling example comes from the recent debt crisis in Greece. In mid-2015, the declared debts owed by Greece to its creditors, based on the face value of those debts, amounted to a massive total of roughly $350 billion, almost twice the size of the Greek GDP; yet, critics contended that, according to alternative calculations guided by International Public Sector Accounting Standards (IPSAS), that figure might be as low as $36 billion.[5]

In American politics, one high-profile numerical benchmark that has come under similar scrutiny is the unemployment rate. Though often presented in popular media as a single, neat datum, the unemployment rate produced by the Bureau of Labor Statistics is an immensely complicated calculation. (It is actually six different calculations, "U1" through "U6," of which "U3" is the one commonly cited as *the* unemployment rate.) Though there has long been considerable agreement on the proper methods for how to carry out this measurement—a "hard-won consensus between social science, economics, journalism, business, and labor"—the complications behind the calculations leave room for controversy. In 2014, for example, critics accused the Obama government of trying to "juke the stats" on unemployment, changing how certain jobs were classified in official labor statistics in order to produce a more optimistic message about unemployment and the health of the national economy. More recently, during the 2016 Presidential Election, then candidate Donald Trump went as far as to call the unemployment rate, then around a modest 5 percent, "a phony number to make the politicians look good," exploiting the technical intricacies of the figure to dismiss its veracity altogether.[6]

Scanning the headlines, it is easy to find many examples in political life in which quantitative reasoning seems to fail to live up to its esteemed reputation. Instead of being a distinctly stubborn, honest, and objective means of thinking through public problems, numbers prove themselves to be remarkably fragile, malleable, and even deceptive. This is hardly a novel observation, of course. Being disappointed in the inadequacies of numbers and calculations is one of the constitutive features of our quantitative age. Alongside the frequent refrain that "numbers never lie" is the arguably even more popular jibe that "there are three kinds of lies: lies, damned lies, and statistics," a late nineteenth-century coinage popularized by Mark Twain, who attributed it (probably incorrectly) to former British Prime Minister Benjamin Disraeli. As it happens, quantitative practitioners have often been among the first to call attention to the frustrations and failures of quantification in public life. For example, the most popular textbook ever written on statistics, selling over a half-million copies in English and translated into numerous other languages, was a brief introductory volume written in 1954 by Darrell Huff entitled *How to Lie with Statistics.* Sometime in the early 1960s, the future Nobel Prize–winning economist Ronald Coase offered a much-repeated quip about how "if you torture the data enough, nature will always confess." More recently, statistical and mathematical experts like Gary Smith and Cathy O'Neil have issued stern warnings about the dangers of misusing quantitative tools—about how they can become, in O'Neil's memorable phrase, "weapons of math destruction."[7]

For quantitative reformers like Gary Smith and Nate Silver, the problems of numerical deception and delusion are essentially technical ones: the solution is to promote better, more responsible quantitative practices—in Silver's case, a Bayesian outlook on statistical forecasting—and to teach the public to read numbers more critically.[8] Other quantitative skeptics, though, have been less sanguine about whether such problems with numbers might be reformed simply by technical improvements. For many, the problem seems to lie with the very numbers and calculations themselves. In the past five years, numerous studies and articles have called attention to a rising "antipathy to statistics" among citizens in the United States and United Kingdom. For example, a study in 2013 found that three-quarters of Americans suspect that political polls are biased—and that was before what was widely seen as a major failure on the part of pollsters to accurately predict the 2016 Presidential Election. Shortly before that 2016 election, another U.S. study found that nearly 44 percent of Americans, and over 68 percent of likely supporters of Donald Trump, distrusted the data about the economy reported by the federal government. In the United Kingdom, a 2015 study by YouGov and Cambridge University found that 55 percent believed that the government was "hiding the truth" when it came to data on immigration into the country. Alongside the buoyant confidence in numerical data and quantitative algorithms evident across many spheres of public life and the private sector, there thus appears to be a powerful countervailing sentiment: that political numbers are in fact distinctly biased, untrustworthy, and deceptive.[9]

While it has taken on particular political salience in recent years, such doubtful and disparaging sentiments about calculative thinking are not new either, any more than their optimistic counterpoints. This brings us to the third epigraph that begins this book. The author, rather more famous than William Pulteney or John Crookshanks, was the Scottish philosopher David Hume. An astute observer of the habits of political reasoning in his own culture, Hume was one of the first to observe the distinctive authority that numerical "facts and calculations" had come to assume in Britain by the middle of the eighteenth century—and to diagnose some of the new dangers that came with it. In 1752, he published an influential essay disputing what had become one of the most revered calculations in eighteenth-century political life: the so-called "balance of trade," a calculation which, according to contemporary thinking, provided a precise barometer of a nation's commercial health. Britons had been arguing about trade balances for decades, convinced that better calculations could settle contested questions about trade policy. Hume felt that all this computational fervor was for naught, a result of the misguided belief that irregular numerical evidence could provide certain political knowledge. After all, Hume explained: "Every man, who has ever reason'd on this subject, has always prov'd his theory, whatever it was, by facts and calculations." You could make the numbers say whatever you wanted them to.

Calculated Values tries to make some sense of this tension between the quantitative confidence expressed by Pulteney and Crookshanks—the notion that truth and numbers are always the same—and Hume's observation that, far from being resolutely stubborn and unflinchingly honest, numerical calculation is a notably flexible mode of reasoning that can be made to support many political positions. The inspiration for this book lies, at least in part, in my own experience of quantitative discontent—or perhaps more appropriately, quantitative disenchantment. Prior to beginning my graduate education in the history of science, I worked as an investment banking analyst at a boutique firm based in New York. My position was in the "restructuring" practice, which focused on advising companies in financial distress, their creditors, and other parties somehow entangled in such "special situations." As an analyst, my primary responsibility was to calculate. Using Microsoft Excel spreadsheets and a battery of hard-won keyboard shortcuts, I measured the total enterprise value of companies, modeled future business performance, and projected the growing costs of retiree healthcare. One time, I tried to calculate the likelihood that a sovereign government would pay an extremely dubious debt. These calculations were often extremely complicated and always labor-intensive.

But what struck me most as a novice analyst was that they seemed disingenuous. For all the numerical effort that I put into getting the right answer, my superiors seemed largely unconcerned with getting to the numerical truth. I was regularly told to adjust inputs and assumptions within the models I built because the answer did not "feel" right, or because the model needed to tell a different story. In one deal, my firm represented a research-based company that was in a dubious financial condition but was also very optimistic about key products in its R&D pipeline. Part of my task as an analyst was to help craft a model of the company's future business prospects that made the company look as good as I plausibly could, in order to attract outside investment money that might give the company enough time to bring its promising research to market. In another project, representing the creditors of a bankrupt Midwestern manufacturer, I was instructed to show computationally that the valuation of that company's business fell within a specific numeric range—not too big, and not too small. If we could convince a bankruptcy judge of that valuation, then our clients— instead of others in the hierarchy of creditors—would receive the coveted equity in the company and thereby hold control of the business in the future. In one especially perplexing project, I found myself building a model showing that our client, a legitimate giant of American industry, was actually in an incredibly dire financial condition.[10]

At first, I found this disconcerting. None of my calculations seemed especially concerned with finding what seemed to me to be the *true* numbers—the correct valuation, the best estimate, the most realistic future outcome. With time, though, I came to realize something that my much more experienced colleagues had long known: the numbers were not supposed to do what I thought they were. The numbers were not supposed to find *the answer;* they were supposed to make *an argument.* Our job was to make a case, and there was invariably another set of calculations making an opposing case on the other side. The problem of sorting out the answer took place through other means: an adversarial negotiation, a market auction, or the judgment of a bankruptcy court. Yet this did not mean that our calculations were cynical exercises in data manipulation or useless numerical window-dressing. They did have a key role to play in the collective activity of thinking through these financial problems. It was just not the one I expected them to have.

A few months later, in my first year as a graduate student, I began looking through a cache of somewhat mystifying pamphlets from the year 1720, recently digitized and available online. I knew those pamphlets pertained somehow to one of the most infamous events in financial history: the "South Sea Bubble," long cited as Britain's first modern financial crisis. Inside those pamphlets were pages of numerical calculations—at first almost entirely inscrutable to me—principally about the financial value of the South Sea Company and its stock. Those pamphlets became the germ of this book; they make up the subject of Chapter 5. One of the things that puzzled and intrigued me most about those pamphlets is that they made an incredibly wide range of different arguments about the value of South Sea Company stock, nearly all of which seemed like they might have been plausible at the time. Some of them strained to show that the South Sea Company was a triumph of financial engineering, whose stock promised unimagined riches; yet others put just as much effort into showing that the stock was worth little more than the paper it was written on. I could not decide whether to think of these calculations as remarkable feats of analytical ingenuity or distasteful examples of numerical deceit. I wondered how contemporaries could possibly have known what calculations to believe.

Over time, I came to see that those eighteenth-century calculations were very much like the financial models I had crafted in my earlier career in banking. In other words, they were arguments, not answers. As I continued to explore the quantitative archives of Britain's long eighteenth century, I recognized that understanding the distinctive role that calculations came to play in that political community required thinking about them first and foremost as *instruments of dispute*—tools for conducting arguments, more than tools for finding answers.

I want to clarify one crucial thing before we continue: this book is not an argument either for or against quantification. It is not intended to be a triumphant tale about the rise of quantitative rationality or a celebration of how better calculations can make a better society. But neither is it a critical screed against quantification or a deconstructive take on how numbers can never say anything "true." I believe—along with Pulteney, Crookshanks, and many of the early modern calculators who occupy this book—that numerical data and intelligent calculation offer extraordinary promise for illuminating complicated social problems, guiding more beneficial public policies, and holding those in positions of public trust accountable. The fact that we use calculations so often to make arguments rather than to find answers does not mean that there *are no* correct answers or that calculations cannot sometimes identify them. But I also recognize that trusting in numbers comes with its own set of problems. Numbers often let us down. Many of the frustrations that arise in our own twenty-first-century quantitative age are frustrations that have existed for a very long time. These are frustrations that cannot be solved through purely technical means, by coming up with better statistical techniques, economic models, or computational technologies. But they are frustrations that we will deal with far more successfully if we understand them better. My hope is that examining the politics of calculation in the past can help inform how we think about numerical thinking in political life—its virtues and vices, what is expected of it and what it actually does—and therefore help us use numbers more effectively in the future.

One final note on terminology. Broadly speaking, I describe this book as a history of *calculation:* namely, the activity of reasoning with numbers using mathematical operations. The calculations discussed in this book comprise a rather broad category, ranging from basic arithmetic to comparatively advanced algebra and the construction of elaborate financial models. While I will use other, related concepts for stylistic variety and clarity in different settings— numbers, arithmetic, quantitative thinking—calculation is the organizing concept. There are three reasons for this choice. First, the focus of this book is on calculation—perhaps *calculating* is even better—as a practice, something people do.[11] My interest is in calculation as a way of thinking and knowing, arguing and persuading, fighting and politicking. Focusing on calculation as a practice offers a different way to think about the intellectual history of political economy, as a dynamic and ongoing activity rather than a punctuated accretion of ideas, as *thinking* rather than *thought*.

Second, focusing on calculation helps to clarify what this book is, and especially is not, about. This history of calculation is not exactly a history of *mathematics,* considered as an academic enterprise and field of abstract knowledge; nor a history of *numeracy,* understood as a social phenomenon, like literacy, describing the extension of numerical competence and arithmetical skill within a society; nor a history of *statistics,* a more narrow and technical category that only emerged in Britain around 1800, after the period discussed in this book—though it intersects with all of those histories. By focusing on calculation, I also wish to distinguish this study from histories of *information* or *data,* topics which have received relatively more emphasis in recent scholarship.[12] The chapters that follow are interested less in the gathering of empirical data about political and economic phenomena, and more on the ways that such data were interpreted, operated upon, and manipulated to formulate assertions, perform arguments, and guide actions. Further, one of the effective arguments this book makes is that all numerical data are, in some sense, *calculated,* the product of active human effort and discretion. (As one recent study has put it: "raw data is an oxymoron.")[13] Focusing on calculation foregrounds that human element.

The third, and most important, reason for organizing this study around calculation is the tension that inheres in the word *calculate*. On the one hand, *calculating* refers to the seemingly neutral activity of "arithmetical or mathematical reckoning"; on the other, *calculating* describes a person who "shrewdly or selfishly reckons the chances of gain or advantage."[14] This tension lies at the very heart of this book, and the title *Calculated Values* is chosen to evoke it. This story is both about *calculation*, as in arithmetical reckoning, and *calculation*, as in the shrewd reckoning of gain or advantage. It is both about numerical *values*, as in the actual numbers that swirled around eighteenth-century political life, and numerical *values*, as in all of the personal, cultural, and political ideals with which numbers were invested during that same period. We continue to struggle with these tensions.

Notes on Style

I have tried to reproduce the original language, spelling, typography, and notational conventions used in primary manuscript and printed sources quoted in the pages that follow. Spelling and punctuation have not been updated, except where it has been necessary to convey the author's meaning clearly. Generally, "*sic*" is not used to identify odd or irregular forms or spellings, except in cases where it is necessary to clarify an ambiguous or important usage. Similarly, I have tried to maintain original *italics*, CAPITALIZATION, and other emphatic typography from original sources (except in the titles of printed texts). Therefore, all such emphases in quoted material should be assumed to come from the original text, except where I have noted "emphasis mine."

Until 1752, England, Scotland, and the British colonies used "Old Style" (O.S.) dates, based on the Julian calendar, with the New Year beginning on March 25 instead of January 1. Thereafter, the British calendar was reformed to align with the Gregorian calendar, resulting in a shift of the calendar forward eleven days to begin the year on January 1. This yielded so-called New Style (N.S.) dates. For events occurring before 1752, I use the Old Style dates used by contemporaries, except that I treat January 1 as the start of the New Year. When necessary, I have attempted to clarify dates falling in the ambiguous period between January 1 and March 24 with a " / ". For example, a date indicated in the text or Notes as "February 28, 1719/20" is one that contemporaries would have considered part of the year 1719, but which falls in 1720 in modern terms.

During the period, the English currency, the pound sterling, was denoted in various ways, including versions of the modern symbol (£), the italicized letter *"L"* or *"l.,"* and abbreviations like "lib." The pound was divided into twenty shillings, usually denoted with *"s.,"* and shillings were divided into twelve pence, denoted *"d."*

CALCULATED VALUES

Introduction:
Political Calculations

THE PHRASE "facts and figures" is a talisman of modern quantitative culture. It adorns brochures and websites and call-out boxes in newspaper articles, identifying the numbers most essential to understanding whatever topic is in question. Its presence evokes deep-seated attitudes about numerical knowledge and its proper place in public life. "Facts and figures" are supposed to be important, trusted, and respected. They are the objective foundation of sound decisions and reasoned debates. They are also dull, mechanical, and emotionless. This image of numbers and calculations as things indispensable but uninspiring has deep historical roots. In 1841, for example, the editors of a new British magazine entitled *Facts and Figures, a Periodical Record of Statistics,* began by admitting that "we are perfectly aware that Statistics are dry food." (They promised to try their best to enliven those dry numbers by "blending them with matters of stirring interest.")[1] To a great degree, what makes facts and figures tedious also makes them trustworthy. They are imagined to be free of opinion and artifice, to be impersonal, impartial, and indisputable.

The expression "facts and figures" first came into use in English well before *Facts and Figures* magazine. The first usage in print likely comes from the year 1727. Its author was William Pulteney (born 1684), a genteel English politician who had represented the Yorkshire borough of Hedon in the House of Commons since 1705. For his first twenty years in Parliament, Pulteney had been a devoted member of the Whig party and a loyal ally to Robert Walpole, Britain's first "Prime Minister." But the alliance between the two had deteriorated, and by 1727 Pulteney had become one of Walpole's fiercest critics and the leader of an opposing faction. He directed much of his hostility toward Walpole's financial policies, especially regarding Britain's mounting public debts. Pulteney excoriated Walpole's fiscal record in a string of printed pamphlets, including a 1727 offering entitled *A State of the National Debt*. Shortly before that publication, Whig advocates had triumphantly asserted that Walpole's wise policies had reduced the national debt by £2 million. Pulteney stridently disputed this, and he believed he had the decisive evidence on his side. He had numbers. His pamphlet presented a detailed numerical analysis tracing all government debts incurred between 1716 and 1725, using careful calculations to clarify these complex data and reveal the true impact of Walpole's policies. (For a portion of Pulteney's calculations, see Figure 0.1.) His conclusion was that over those nine years, the national debt had actually increased by £7.8 million under Walpole's watch, up to an effective total of more than £56 million. The numbers, Pulteney suggested, revealed dire truths that Walpole and his supporters had tried to obscure. "*Facts* and *Figures* are the most stubborn Evidences," he declared, "they neither yield to the most persuasive *Eloquence,* nor bend to the most imperious *Authority*."[2]

In some ways, Pulteney's ode to "*Facts* and *Figures*" seems like a familiarly modern notion—that quantitative data constitute an especially strong form of evidence, more reliable than words alone, no matter how eloquent. But in many other ways, William Pulteney's use of facts and figures bears surprisingly little resemblance to the standard image that had come to prevail by the age of *Facts and Figures* magazine, and which remains dominant today. The facts and figures that filled Pulteney's pamphlet were not dull, mechanical, or unimaginative; rather, they were the product of deeply creative computational work by the author. Nor were they impartial or impersonal exactly, but unabashedly political, intended to score points in a partisan dispute and undermine a personal enemy. Nor were they "cold" and "dry," but a matter for heated dispute. On the opening page of his pamphlet, Pulteney explained that he felt moved to craft his calculations on the subject of the national debt because he had been "met with such warm and angry Contentions about it, as have very much disturbed the Peace and Quiet of the Neighbourhood where I live."[3]

Pulteney may have been the first to label his calculating efforts as "*Facts* and *Figures.*"[4] But he was just one participant in a much broader transformation in British political practice and culture that transpired over his lifetime—what

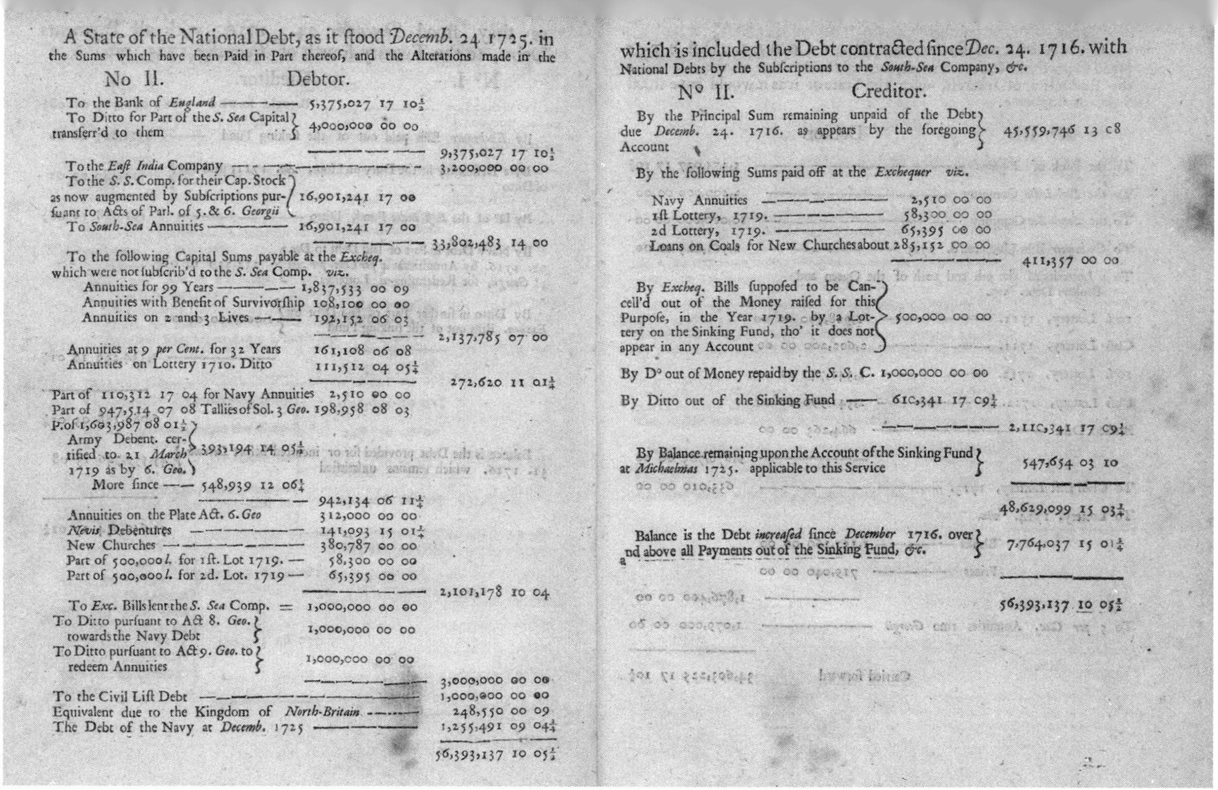
A State of the National Debt, as it stood Decemb. 24 1725. in the Sums which have been Paid in Part thereof, and the Alterations made in the

No II. Debtor.

To the Bank of *England*	5,375,027 17 10½	
To Ditto for Part of the *S. Sea* Capital transferr'd to them	4,000,000 00 00	
		9,375,027 17 10½
To the *East India* Company		3,200,000 00 00
To the *S. S.* Comp. for their Cap. Stock as now augmented by Subscriptions pursuant to Acts of Parl. of 5. & 6. *Georgii*	16,901,241 17 00	
To *South-Sea* Annuities	16,901,241 17 00	
		33,802,483 14 00
To the following Capital Sums payable at the *Excheq.* which were not subscrib'd to the *S. Sea* Comp. *viz.*		
Annuities for 99 Years	1,837,533 00 09	
Annuities with Benefit of Survivorship	108,100 00 00	
Annuities on 2 and 3 Lives	192,152 06 03	
		2,137,785 07 00
Annuities at 9 *per Cent.* for 32 Years	161,108 06 08	
Annuities on Lottery 1710. Ditto	111,512 04 05¼	
		272,620 11 01¼
Part of 110,312 17 04 for Navy Annuities	2,510 00 00	
Part of 947,514 07 08 Tallies of Sol. 3 *Geo.*	198,958 08 03	
P. of 1,603,987 08 01½ Army Debent. certified to 21 *March* 1719 as by 6. *Geo.* 393,194 14 05¼; More since 548,939 12 06¼		
	942,134 06 11¼	
Annuities on the Plate Act. 6. *Geo*	312,000 00 00	
Nevis Debentures	141,093 15 01¼	
New Churches	380,787 00 00	
Part of 500,000*l.* for 1st. Lot 1719.	58,300 00 00	
Part of 500,000*l.* for 2d. Lot. 1719	65,395 00 00	
		2,101,178 10 04
To *Exc.* Bills lent the *S. Sea* Comp.	1,000,000 00 00	
To Ditto pursuant to Act 8. *Geo.* towards the Navy Debt	1,000,000 00 00	
To Ditto pursuant to Act 9. *Geo.* to redeem Annuities	1,000,000 00 00	
		3,000,000 00 00
To the Civil List Debt		1,000,000 00 00
Equivalent due to the Kingdom of *North-Britain*		248,550 00 09
The Debt of the Navy at *Decemb.* 1725		1,255,491 09 04¾
		56,393,137 10 05½

which is included the Debt contracted since Dec. 24. 1716. with National Debts by the Subscriptions to the *South-Sea* Company, *&c.*

Nº II. Creditor.

By the Principal Sum remaining unpaid of the Debt due *Decemb.* 24. 1716. as appears by the foregoing Account		45,559,746 13 08
By the following Sums paid off at the *Exchequer viz.*		
Navy Annuities	2,510 00 00	
1st Lottery, 1719.	58,300 00 00	
2d Lottery, 1719.	65,395 00 00	
Loans on Coals for New Churches about	285,152 00 00	
		411,357 00 00
By *Excheq.* Bills supposed to be Cancell'd out of the Money raised for this Purpose, in the Year 1719. by a Lottery on the Sinking Fund, tho' it does not appear in any Account	500,000 00 00	
By Dº out of Money repaid by the *S. S.* C.	1,000,000 00 00	
By Ditto out of the Sinking Fund	610,341 17 09¼	
		2,110,341 17 09¼
By Balance remaining upon the Account of the Sinking Fund at *Michaelmas* 1725. applicable to this Service		547,654 03 10
		48,629,099 15 03¾
Balance is the Debt *increased* since *December* 1716. over and above all Payments out of the Sinking Fund, *&c.*		7,764,037 15 01¾
		56,393,137 10 05½

Figure 0.1 Part of William Pulteney's calculation of the size of the national debt as of December 24, 1725, and its change since December 1716. From *A State of the National Debt, As It Stood December the 24th, 1716. With the Payments Made towards the Discharge of It out of the Sinking Fund, &c. Compared with the Debt at Michaelmas, 1725* (London: Printed for R. Francklin, 1727), unpaginated appendix. Courtesy of the Kress Collection of Business and Economics, Baker Library, Harvard Business School. (HOLLIS: 007392800).

amounted to the emergence of a new quantitative age. Beginning in the late seventeenth century, calculations like Pulteney's account of the national debt became a fixture of British political life. Complicated calculations appeared in Parliamentary debates, in the internal deliberation of statesmen, and especially in mounting piles of printed pamphlets and newspapers. As in Pulteney's case, these figures were almost invariably related to questions about money, particularly public money. They concerned taxation, government budgeting, import-export statistics, the stock market, and especially the national debt. And, again like Pulteney's case, what inspired these calculations was political antagonism. Britain's new quantitative age was not fashioned by dispassionate scientific practitioners seeking "objective" knowledge about society or the economy, nor by diligent bureaucrats trying to advance the interests of the state. Rather, political actors of various stripes, from eminent ministers and members of Parliament to hack writers and out-of-work accountants, found that numerical calculation offered an especially useful tool for carrying out political arguments. They used numbers to critique and defend public policies, interrogate leaders, embarrass opponents, and uncover secrets. This motley collection of calculators made fighting with numbers a regular part of British politics.

These calculative conflicts were angry, uncivil, and fiercely partisan. They were driven by the ever-shifting demands of national politics, and rarely yielded conclusive answers to the issues at hand. They transpired largely through cheap, ephemeral texts, not the grand, abstract treatises that usually garner most attention as contributions to intellectual history. Consequently, later observers have rarely seen the unprecedented volume of political calculations produced during Pulteney's lifetime as a meaningful chapter in the history of political economy, social science, or statistics. In fact, this period has often been cited as a kind of dark age in the history of quantitative methods in government, politics, and social inquiry, separating the grand empirical ambitions of the Scientific Revolution—most famously the "political arithmetick" of William Petty (1620–1687)—from the birth of modern statistics in the nineteenth century. In 1837, for instance, Scottish political economist and statistician J. R. McCulloch asserted that "during the long interval between Sir William Petty and Dr. [Henry] Beeke"—who published an important quantitative study on the income tax in 1799—"statistical science could hardly be said to exist."[5] Modern scholars commonly claim that quantitative thinking first gained the political prominence it holds today in the early nineteenth century—more specifically in the 1820s–1830s, when new statistical bureaucracies produced what Ian Hacking has called an "avalanche of printed numbers" about social phenomena.[6]

Yet the novel calculations forged by Pulteney and his contemporaries constituted a vital chapter in the history of quantitative thinking, with significant and lasting effects on financial practices, economic ideas, and political culture in Britain.[7] Those calculations were empirically rich and technically inventive, even if they were not recognizable as "statistical science." Calculators like Pulteney put tools of mathematical calculation—mostly arithmetic, along with a few more advanced techniques like logarithms and occasional algebra—to work in many new and clever ways. They synthesized information from scattered sources, used estimates and interpolations to fill in gaps in irregular data, and crafted elaborate arithmetical models to explore political uncertainties. They fashioned new ways to think about change over time, not only tracing changes in the past, but projecting possibilities for the future. Through their quantitative conflicts, calculators fashioned new ideas about what *could be* calculated, from the "intrinsic value" of a stock to the fair compensation one nation (England) owed to another (Scotland) for entering into a constitutional union.

Perhaps the most important consequence of these lively political calculations was not technical, but cultural. Pulteney and his fellow calculators presented an unprecedented volume of numbers before the British public, and their numerical controversies became a powerful engine for publicizing the virtues of numerical thinking. Partisan calculators disagreed fervently with one another about what the right numbers were on any given question. Yet their relentless calculations—their efforts to get slightly closer to the correct numbers than their opponents—fueled a common sense that a domain of numerical truth did exist somewhere, which ought to serve as an ultimate arbiter of political knowledge. Politically engaged Britons, most of whom had at best basic numerical skills, became increasingly familiar with seeing political numbers printed in pamphlets and newspapers, reading about numerical arguments secondhand in magazines, perhaps even hearing them discussed in pubs and coffeehouses. More and more, they came to share in Pulteney's belief that quantitative knowledge was deserving of special epistemological esteem—that "*Facts* and *Figures*" constituted "the most stubborn Evidences." By the middle of the eighteenth century, calculators could take it for granted that numerical assertions would garner special weight with public audiences. In fact, some observers began to worry that Britons were too willing to believe in claims that were "cloath'd in the Dress and Appearance of Calculations."[8] Numerical thinking had come to garner such authority in

British public culture that some Britons began to worry that their countrymen trusted numbers too much.

There were limits to how far this new trust in numbers extended, to be sure. It was never universal. There were always hold-outs who rejected their contemporaries' fondness for the quantitative, as there are today. As numbers and calculations became increasingly prominent in public life, they provoked new suspicions and new criticisms. There were also many boundaries shaping who was able to partake in Britain's emerging quantitative culture. Participants generally needed sufficient property to vote in Parliamentary elections, sufficient education to do basic arithmetic, and sufficient leisure to read newspapers and engage in political conversation. This was not an inconsequential number of people, but many—including women and most working men—would have been excluded in formal or informal ways. The mounting authority granted to quantitative knowledge was never hegemonic or dogmatic, either. It did not completely displace other attitudes about what constituted reliable knowledge, nor entirely vanquish older prejudices against numerical thinking.[9] Those Britons who did show special respect for quantitative thinking did not automatically believe every politically salient number put in front of them, any more than people do today.

Even with these limitations, the new authority granted to numerical calculation marked a decisive transformation in Britons' collective *civic epistemology*. By the middle of the eighteenth century, there existed a powerful and relatively pervasive belief within the British political community that numerical facts and figures constituted an especially valuable and virtuous form of public reasoning. This was a significant departure from previous eras. Numerical calculation had long played some part in various governmental activities, of course, for example in keeping track of Crown funds and in maritime navigation. Throughout the seventeenth century, enthusiastic calculators like William Petty had made bold proclamations about how "political arithmetick" could advance the public good. But overall, up through most of the seventeenth century, quantitative thinking held no special prominence in political practice or public culture. Serious quantitative discussion about political questions was relatively rare, and public audiences did not accord quantitative assertions any special weight. In fact, many early modern people had looked upon calculation with a measure of suspicion, seeing calculations as the province of pedantic academics, covetous merchants, even sinister astrologers, not honorable gentlemen and honest citizens. Around the time of the Glorious Revolution of 1688, though, this began to change. Numerical calculations became more frequent, technically involved, and rhetorically compelling. Cultural prejudices against numerical thinking receded, while a new confidence in calculation rose. The new calculative political culture that coalesced would prove highly durable, not only in Britain but elsewhere in the Anglophone world.[10]

Calculated Values tells the story of this new era of calculation.[11] It may seem like a counterintuitive story. Britain's first age of political "*Facts* and *Figures*" helped lay the foundations for the "trust in numbers" that has become such a critical feature of political culture in many modern polities.[12] Yet the numbers and calculations that fill this book seem to bear few of the virtues that are often associated with "objective" quantitative thinking. Calculators did not produce page upon page of numbers to depoliticize certain contentious questions and remove them from the realm of debate. Nor was the period's numerical culture one in which cynical political actors took otherwise honest, objective numbers and corrupted them to partisan ends. Rather, the calculations that flourished in this era were born of politics, and they remained political, through and through. To put it another way, the era's political *calculations*—as in, intensive numerical analyses of political questions—were invariably driven by *political* calculations—as in, the strategic pursuit of partisan objectives. Yet, the fact that numbers were used in such overtly political ways in the eighteenth century did not hinder Britons' general confidence in quantitative knowledge, but rather the opposite. It was the deliberate use of calculation toward political ends that generated calculation's public authority. That irony is at the heart of the story that follows.

Public Calculation and Civic Epistemology

The emergence of Britain's new calculating age can be seen as a crucial episode in the history of what Sheila Jasanoff and other scholars in Science and Technology Studies (STS) call *civic epistemology:* "the social and institutional practices by which political communities construct, review, validate, and deliberate politically relevant knowledge." STS scholars have shown that different political communities often think through problems and craft collective knowledge in quite different ways. This is particularly true when it comes to political problems involving specialized technical knowledge, like environmental protection, defense strategy, or regulating hazardous technologies. Some polities might favor contentious argumentation rather than the pursuit of consensus; might grant varying levels of trust to certain kinds of experts or identify experts in different ways; might place different amounts of weight on different kinds of evidence, like numerical data, visual images, or personal testimony by citizens; or might privilege certain values over others in the decision-making process, like transparency, community participation, or the independence of experts. These different civic epistemologies structure how political decisions are made, what arguments "count," and what policies get enacted.[13]

Civic epistemologies are historical and cultural products, grounded not only in formal constitutional structures, but in informal beliefs, values, and traditions. Of course, attitudes about public knowledge vary from citizen to citizen, and institution to institution, within a given polity; they are always under contention and subject to change. Yet, certain epistemological habits, routines, and presumptions gain widespread purchase across a polity, spanning different institutional sites and partisan positions. Some of these elements also prove highly durable over time, becoming an ingrained feature of a community's political culture. By comparing how different political communities reckon with certain problems of knowledge—like making economic policy, planning public investment projects, or regulating biotechnology—researchers have been able to identify certain key features that distinguish the civic epistemologies of different nations.[14] For example, in contemporary Germany, public knowledge-making procedures rely heavily on institutional committees aimed at achieving a "communally crafted" consensus. In France, responsibility for making public knowledge has long been deferred to state experts, whose authority and prestige derives in large part from their selection through the nation's hierarchical system of elite education. By contrast, experts bear less esteem and autonomy in the civic epistemology of the United States, where there is a much greater emphasis on transparency, accountability, and objectivity—and where quantitative procedures and forms of evidence are given particular weight.[15] (The prominence of quantification in American civic epistemology is a point to which I will return in the Conclusion.)

Though STS scholars have recognized that civic epistemologies often have deep historical roots, most of the research on the subject has focused on synchronic and comparative studies of the recent past. *Calculated Values* seeks to add a new dimension to this scholarship, offering a historical, diachronic study of civic epistemology in the making. Over the decades that followed the dramatic political changes of 1688, a new reliance upon, and reverence for, numerical calculation became part of the fabric of political life in Britain, independent of any specific individual, issue, or ideology. This calculative habit came to prevail across many sites of political activity: the formal deliberations of Parliament and various Parliamentary committees, the internal reckoning of ministers, and especially print debates in the public sphere. It also came to extend across the ideological spectrum. Calculators could be found among Tories and Whigs, "Court" and "Country," moderates and radicals. By tracing how calculation gained such prominence as a tool of political thinking and knowing in post-1688 Britain, we can gain critical insight into how civic epistemologies come into being.

So how did this transformation in civic epistemology transpire? What first inspired political actors to turn to complex calculations as a mode of political analysis and argument, given that it was a recondite form of argumentation long viewed with suspicion? What institutional and cultural features of British politics after 1688 proved hospitable to calculation as a form of public reason? What made calculation "stick," as it had not before, as a regular and enduring feature of political life in the eighteenth century? Who were the individuals who carried out these calculations, how did they achieve that role, and what status did they hold in British politics? What political values and imaginaries—what images of a desired political future—were encoded in practices of political calculation? These are the questions that occupy the chapters that follow.

At its broadest, the argument of this book is that numerical calculation gained new and enduring authority in post-1688 Britain because of its usefulness as an *instrument of dispute* within an intensely antagonistic, partisan, and public political culture. Specifically, Britons found calculation a vital tool for arguing about issues related to commerce and finance—about the merits of a new excise tax, or about whether to pose trade restrictions on France, or about the best strategy for paying off the nation's public debts. Entangled in these technical questions, though, were deeper political controversies and concerns—about the relative power of the Crown and Parliament, or about whether the nation's political representatives were acting in the public good, or about how to govern the nation in a way that was fair to its different regions and interests. Calculation gained new authority not because it solved these problems, but because it turned out to be a compelling means to conduct these arguments.

In claiming that numerical calculations gained authority because of their value as instruments of dispute, I do not mean just that calculations proved useful as a form of rhetoric, a means to persuade.[16] This is certainly true; calculation did become a potent tool of rhetorical persuasion in eighteenth-century Britain. But persuasion is only part of the story. Calculations also proved useful for a whole range of other argumentative tasks: to interrogate, measure, judge, flatter, needle, undermine, and insult. They became a medium not just for making arguments, but for *having* arguments, a way for political adversaries to fight out problems. As this happened, numerical calculations also gained new power as a way to make knowledge about the social world. They gained not only rhetorical authority, but epistemological authority.

What made numerical calculation so well adapted to that particular political environment? To begin, I argue that the most important driver of new quantitative thinking in British politics during this formative period was an impulse for political critique. Numerical thinking gained a foothold in politics through the efforts of relative outsiders, who used calculations as a forensic tool to interrogate and disparage more powerful adversaries, particularly agents of the government and government-supported companies. Many of the era's most inventive and influential calculators—Charles Davenant, Archibald Hutcheson, Richard Price—fostered a "Country," republican political philosophy that prized transparency, worried deeply about political corruption, and viewed state and financial power with great suspicion. These pioneering calculators used arithmetic to draw conclusions from the imperfect data made available by the government, to speculate about what was going on behind the veils of power, and to pressure insiders into disclosing better information.[17] Often, it turned out that the numerical information critics demanded did not yet exist, because government insiders did not consider it especially important. It was the outsiders who assigned that information a political value. By analyzing key commercial and fiscal metrics, like tax yields or the size of the national debt, such calculating critics also fashioned new methods for assessing the efficacy of government and the ethics of government leaders. Numbers provided a powerful tool for comparing political ideals to political realities.[18] Political criticism was the spark that ignited Britain's quantitative age.

What kept the nation's calculating culture burning was partisan conflict. As critics marshaled calculations to attack government insiders and other partisan adversaries, those attacked increasingly answered back with figures of their own. Dueling calculations became a common feature of public politics. This calculative combat was fueled by the distinctive political environment that arose after the Revolution of 1688, characterized by extreme partisan polarization and a fast-moving, relatively unrestricted public press. As an instrument for political dispute, numerical calculation proved to be especially well-suited to this adversarial context. For one thing, calculation was incisive. It offered politicians and their public defenders a way to render arcane, diffuse, and technical policy disputes—about international trade policy, for example, or the merits of different public borrowing schemes—into more pointed public arguments (albeit arguments that were often still highly complicated). Public calculators often sought to translate such muddy problems into a headline number that delivered a clear political message, as when William Pulteney sought to demonstrate the failures of Walpole's fiscal policy by calculating that the national debt had risen £7.8 million under his management. In order to do so, calculators had to be creative. In matters of public finance and commercial policy, it took care and ingenuity to make sense of the sparse and irregular information that was available and turn those messy data into meaningful political claims.

In this unruly context, another feature of calculation proved vital: its flexibility. Using the same public data, different calculators could deploy the tools of arithmetic to construct a range of analyses supporting different partisan positions. This meant that opposing sides of a particular dispute could usually marshal plausible numbers supporting their cause. Calculation was also flexible in a deeper, ideological sense. Calculators fostered a range of different attitudes about what defined a good calculation, and what made calculation itself a good form of political reasoning. Britons identified several virtues in quantitative thinking: its illuminating ability to uncover secrets and deceptions; its association with the rigor of mathematics; the way it clearly articulated the shrewdness of merchants; its plain-spoken, "common sense" clarity. As sociologists Wendy Nelson Espeland and Mitchell Stevens have observed, the authority of quantification is "polyvalent."[19] Because calculators had different ideas about what made calculation a virtuous mode of thinking, it could be adapted for use in the service of a range of different political perspectives. For example, Whig calculators might offer numerical arguments that purported to express the common experience of the nation's merchants, while Tory calculators might offer numbers of their own that promised to protect the public against misrepresentations advanced by commercial special interests. These different values were manifest in subtly different calculating styles, shaping how different calculators treated available data, what assumptions they made in their analyses, and how they structured their calculations. These different styles made the era's numerical disagreements all the more heated.

The flexibility of calculation did have limits, though. Calculative combat was not an entirely lawless affair, and not all calculations were equally credible. Publicly available data, emerging norms of calculative practice, even a calculator's previously published figures—all of these placed bounds on how far a calculator could stretch the figures. Partisanship itself imposed the most significant limitations. Partisan interest drove calculators to spin the numbers to their party's advantage, but it also drove calculators to ruthlessly police one another. All public calculations were subject to intense scrutiny from unforgiving adversaries, who were quick to identify errors and unjustifiable claims.

These calculative battles rarely reached conclusive answers. Describing the challenges faced by economic reformers around the turn of the eighteenth century, historian Paul Slack writes that "knowledge was rarely clear, never complete, and could always be manipulated to suit the case of contending parties and factions."[20] If a question were important enough, calculators could almost invariably find another calculative move to play—a new assumption to critique, a new analysis to conduct, a new data source to incorporate. Often, it seemed like dueling partisan calculators could keep trading blows back and forth indefinitely without reaching a conclusion. This gave computational controversies an adhesive quality; people seemed to get stuck in them. As certain political arguments became fixated upon a specific numerical dispute, space for other kinds of arguments and analyses might get crowded out. Yet these ugly, open-ended disagreements were also often highly generative. Calculation facilitated meaningful disagreement. Calculative conflicts generated an unprecedented volume of printed numbers, which helped familiarize a broader public with the virtues (and vices) of quantitative thinking. They also generated new computational techniques, encouraged calculators to track down new sources of knowledge, and forced them to reckon with the methodological and epistemological complications of their calculations. Whereas in modern political life quantitative techniques are often criticized for flattening out complexity and obscuring context, in eighteenth-century Britain, quantification often tended to make political discussions more subtle, searching, and sensitive to contextual factors.

Instruments of Dispute

Over time, this calculative combat effected a gradual transformation in Britain's civic epistemology, winning a newfound authority for numbers in public life. Fighting with numbers made them more powerful. Especially striking about this story is that it does not conform neatly to either of the most prominent explanations scholars have offered for the authority of quantification in modern political life: the success of calculations as instruments of control or as instruments of trust. Instead, *Calculated Values* contends that the authority of calculation in British politics and public culture derived from its use as an instrument of dispute. To put it another way: previous studies link the authority of quantification to processes of objectification or to the pursuit of objectivity; this book links it to argumentative acts of objection.

First, much of the literature on the political uses of quantification has focused on the ability of numbers to impose order on unruly social phenomena. Counting, calculating, and measuring promises to make things legible, comparable, and thus manageable. Specifically, many studies have explained the historical empowerment of numbers and calculations as tools of political reasoning in terms of the imperatives of state control—as leading historian of statistics Alain Desrosières put it, "the need to know a nation in order to govern it."[21] Quantitative tools like censuses, land registers, and statistical surveys have been central to how modern states envision the people and territories they govern, casting them as objectified things susceptible to rational management.[22] This is especially evident in colonial states.[23] Of course, in practice, quantitative methods often fail to provide the kind of clear vision and administrative order that state agents imagine they impart. Nonetheless, scholars have argued that the "illusion of bureaucratic control" has been a crucial driver of the authority numbers hold within modern governmental rationalities.[24]

The story of post-1688 Britain follows a different script. The increasing intensity and inventiveness of political calculation in that era cannot be explained entirely—or even primarily—in terms of the internal needs and actions of the British state. Relative political outsiders, not direct agents of the executive government, were the foremost computational innovators and the leading advocates for the virtues of calculation as a mode of political thinking (at least until the ministry of Robert Walpole in the 1720s). This is not to say that state actors or institutions had no interest in numbers, or that the state played no role in the cultivation of the nation's rising quantitative culture. Some state agents definitely did place great value upon numerical information and calculative thinking, like the famed administrative reformer William Lowndes; some state institutions, notably the excise tax administration and the navy, seized upon quantitative techniques of surveillance and measurement as means to impose control over unruly markets and intransigent citizens. Yet the stories told in this book trouble the notion that state imperatives were the main stimulus generating enthusiasm for numerical calculation in the political realm. Generally speaking, it was the demands of public political argumentation, rather than the state's instrumental desire for numerical information, that was the greatest source of calculative energy during Britain's early quantitative age. When leading statesmen began to incorporate calculation more deeply into governmental reasoning, as Robert Walpole did in the 1720s, their use of calculation was fundamentally shaped by this environment of public quantitative combat.[25]

A second crucial explanation "for the prestige and power of quantitative methods in the modern world," articulated most thoroughly by Theodore Porter, is that numerical techniques are powerful instruments for engendering trust.[26] Making knowledge that can be denominated in numbers frequently requires following strict, mechanical procedures, which limit the effect of individual interests or biases. Consequently, numerical facts and figures are perceived to be impersonal and objective; they seem to represent the objects they measure rather than the subjective attitudes of those who make them. Porter contends that this makes numbers especially useful as a "technology of distance," well-suited for communicating among people who do not naturally trust one another due to geographic separation, social difference, or political rivalry.

One key example of a numerical instrument of trust in the political realm, Porter observes, are the mathematical cost-benefit analysis techniques employed in governmental decisions about public investment, most notably in the twentieth-century United States.[27] An older example are the double-entry bookkeeping techniques first popularized in the late medieval Mediterranean. Scholars have argued that the success of such formal methods of commercial accounting was attributable largely to their function in generating trust among merchants. By adhering to the exacting methods of double-entry, merchants could display their personal discipline and ensure their counterparties that they were not manipulating their records. This was not only true in commerce, but government as well, where transparent, rigorous accounting has helped states generate credibility with citizens. "Over and over again," writes historian Jacob Soll, "good accounting practices have produced the levels of trust necessary to found stable governments and vital capitalist societies."[28]

In a sense, the political calculations that flourished during Britain's long eighteenth century were also a product of the challenges of political distrust. But they did not function as technologies of trust in the way Porter and others have described, for several reasons. First, numbers did not really "travel well" at all in the period, and they were not a particularly effective "technology of distance."[29] Public calculations were often riddled with computational and even typographical errors. Once in circulation, calculations were often read and interpreted in ways that were very different from what their calculators intended. Second, public political calculations in this era were not rigid, formal, or mechanical, designed to limit the subjective discretion of the calculators who produced them. In fact, those eighteenth-century calculations were remarkably flexible and creative modes of political expression. This flexibility was a key reason why calculation proved to be well-suited to two-sided, adversarial dispute. Rather than being an evidently impersonal or impartial mode of thinking, calculation was entangled with individual personalities and partisan interests in complex ways. Third, Porter emphasizes that quantitative techniques are often seized upon by actors and communities in positions of political weakness, due to lack of trust or authority. In eighteenth-century Britain, calculation was also frequently an instrument prized by weaker actors, operating in the political minority or outside the center of political power. But whereas the practices of numerical discipline described by Porter are frequently a defensive tool—used by actors who are targets of distrust as a "defense against meddlesome outsiders"—numerical calculation in the eighteenth century was foremost an offensive tool—a way to express distrust rather than defend against it.[30]

Scholarship on quantification is rich and varied; organizing that research into only two categories, control and trust, necessarily leaves out other important themes and elides crucial subtleties—including the fact that the imperatives of trust and control often intersect with one another.[31] But it does help to identify what makes the calculative age in post-1688 Britain unexpected and illuminating. While the drive for control or trust may serve to explain much of the power quantification has attained in various modern political contexts, neither can fully explain the remarkable efflorescence of public calculation during that historical moment. To understand that, we must look elsewhere: to calculation's distinctive affordances as an instrument of dispute.

Trivial, Dishonorable, and Unbecoming? Calculation Before 1688

The power that numerical calculation attained in the decades following 1688 was unlike anything before. This did not mean that this was the first time anyone had ever put calculations to political use, or engaged in public disputes over numbers, or touted the special virtues of numerical thinking. All of these had earlier precedents. But it does mean that, overall, there was a decisive shift in Britons' civic epistemology beginning around 1688. Calculations went from being an occasional, temporary, and marginal part of political practice to being a regular, durable, and expected one. Whereas earlier generations, including many among the social and political elite, had commonly looked upon numerical calculation as a specialized and even suspicious way of thinking, eighteenth-century Britons came to see it as indispensable and worthy of special esteem.

To understand the extent of this change, it is worth looking back at the different status calculation held in politics and culture in earlier centuries.

To begin, it is important to recognize that there were many points in the preceding centuries when people had put numbers and calculations to work to improve government, shape political argument, and advance social understanding. Monarchs had long been interested in keeping careful watch over the treasure and territories they commanded. The year 1086, for example, saw the completion of a "Great Survey" of English and Welsh lands, ordered by King William the Conqueror, in order to determine the appropriate taxes and fees owed to the king by his subjects. The resulting records, the famed Domesday Book, contained extensive numerical information on thousands of different localities, including the number of households, extent of arable land, and assessed taxes. In the ensuing centuries, the English monarchy would develop orderly procedures for recording certain kinds of fiscal information, like the sophisticated accounting of customs tax revenues evident during the reign of Edward III (1327–1377). During the tumultuous reign of Henry VIII (1509–1547), key advisers initiated several creative endeavors to gather systematic numerical information for governmental use. Cardinal Wolsey, for instance, surveyed agricultural and military resources during the 1520s, while Thomas Cromwell undertook an extensive accounting of Church wealth. During that period, administrators began to compile crucial archives of numerical data, including the parish registers (initiated in 1538) and the London bills of mortality, which began as lists of plague deaths under Wolsey. Occasionally during this period, enterprising political thinkers used arithmetic to formulate political arguments, like in 1529, when religious agitator Simon Fish produced numerical estimates on the extent of Church property and the cost of supporting friars in order to critique the Roman Catholic Church.[32]

Throughout the following century, there were several moments when numerical information and calculative thinking briefly attracted heightened attention in political practice. In the 1620s, for instance, royal advisers and newly vocal merchants generated an abundance of new and creative ideas for how to remedy the nation's dire economic conditions at the time. Thinkers like Thomas Mun turned to calculative concepts like the "balance of trade" in order to make sense of the nation's commercial challenges. During the crisis of the Civil Wars in the 1640s–1650s, fiscal pressures encouraged political leaders on various sides of the conflict to experiment with new public-financial practices and to explore quantitative approaches to improving government accountability.[33] Some of this enthusiasm for quantitative information and public accountability carried over into the Restoration period after 1660, as evidenced by a short-lived Parliamentary project to audit the public accounts, the Brooke House Commission. Across the seventeenth century, a handful of numerically inclined thinkers formulated quantitative approaches to addressing problems of political economy: for example, Mun in the 1620s, Rice Vaughn in the 1630s, and Samuel Fortrey in the 1660s. (Computational innovation was not limited to cities or national politics, either. Consider Richard Loder, a yeoman farmer from Berkshire, whose 1610s account books contained a variety of creative calculations, for example, analyzing the relative profitability of barley versus wheat.)[34] The Restoration period also gave rise to the most famous early project in applying "Number, Weight, and Measure" to political questions: William Petty's "political arithmetick."[35]

These early numerical efforts laid important foundations for the calculative culture that would take hold in the long eighteenth century. Earlier calculators crafted new computational techniques, administrative routines, and empirical archives, and helped to cultivate an aspirational sense that quantitative thinking might make a better polity. Nonetheless, the quantitative age that arose after 1688 was substantially different from what came before. Throughout the seventeenth century, visible political calculation was mostly a sporadic and unsustained activity; the flames of calculative enthusiasm that did get ignited usually flickered out quickly. Speculations about trade balances during the crisis of the 1620s, for example, did not lead to sustained inquiry into the nation's import-export data—that would not begin until the 1690s, accelerating rapidly thereafter. The Brooke House Commission proved a temporary exercise—such a full-scale audit of public accounts would not begin until 1691. Petty's ambitious vision for political arithmetic went largely unfulfilled—only to be revived in a new form in the mid-1690s by Charles Davenant and others. Each of these early quantitative projects had its own immediate reasons for failing to catch on, but the series suggests a deeper trend. Through most of the seventeenth century, British political culture was generally inhospitable to quantitative thinking in a variety of ways.

For one, most medieval and early modern political leaders, from kings to leading ministers to MPs, simply did not place much value on numerical information as a form of political knowledge and guide to political action. This was true even as growing administrative capacities generated quite a lot of potentially illuminating numbers. Historian Olive Coleman observes that, during the medieval period, "a complex and sophisticated administration, whose very existence belies many assertions about the simplicity of medieval men, remained strangely separated from political decisions." Paul Slack has similarly shown that numerical thinking remained on the periphery of elite politics into the seventeenth century, noting a "disjuncture between the intoxication of new computations"—like those offered by William Petty—"and hard-headed calculation of their utility in the everyday practice of politics." Of course, some statesmen did appreciate the value of quantitative information as a political and governmental tool, particularly toward the end of the century. Notable was Thomas Osborne, known as Lord Danby (later Marquess of Carmarthen and Duke of Leeds), Lord High Treasurer under Charles II from 1673 to 1679. Danby organized the 1676 "Compton Census," a parish-by-parish survey of conforming Anglicans, dissenting Protestants, and Catholics carried out by local clergy, which revealed that the vast majority of the country was devoted to the Church of England. In an indication of how little weight seventeenth-century monarchs placed on such quantitative reckoning, Charles's brother James II disregarded the census results and initiated dramatic religious reforms that alienated the nation's Anglican majority, a key factor in his fall from power in 1688.[36]

The fate of "political arithmetick" is especially illustrative of the limited esteem numbers attracted among England's political elite. Histories of statistics and economics have often described the work of Petty and fellow arithmetician John Graunt as a "brilliant flowering" of quantitative methodology carried out by thinkers "ahead of their time."[37] Graunt's use of numerical tables and "shop arithmetic" to study the London Bills of Mortality (1663) was a landmark in quantitative demography and epidemiology. Petty's innovative calculations sought to quantify a remarkable array of social, political, and economic phenomena, including the aggregate wealth of the nation, the distribution of wealth among different social classes, and even the monetary value of individual laborers. But the significance of these technical advances was rather lost on their contemporaries. This was particularly frustrating for Petty. In his definitive biography, Ted McCormick argues that Petty envisioned political arithmetic as not just a battery of calculative techniques but an ambitious program in social engineering designed, in part, to "transmute" the fractious Irish into English citizens. This program depended on the support of a strong sovereign government, yet Petty failed to attract almost any attention from the Crown. "Political arithmetick" would eventually play a pivotal role in the rise of Britain's quantitative age, but only after its "radical reinterpretation" in the 1690s—a story for Chapter 1.[38]

Political leaders' indifference toward numbers echoed broader social prejudices against calculation, which many saw as a form of thinking unbefitting a gentleman. In the sixteenth century, humanist educators like Elizabeth I's tutor

Roger Ascham worried that students' wits would be "marred by over-much study and use of some sciences, namely, music, arithmetic, and geometry," dulling their conversational skills and distracting from proper learning of the classics. "Mark all mathematical heads, which be only and wholly bent to those sciences," Ascham wrote in 1570, "how solitary they be themselves, how unfit to live with others, and how unapt to serve in the world."[39] Later commentators echoed Ascham's concern that numerically minded people were unworldly. One famous voice for this view is Iago in Shakespeare's *Othello* (1604). In the opening scene, Iago rails against Othello's recent choice of lieutenant, "a great arithmetician, one Michael Cassio," whom Iago dismisses as a mere accountant, not a proper military officer. The insult typified contemporary suspicions that numerical skill was incommensurate with manly leadership. More subtly, Iago's pejorative reference to "arithmetic" may have evoked contemporary prejudices against the incursion of Hindu-Arabic numerals—pertinent prejudices given that Othello is a "Moor." Arabic numerals had only begun to attract a significant following in England in the first half of the sixteenth century, and many in Shakespeare's time still rejected this new style of "cyphering" in favor of older reckoning techniques like roman numerals or physical counting boards. (Especially unsettling was the zero digit, the "cypher." Shakespeare scholar Patricia Parker argues that Othello's name—"O! O! O!"—may have evoked a numerical pun, playing on contemporary anxieties about the "infidel o"!)[40]

Skepticism and even hostility toward numerical thinking was especially common among the nation's landowning elite, who often felt that only grubby artisans and acquisitive merchants needed calculating skills, which were thus unconducive to gentility. This social prejudice frustrated those Englishmen who did see value in numerical thinking, like Francis Osborne, a popular writer of educational advice literature who stressed the need for young gentlemen to learn arithmetic and accounting. In 1658, Osborne lamented how those calculating skills—what he called "the *Œconomickes*"—were "the least esteemed with our Gallant: Looked upon by some as triviall, by others as dishonourable and unbecoming a Masculine Imployment."[41]

One place where the tension between quantitative thinking and genteel manners became especially evident was, oddly enough, in natural science. During the mid-seventeenth century, a coterie of natural philosophers fashioned a new, experimental approach to investigating nature, given institutional form in the Royal Society of London in 1660. Leading this experimental project was Robert Boyle, famed for his investigations into the properties of gases. Steven Shapin has argued that the genteel Boyle, son of an earl, wished to create a community of gentlemen-philosophers who could be counted on to report their experimental findings honestly and bear faithful witness to one another's investigations. This required limiting reliance on numbers and mathematics, which Boyle considered "an abstract, esoteric, and private form of culture" incompatible with the collective enterprise of gentlemanly science. Boyle worried that too much mathematical calculation would exclude many potential gentlemen from participating, and that seeking high numerical precision or absolute mathematical certainty was likely to heighten disagreement and incivility. Boyle preferred plain-language accounts of experimental results; when numerical measurements seemed unavoidable, he emphasized that his numbers were inexact.[42] The resistance of Boyle and his gentlemanly colleagues to quantitative methods highlights a crucial point: the emerging authority of numerical calculation in public life in eighteenth-century Britain was not simply a side effect of an earlier "scientific revolution."[43]

For some in the seventeenth century, the problem with "Number and Measure" was graver than just being ungentlemanly. It was a veritable dark art. "My memory reacheth the time," wrote Francis Osborne in 1655, "when the Generality of People thought her most usefull branches, spel[l]s, and her Professors, Limbs of the Devill."[44] Historians have documented a deeply rooted belief in English culture that mathematical calculation amounted to a form of conjuring, especially visible in the sixteenth century but lingering well into the eighteenth century and even beyond. There seem to have been several reasons for this association, including mathematicians' cryptic symbols and the longstanding use of numerical calculations in astrology.[45] In fairness, some leading mathematical thinkers only strengthened the linkage between mathematics and the occult, most famously the eclectic John Dee. Famed for his stirring defense of mathematical thinking in the preface to the first English translation of Euclid's *Elements* (published in 1570), Dee was also unabashedly interested in the application of arithmetic and geometry to magic, astrology, and "skrying," a method for communicating with angels. Dee and some mathematical associates were in fact arrested in May 1555 for illegally calculating horoscopes for Queen Mary, Princess Elizabeth, and others.[46] Throughout the early modern period, mathematical promoters (Dee included) attempted to assuage fears about mathematics, emphasizing its usefulness in domains like carpentry, navigation, and surveying. But as Osborne's recollection shows, the notion that numbers and calculations were weird or wicked persisted.

Complicating matters was the fact that, well into the mid-seventeenth century, one of the most visible forms of quantitative thinking in English culture was in fact astrology. Printed astrological almanacs were wildly popular. As historian

Benjamin Wardhaugh writes, they were "for many people, their only contact with mathematics and calculation." The astrological legacy offers an important point of comparison for the subject of this book. For one, astrology was a number-dense practice that aesthetically resembled the financial calculations that proliferated in the next century; astrological tables and financial tables often looked a lot alike. Further, astrology offered a crucial early example of calculation being put to explicitly public, political ends. Many almanacs contained not only "natural" astrology—solar and lunar data, weather forecasts, agricultural tips—but "judicial" astrology as well, which aimed to project the course of human affairs. The eruption of civil war in 1642 was a particular boon to judicial astrologers like William Lilly, who printed popular astrological texts forecasting political events. A supporter of the Parliamentarian cause, Lilly calculated prophecies in the 1640s warning King Charles I and his Royalist supporters of their impending fall. Leading generals of the Parliamentary New Model Army consulted astrological predictions in formulating strategies. As a technical enterprise, judicial astrology was even more sophisticated than its "natural" counterpart, "embodying centuries of accreted methodology and tradition." According to historian Bernard Capp, it also played a crucial "role in fostering political awareness among the reading public."[47]

But astrological prognostication was always extremely controversial, both politically and religiously. After 1660, it fell into particular disrepute. Though participants on both sides of the Civil Wars had turned to astrology for guidance, Restoration critics like satirist Samuel Butler cast judicial astrology as a folly of the losing side, alongside other misguided notions like anti-monarchy and Puritanism. Wardhaugh argues that, in the wake of such astrological excesses, the term "mathematician" gained a sarcastic, even sinister, connotation. Almanacs continued to be a booming business—400,000 were printed between 1660 and 1688—but they jettisoned most political content and openly parodied their astrological predecessors. In the long term, judicial astrology likely did more to harm than help the status of numerical calculation in British civic epistemology.

Prior to 1688, there was ample evidence to suggest that, in fact, numerical calculation was unlikely to become a prominent part of political practice or attain distinct authority in the nation's civic epistemology. When it was not dismissed as an occult practice, calculation was seen variously as trivial, dull, pedantic, miserly, uncivil, and unmanly. This antipathy did not suddenly disappear in 1688, of course. As we will see, other early quantitative boosters like Charles Davenant, John Arbuthnot, and William Paterson worked hard to win over contemporaries who were skeptical of calculation as a mode of public reasoning. Long-standing cultural prejudices against calculation lived on into the eighteenth century (and beyond), and it was not uncommon for critics to try to discredit calculators by likening them to conjurers.[48] But it was not long before these anti-numerical forces would, for the most part, yield to a rising confidence in the quantitative in public life. So what changed?

The Constitution of a Quantitative Age

The rise of calculation in British civic epistemology cannot be reduced to any single factor, of course, nor did it occur in isolation from broader social and cultural changes. Calculative thinking attained new influence across many areas of British life over the eighteenth century. One evocative, though imperfect, way to visualize this transformation is the analysis of linguistic usage. Beginning around 1700, the term *calculate* and its various forms—*calculated, calculator, calculable,* and so on—started to become dramatically more common in English printed texts. While the frequency of *calculate* and its variants had held steady throughout the seventeenth century (particularly after about 1640), the term began to grow more frequent beginning precisely at the turn of the eighteenth century, becoming more than twice as common by 1750 and more than three times as common by 1800—depending on how you measure, of course.[49]

The rising prominence of calculation, both as an activity and a word, in British life in this period was a broad phenomenon that extended well beyond the sphere of politics. Across British society, numeracy had been on the rise throughout the early modern period, as more and more Britons were learning how to work with numbers and arithmetic in their private lives.[50] Eighteenth-century Britons came to calculate about a great many things.[51] Consider, as one sample, the body of texts printed between 1725 and 1729—two years on either side of William Pulteney's *"Facts* and *Figures"* pamphlet. Among just those that contain some version of the word *calculate* in the title, you can find introductory arithmetic textbooks, several almanacs (ranging from practical agricultural guides to esoteric astrological ones), texts on various commercial and financial topics (foreign exchange, taxation, stock trading), on nautical navigation, carpentry, Newtonian astronomy, the study of magnets ("loadstones"), gambling strategy, and religious prophecy ("calculations . . . pointing out the Introduction of the *Blessed Age*.")[52] As this list suggests, there were many entangled historical trends encouraging Britons' enthusiasm for calculation. Broader economic transformations were certainly crucial: the expansion of domestic and international trade, the increasing monetization of economic activity, the development of new financial technologies and markets. So too were trends in scientific culture, particularly the rising popularity of a Newtonian style of "mixed mathematics" inspired by the successes of Newton's *Principia* (published 1687).[53] Scholars have recently shown that, contrary to the assumption that calculative thinking was a "secularizing force," religion was also a significant driver of quantitative ingenuity, notably in areas like sacred history and historical demography.[54] So too was a broader recognition that "mathematics could be fun."[55]

At a low level of historical resolution, the rising intensity and importance of calculation in political practice might be swept together with this broader surge of calculative thinking across British society. Certainly, new calculative activity in politics was enabled by higher levels of numeracy and was closely linked to quantitative developments in many other areas, including mathematics, philosophy, commerce, finance, and even religion. These entanglements are evident in the *dramatis personae* that occupy the ensuing chapters, which include a failed playwright turned civil servant (Charles Davenant), a banker (William Paterson), an eminent university mathematician (David Gregory), a lawyer (Archibald Hutcheson), a commercial accountant (John Crookshanks), a couple lifetime politicians from genteel families (Robert Walpole, William Pulteney), and a Unitarian minister and political radical (Richard Price).

Yet Britain's emerging calculative political culture was not an automatic result of rising numeracy. Even with numerical education on the rise, very few Britons would have gained the kind of elite calculating skills needed to follow every detail of a calculative text like Pulteney's *State of the National Debt,* let alone execute the calculations themselves. Britain's new politics of calculation was not a consequence of all political actors becoming expert calculators; rather, it emerged because politically minded Britons were increasingly willing to pay attention, and grant credence, to calculations whose complexities they could not fully grasp. Nor was Britain's new calculative political culture a direct corollary of increasing enthusiasm for calculation in other domains, like commerce or science. The fact that an increasing number of Britons had developed calculating skills for their instrumental use in commerce, or that some elite men of science had seized upon mathematics as a means of understanding nature, did not guarantee that intricate arithmetical calculations would become a trusted form of public knowledge in the political realm. But they did—through a process that was largely endogenous to politics itself. As quantitative thinking became increasingly empowered in political life, that process reinforced the growing prestige of calculation in British culture at large.

This political process, of course, had many deep historical roots, stretching back into the seventeenth century and earlier. The quantitative age of the long eighteenth century built upon many preexisting materials in British intellectual, economic, and political life. As discussed in the previous section, earlier exercises in political arithmetic and public accounting provided crucial precedents. The mid-seventeenth century also witnessed foundational developments in public and private finance, which would prove to be the key object of calculative controversy in the long eighteenth century. Established features of early modern English political culture also provided key resources for the development of calculative politics, notably a deep reserve of republican political sentiments, a tradition of text-based argumentation as a key political practice, and a pattern of organized factional antagonism.

The dawn of the quantitative age in British politics was not simply a result of gradual change, though. It took the revolutionary political events of 1688 to activate these existing materials and trigger the development of an enduring culture of calculation. In November 1688, the stadtholder of the Dutch Republic, William of Orange-Nassau, landed an invasion force at Torbay on the coast of southwestern England. William had been summoned by a group of English Parliamentary representatives outraged by reigning King James II, particularly his Roman Catholic faith and his pretensions to refashion the government into an absolutist state on a French model. James soon fled his throne. In February 1689, a Parliamentary Convention made William and his wife Mary, the eldest daughter of James II, England's new co-monarchs, though with various constitutional restrictions designed to bolster the authority of Parliament and prevent future abuses of monarchical power.

Though historians long portrayed 1688 as a rather moderate, even conservative affair, recent studies have brought renewed attention to the sweeping and violent consequences of that first "modern" revolution.[56] This book suggests another sense in which 1688 was revolutionary: it marked a significant turning-point in the nation's civic epistemology. The Revolution transformed *how* political leaders, representatives, and citizens thought through political problems. The Revolution reconfigured the structures of political deliberation in England, creating an environment remarkably hospitable to numerical calculation as a mode of political thinking and arguing. Three such changes were especially crucial: the expansion of Parliamentary power, the rise of party politics, and the unleashing of the public press.

First: Parliament. The "Revolutionary Settlement" of 1689 reshaped the constitutional balance of power between Crown and Parliament. A series of Parliamentary acts—especially the Coronation Oath Act (1689), the Bill of Rights (1689), the Triennial Act (1689), and the Act of Settlement (1701)—explicitly delimited the authority of the monarch. While the respective powers of Crown and Parliament remained open to contestation, these acts restricted the Crown's ability to interfere with the courts or elections and secured vital Parliamentary powers, including the right of free speech within Parliament and the requirement that new Parliamentary elections be held every three years (at least until 1716). A key component of Parliament's expanded constitutional authority was the "financial settlement" that followed the Revolution. Parliament had claimed the exclusive constitutional authority to levy taxes since the thirteenth century. Practically, though, they rarely did much to leverage this "power of the purse." Parliament generally signed off on Crown revenues for the life of a monarch, and so were only consulted on fiscal matters when a new monarch ascended or in fiscal emergencies, notably during wars. After 1688, though, Parliament consciously limited the size and duration of revenue grants, meaning the Crown had to reapply for monetary "supply" repeatedly. William's landing at Torbay also initiated a war between England and France, which put mounting stress on the nation's finances and increased Parliament's relative control.

These new powers reshaped the very nature of Parliament itself. As historian John Brewer puts it: "The permanence of Parliament, the greater length of its sessions, and its much increased legislative activity . . . made it a far more important policy-making body than it had been before 1688."[57] Critically, new powers brought demands for new kinds of knowledge, particularly in the House of Commons, which held precedence over monetary matters. Instead of being resolved entirely by individual ministers or small, closed-door councils, fiscal and commercial policies were often discussed among a relatively large group of elected representatives with diverse skills and objectives. MPs found themselves deliberating regularly over technically complex questions about taxation, public expenditure, trade regulation, the stock market, and the national debt. This constituted a shock to existing patterns of civic epistemology, which opened the way for new practices of thinking and knowing—like numerical calculation—to gain influence.

Second: parties. In 1688, people of various political leanings came to a consensus that James II had to go. But many disagreed starkly about what the nation's future ought to look like beyond that. Frequent Parliamentary elections and regular Parliamentary sessions offered ample space for competing political visions to clash with one another. Post-1688 politics became a highly antagonistic and partisan affair. Political combatants arrayed themselves along a variety of axes—"Country" versus "Court" in the 1690s or establishment versus opposition Whigs in the 1730s. But one line of conflict dominated all others: the contest between Whig and Tory. These dueling party identities dated to the 1670s, but Whig-Tory animosity grew especially heated after 1688, peaking during the reign of Queen Anne (1702–1714). As Geoffrey Holmes has definitively shown, two-party politics became "the most dominating, inescapable fact of political life at all levels" during that period. Party affiliations shaped what newspapers Britons read, where they drank their coffee, what stocks they invested in, and which doctors they visited. The Whigs and Tories had deep disagreements about what post-Revolutionary Britain ought to look like: Whigs guarded the hard-won powers of Parliament, while Tories supported a stronger monarchy; Whigs pushed broader toleration for "dissenting" Protestants, while Tories defended the exclusive authority of the Church of England; Whigs advanced an aggressive, interventionist, and anti-French foreign policy, while the Tories favored an insular, naval-based one. The parties also disagreed on countless points of political economy, including taxation, banking schemes, commercial tariffs, trade monopolies, interest rates, and public debt.[58]

The composition, agendas, and relative strengths of the two parties certainly varied over the period covered in this book. For our story, though, what was most important was the fact of partisanship itself. Throughout much of the period covered here, politically engaged Britons envisioned politics as an oppositional contest of "us versus them," in which victory by the other side threatened devastating consequences to the nation. This polarization created distinctive conditions for how political knowledge was made and evaluated. Assertions of political fact became contentious, adversarial acts. Often, showing that the other side was wrong was more valuable than being right, and usually easier to accomplish. But this relentless antagonism brought some measure of order to collective political thinking, and in many ways energized the pursuit of political knowledge.[59] Political claims were subjected to an extremely high level of scrutiny, as partisan combatants jumped on any conceivable error or irregularity in their opponents' arguments. Questions about evidence, methodology, and indeed epistemology became a regular part of political practice. This ruthless, adversarial context became fertile ground for quantitative arguments.

Third: the press. One of the critical conditions for the dawn of the quantitative age was the development of a vibrant print culture, in which short, cheap texts were used as a primary vehicle for political argumentation. This distinctive media environment reflected many longer historical changes. Political writing in various media—pamphlets and news reports, sermons and satires, print and manuscript—had long played a central role in British political practice. At no point was this clearer than during the Civil Wars, as evidenced by the extraordinary collection of tens of thousands of tracts gathered by publisher George Thomason. Over the seventeenth century, printed texts took on various new functions in political practice. Parliament, for example, increasingly dispensed with the traditional notion that what happened in its chambers were *arcana imperii* best kept secret and turned to print to publicize information on the legislative process.[60]

The development of political print culture accelerated notably at the end of the seventeenth century. In 1695, Parliament refused to renew the Licensing Act, a collection of restrictions initiated in the 1630s establishing strict oversight of printing presses and requiring texts be approved prior to printing by the official Licenser of the Press. In the later part of the century, that job had been held by the steadfast Royalist Roger L'Estrange, who carefully monitored texts for politically subversive ideas. After 1695, publishers seized upon the new commercial opportunities presented by the end of pre-publication censorship, producing an abundance of cheap tracts on political issues, often with the authors' identity concealed. These new media outlets proved vital in the contentious partisan politics of the post-1688 period. Party leaders relied on newspapers and pamphlets to influence citizens, sway undecided representatives, and discredit political enemies, producing a sustained pattern of what historian Mark Knights has dubbed "paper warfare."[61] Leading ministers like Robert Harley employed teams of illustrious writers—the chief polemicists for Harley and the Tories included Charles Davenant and the tireless Daniel Defoe—to craft print arguments in support of partisan causes.[62] This new media environment had a dramatic effect on the practices of political thinking in Britain. Disputes about technical policy questions, like public finance or trade policy, were not just a matter for expert councils or even elected representatives. Instead, they were played out in pamphlets, broadsides, and newspaper articles, which were discussed in new spaces of public discussion like coffeehouses.

In his famous account, social theorist Jürgen Habermas cited precisely this moment in British history as the origin of a new form of political life: a *public sphere* characterized by private individuals coming together in open, rational conversation, unsupervised by state authority.[63] Habermas's depiction has energized a great deal of scholarship, and criticism, from British historians.[64]

Recent studies have demonstrated that, while changing modes of publicity critically shaped the character of political life around the turn of the eighteenth century, public conservations rarely lived up to the orderly description Habermas offered. Far from being its own, cordoned-off realm of "civil society," the world of the coffeehouses was deeply entangled with the machinations of courtly politics and the messiness of party conflict.[65] Though cultural commentators stressed a new ethos of "politeness," the political conversations that occupied public attention in this era were often nasty, offensive, and deeply personal, anything but polite.[66] This was the public sphere—uncivil, impolite, sometimes irrational—in which Britons' new appreciation for numbers would emerge. Britain's new quantitative age was a *public* phenomenon, to be sure, but it cannot be simply explained in terms of what Habermas called "people's public use of their reason."[67]

In fact, the public sphere that nurtured Britons' newfound confidence in calculation was shaped as much by fear as by reason—particularly a fear that things were not as they seemed.[68] Britons in the post-1688 era were constantly on guard against being deceived, especially by the people and institutions who asked their trust: political ministers, government administrators, corporate leaders, inventors, journalists, printers, lawyers. In political life, some Britons lived in fear of conspiratorial plots by Jacobites, traitorous Britons who remained loyal to King James II and his heirs. Others were haunted by the specter of corrupt politicians co-opting the Revolutionary cause for personal gain.[69] In economic life, they feared forged credit instruments, the lies of the "stock-jobber," and the false promises of the "projector."[70] The press exacerbated these anxieties, as the reading public was hit with a torrent of printed texts, often unnamed, trafficking in sensational claims, accusations, and innuendos.[71]

During that contentious period, Britons struggled with anxieties about political knowledge itself—about trust and truth, fact and falsehood, representation and misrepresentation. As Knights has written, "Knowledge and truth were thus political issues."[72] It was within this environment of fear, animosity, and distrust that Britons began to make numerical calculation into an integral part of the practice of politics.[73] Calculators trumpeted the power of numbers to "undeceive" the public—to catch lies, pierce the veils of power, and force secrets out into the open. But this was not simply a matter of some high-minded Britons reaching for a more rational form of public reasoning as an escape from the partisan muck. Calculators were in that muck themselves. Their numerical arguments were unscrupulous, impolite, and unreasonable in many ways, and they rarely yielded conclusive answers. Yet numbers proved distinctly well adapted to political discussion and deliberation in a highly disagreeable time. The numerical disagreements that arose helped imbue numerical thinking with new authority.

Money Problems

The rise of calculation within British civic epistemology was triggered by changing political structures and practices, but also by new political problems. These problems were predominantly about money—about taxation, government expenditure, public debt, interest rates, international trade, and the stock market. It is hardly surprising that financial matters were a source of political contestation for Britons in the long eighteenth century; public resources have forever been a source of political conflict. Nor is it surprising to say that financial questions provoked quantitative answers; money seems an inherently numerical thing. But not all polities have featured high-profile, public fights about the numerical details of tax yields and public debt restructuring as a prominent feature of political life. Political conflicts over money do not always take such intensely quantitative form, and elaborate calculation is not the only way to think financially. The particular form of financial politics—adversarial, public, technical, and, above all, calculative—that came to prevail in post-1688 Britain demands historical explanation.

British calculators around the turn of the eighteenth century had to reckon with dramatic changes in the nation's economic circumstances. Chief among these was a "financial revolution" that brought a massive expansion in Britain's "fiscal-military state" and its private financial markets. This transformation began with a variety of minor, often uncoordinated developments, dating at least to the 1640s. A crucial early change was the implementation of new excise taxes, notably on beer and ale, during the period of the Civil Wars. Further advances were made after the Restoration, including the development of new paper instruments for government borrowing ("Exchequer orders") in the 1660s and the movement of tax collection from a semi-private "farming" system to centralized administration (customs in 1671, excise in 1683). These earlier developments laid the groundwork for an even more dramatic expansion of the government's fiscal capacities after 1688, especially in public borrowing. Prior to 1688, the government's credit needs had been serviced primarily through relationships with a few dozen individual bankers, merchants, and "undertakers." After 1688, though, Parliament began raising large public loans by selling financial instruments to a broader, comparatively anonymous base of public investors. The professionalization of customs and excise administration before 1688, plus the introduction of new revenue sources like the land tax thereafter, made it possible for Parliament to guarantee interest payments on the proceeds of specific taxes. New institutions were created to help facilitate government borrowing, most notably the Bank of England in 1694. The state's new "public credit" proved an essential weapon in the series of expensive wars with France that began in 1689. Innovations in public finance developed in tandem with private finance. New advances in government borrowing capitalized on the practical knowledge developed by scriveners, brokers, and bankers in the City of London. Government finance projects produced large and exciting new objects for private investment, which in turn fueled the rapid expansion of secondary financial markets, London's so-called "Exchange Alley."[74]

Economic historians continue to debate the chronology, causes, and consequences of Britain's financial revolution. *Calculated Values* seeks to approach that critical moment from a new angle. It considers the financial revolution as a moment of unsettling economic and indeed technological change, in which unfamiliar, risky innovations produced troubling problems of knowledge for the British polity. To put it another way, this book examines the financial revolution as a trial for Britons' civic epistemology. During the late seventeenth and early eighteenth century, Britain's financial system became dramatically larger, more varied, and more technically complex. Politicians, corporate leaders, and projectors engineered an array of untested financial ventures. These included new market instruments (annuities, lotteries, tontines), public finance techniques ("equivalents," "engraftments," "sinking funds"), and regulatory schemes (international trade liberalization, interest rate reductions). These financial innovations were often deeply *problematic;* they generated technical, conceptual, and political problems.

On a technical level, there was often widespread disagreement about how and whether certain inventions worked, how best to design them, and how to evaluate their success. There was little in the way of an established body of financial knowledge to refer to in arbitrating such questions, and few trustworthy financial "experts" to whom the public could turn. Instead, arcane questions about financial technicalities—for example, about how a change in maximum "usury" rates would affect outstanding government debts, or about how expanding the amount of capital held by a joint-stock company would influence various stockholders—became subjects of public debate.[75] In many cases, even the most financially savvy and experienced Britons found themselves perplexed.

James Bateman, a leading financier and a founding director of the Bank of England, admitted as much in an April 1697 letter. Reflecting on a novel and recently approved plan for the Bank to refinance, or "engraft," a large quantity of short-term government debt instruments ("Exchequer bills"), he wrote: "So we shall now soon see w[hat] alteration it will make on the Bank. I cannot yet make any judgm^t of it. and I find others of various oppinions."[76] Such comments were common in the era. During Britain's financial revolution, debates about technical arcana were often bound up with far deeper conceptual, even philosophical, questions about finance itself.[77] During debates about a potential union between England and Scotland in 1706, for example, a question about compensating Scottish taxpayers for accepting higher tax rates elided into a far more profound dispute about the value of the distant future. During the South Sea Bubble in 1720, conversations about the fair price of South Sea Company stock refracted a fundamental debate about the very nature of financial value.[78]

Many of the new financial projects and devices developed during this transformational period were designed, at least in part, to solve public problems, particularly the government's need to raise money and manage its debts. They also had potentially significant, dangerous, and unanticipated consequences for British society. Thus, these financial inventions were politically problematic as well. Many Britons—country landowners, religious conservatives, "classical republicans"—harbored deep concerns about what new financial developments would do to their polity. They worried that rampant government borrowing would impoverish future generations, that easy credit-money would feed a corrupt state and overgrown military, and that new "paper" wealth would enrich the "monied interest" and upend an established social order long based on the equation of political influence with landed riches.[79] In the course of political arguments about public finance, such deep-seated political concerns became bound up with highly technical questions about the productivity of certain taxes, about the fair value of stocks, or about the safest way to restructure the national debt. It was not just that these questions were hard to answer; around 1700, Britons did not even have a good way to go about answering them. Financial knowledge itself became a political problem.

Britain's financial revolution thus offers a particularly rich site for studying a civic epistemology in the making. The chapters that follow reconstruct the contentious and confusion-riddled practices by which Britons actually thought through the public-financial changes they were experiencing: frustrated MPs trying to audit records of public expenditures, partisan pamphleteers arguing about the national debt, and ambitious prime ministers and propagandists debating the technicalities of compound interest.[80] The various entangled problems—technical, conceptual, political—wrought by the new finance forced Britons as a polity to find new, collective mechanisms for making public knowledge. Over time, Britain's civic epistemology was reconfigured, with numerical calculation taking a newly prominent role. Britons would look to numbers to guide them. But that process took time, and it was not a foregone conclusion that Britons would invest their confidence in calculation rather than some other source of authority—like a particular group of financial experts or a specific body of theory that was not heavily dependent on numbers.

In certain ways, many of the analyses and models produced by eighteenth-century financial calculators appear surprisingly modern in their technical complexity and in the financial principles underlying them. Yet in other ways, those calculations do not conform to any standard, modern image of "financial calculation" and its function. The calculations that circulated in print in the eighteenth century were rarely produced by self-interested, "rational" agents looking to maximize their own financial gain within a competitive market. Rather, those calculations were remarkably open-ended, inventive tools for thinking through financial questions, making political arguments, and even critiquing financial innovation itself.[81] This book thus contributes a new perspective to a vibrant discussion within the "social studies of finance" on the role calculations play in financial practices and the development of financial markets.[82]

The strain that new financial development placed on Britons' civic epistemology was most evident in moments of financial crisis, like Britain's notorious South Sea Bubble of 1720. "Manias, panics, and crashes" have, of course, attracted more attention than any other stories in the annals of financial history. Those stories have been told in many ways: as psychological dramas about "the madness of crowds"; as morality plays about greed run amok; as dark thrillers about devious capitalists and their political cronies scamming the unwitting public.[83] Economists have long tried to understand the causes and consequences of crises, whether "to reduce the risk of future crisis and to better handle catastrophes when they happen," or to use past bubbles as historical tests of the rationality of financial markets.[84] This book offers a different way to think about the unprecedented financial crisis Britain experienced in 1720—and perhaps financial crises in general. It was a crisis in collective knowledge about finance. In the early eighteenth century, the pace of financial innovation outstripped Britons' collective ability to understand those financial technologies and their dangers. In 1720, the decisive innovation was a convoluted financial project engineered by the South Sea Company. That scheme stretched the limits of contemporary financial knowledge and elicited an array of different reactions from the various parties—company leaders, politicians, investors, businesspeople, writers, concerned citizens—who encountered it.

When the once-soaring price of South Sea stock collapsed in autumn 1720, inflicting severe losses on investors and badly damaging public credit, Britons were left wondering how they possibly could have been so wrong. But the crash soon proved edifying. One Briton who had identified dangers in the South Sea scheme was the MP and prolific public calculator Archibald Hutcheson, who repeatedly warned that, according to the numbers, the South Sea scheme simply did not add up. After the crash, Britons turned to Hutcheson's calculations to explain what had gone wrong in 1720 and, perhaps even more importantly, to decide who was to blame. Britain's first modern financial crisis turned out to be a decisive moment in the evolution of Britain's civic epistemology, which offered a resounding lesson in why the polity was best off trusting the numbers. A moment that was in many ways the nadir of Britain's financial revolution was also the climax of its quantitative revolution. The South Sea Bubble offers remarkable insight into how problems of knowledge drive financial crises, and into how such crises can transform civic epistemologies.

Britain's Calculative Culture in Context

Calculated Values is an account of a historical change in only one nation (though one which, as I discuss further in the Conclusion, helps to illuminate the development of quantitative culture elsewhere in the Anglophone world, especially the United States). This choice of scale brings critical benefits. It makes it possible to attend to the details of calculative practice and political controversy in higher resolution and to trace change over time more precisely than would be possible in a broader study encompassing multiple nations. But the goal of this book is certainly not to reward Britain for being "first" nor to highlight some way in which Britain was "exceptional." Indeed, it is crucial to acknowledge that Britain was just one among many historical polities where numerical calculation became a valued instrument of political knowledge, both during the eighteenth century and before. A thorough comparative history of quantitative thinking across early modern polities lies beyond the scope of this study. But a preliminary sketch of some uses of political calculation in other national contexts can help to place the British example in a broader frame and to identify the specific insights that example may offer.

A certain reverence for numbers and mathematics has an ancient pedigree, of course. Geometry, as enshrined in the *Elements* of Euclid (c. 300 BCE), was long seen to be the very height of certain knowledge. Ancient philosophers like Pythagoras and Plato speculated that the essential order of the universe lay hidden in numbers. Practical arithmetic was not always accorded the same respect as Euclidean geometry, but ancient commentators did speak approvingly of the practical powers of arithmetical calculation to make better lives and better societies. In constructing his ideal *Republic* (c. 380 BCE), Plato argued that command over numbers was necessary for both warfare and philosophy, and thus ought to be central to the education of the elite "guardian" class. Since "calculation and arithmetic are wholly concerned with numbers," Plato contended, "then evidently they lead us towards truth." (Though Plato was adamant that guardians only use calculation in a noble fashion, "for the sake of knowing rather than trading.")[85] Numbers have also been deployed in the practice of politics since ancient times as well, notably by the famed Roman rhetorician Cicero. Amidst a series of speeches made around 70 BCE against Gaius Verres, a former governor of Sicily accused of corruption, Cicero recited an intricate computation showing the massive sums that one of Verres's lieutenants, Apronius, had extorted through illegal tax collection procedures: 400,000 *modii* of grain, or two months' provisions for the entire Roman population.[86]

Throughout the ensuing centuries, calculators across various global cultures developed innovative quantitative practices and sought to apply them toward public ends. Many of the foundations of practical (and theoretical) mathematics were laid by Indian and Islamic mathematicians, including the modern Hindu-Arabic numeral system and the basics of algebra. In his *Compendious Book on Calculation by Completion and Balancing* (c. 820 CE), Persian mathematician al-Khwārizmī developed algebraic methods for solving linear and quadratic equations and fashioned new mathematical approaches to a range of commercial and legal problems. When it came to using numbers in the practice of government, China arguably led the way until the period of the Song Dynasty (960–1279 CE), developing the world's most sophisticated techniques for state accounting.[87]

Within Europe, many different political communities began to develop their own distinct quantitative cultures in the late medieval and early modern periods. From the twelfth to the sixteenth century at least, the republican city-states of Northern Italy like Genoa, Venice, and Florence were remarkable sites of computational innovation. Northern Italian calculators pioneered new mathematical techniques in finance and accounting, while the Italian republics integrated numerical calculation into government practice as a tool of administration and accountability. As early as 1327, for example, the Republic of Genoa established formal laws governing the keeping of accounts by Genoese businesses; by 1340, the Genoese government itself had begun to employ double-entry bookkeeping in state accounts. In Florence, a 1427 law required that all merchants and landowners keep double-entry books and make them available for a controversial tax census (*catasto*).[88]

Another site where a vibrant quantitative culture developed in the early modern period was the Dutch Republic, particularly during the seventeenth-century "Golden Age." Jacob Soll argues that it was the Dutch who effectively invented "political economy" by applying the "rationalizing tools of accounting" to the problems of state management. Leading proponents for this calculated view of statecraft included the engineer Simon Stevin, the political theorist Pieter de la Court, and the polymath Jan De Witt, "Grand Pensionary" of Holland (1652–1672) as well as the author of a foundational text on financial annuities. Dutch thinkers employed creative numerical calculations in political analysis and argumentation as well. A notable example came in 1644–1645. As leaders of the Dutch West India Company pondered the future of that company's colony in Brazil, multiple calculators fashioned intricate analyses trying to quantify how much the company's New World colonies were worth to the Republic.[89]

At the same time that Britain's quantitative age was beginning, other nations were integrating numbers into political thought and practice in their own distinctive ways. French statesmen, for example, began to recognize that numerical information and calculation could be a great source of power for a monarchical state. Jean-Baptiste Colbert, Controller-General of Finances under King Louis XIV from 1665 until 1683, undertook extensive economic and fiscal reforms and developed an elaborate "secret state intelligence system," in which quantitative information played a central role. Colbert commissioned elaborately decorated ledger-books containing annual summaries of the state accounts for Louis XIV's review. Population was a particular object of quantitative concern in French political reckoning. Later in Louis XIV's reign, the military engineer Sèbastien Le Prestre de Vauban urged the state to gather comprehensive numerical data on the nation's population, famously contending that "the greatness of kings is measured by the numbers of their subjects." Throughout the eighteenth century, state agents and public commentators took great interest in precisely quantifying the French populace, experimenting with new statistical techniques (a "universal multiplier") in order to overcome the administrative challenges inherent in a nationwide census.[90]

German thinkers developed their own, distinctive interests in social and political numbers, which were closely tied to religious and legal concerns and which often took on an academic character. The illustrious mathematician Gottfried Wilhelm Leibniz, for instance, had a keen interest in the application of mathematics and quantitative data to political questions. He wrote an early essay on "juridical mathematics" in 1683 and crafted an elaborate proposal for an official Prussian statistical bureau in 1700, which specified fifty-six different variables necessary for the proper evaluation of a state. Demography was an area where German calculators particularly excelled, often inspired by religious concerns. This was exemplified by the mortality data gathered by Breslau pastor Caspar Neumann around 1690, foundational to early mathematical research on life expectancy, and by the work of Johann Peter Süssmilch, author of *The Divine Order in the Circumstances of the Human Sex, Birth, Death and Reproduction* (1741). It was in Germany that the application of mathematics to law and state administration—what Frankfurt professor Johann Friedrich Polack termed "forensic mathematics" in 1740—first became integrated into university curricula, one component in a broader science of "cameralism." It was also in Germany where the term *Statistik* originated in 1748 (though such "knowledge of the state" was originally seen as a broad domain containing qualitative as much as quantitative information).[91]

Clearly, then, Britain was not the only place where numerical action was happening in political life in the early modern period, nor was Britain's calculative culture necessarily "superior" in any normative sense—older, more empirically astute, more methodologically sophisticated. But the status of calculation was certainly not uniform across these polities, either. Calculation was put to different political uses, carried out by different kinds of actors and in different institutional settings, and energized by different political concerns and epistemological values. My hope is that this study can inspire further inquiry into the comparative history of calculation and political reasoning across different national contexts, especially before 1800. How did differences in the constitution of political power, patterns of political deliberation, and media of political communication relate to variations in calculative cultures? What styles of calculation were favored in different national contexts, and what "epistemic virtues" did different polities celebrate about numbers (if they celebrated numbers at all)?[92] How did these privileged epistemic virtues relate to the political values those nations cherished—in short, did the kinds of numbers people used reflect the kind of nation they wished to be?

If there is one factor that most clearly distinguishes Britain's from other early modern calculative cultures, it was its remarkable publicity. In many other polities—Colbert's secret intelligence system in France being the signal case—the accumulation and interpretation of numerical information were state activities that took place predominantly behind closed doors. Under prevailing "reason of state" theories, politically valuable numbers were seen as *arcana imperii* that, for reasons of state safety, had to be hidden from the prying eyes of the people and rival nations. Alain Desrosières argues that publicity (or lack thereof) constituted an "essential difference" between the early statistical cultures that emerged in eighteenth-century Britain and France. In France, he notes, quantitative "descriptions, whether intended for the king or his administration, were secret and linked to royal prerogative. They were not intended to inform a civil society distinct from the state, nor an autonomous public opinion."[93] A similar secrecy prevailed in the German states. In Brandenburg-Prussia, for example, great strides were made in the enumeration of population and state resources under the reign of Friedrich Wilhelm I (1713–1740); but a 1733 law explicitly forbade publication of such quantitative data out of fear it might reveal vital intelligence to the state's enemies.[94] Even in republican polities that more closely resembled Britain, like the Dutch Republic, calculative discussion and debate were often still confined to relatively private spaces, like the directors' meetings of the great trading companies.

In Britain, though, political calculation became an unabashedly public affair.[95] Plenty of calculations were still conducted in private spaces, of course, and some critical sites of calculative debate—like the Halls of Parliament—were shielded, albeit porously, from public view. But many of the most creative and contentious calculating throughout the period took place in cheap printed texts, designed to engage public audiences and open to public scrutiny. Against reason of state arguments for keeping public numbers secret, Britons demanded that numerical knowledge be public. Not only did they take place *in* public, political calculations in Britain were fundamentally concerned with questions *about publicity*—about struggles for control over information between citizens and state. Post-1688 Britain thus offers an especially valuable site for thinking through the possibilities and problems that come with using numbers to make public knowledge.

The Course of the Narrative

The chapters that follow constitute a punctuated narrative. Rather than offering a comprehensive accounting of all of the calculative activity in British politics across the period, it focuses on particular moments of transition and controversy. It seeks to explain those quantitative moments in detail, often through the "close reading" of historical calculations. One trade-off is that not every pertinent moment can be covered. There is much more that could be said, for example, about the recoinage controversy of the mid-1690s, about debates over lowering legal interest rates in 1714 and 1737, about Walpole's notorious "excise scheme" in 1732–1733, and so on. But there is good reason for this trade-off—what amounts to a methodological argument about how we ought to study the history of quantitative cultures. I contend that numerical calculations cannot be treated as "black boxes," whose operation in the broader world can be understood separately from their inner workings. The details of the calculations matter: how they were built, the assumptions behind them, the objectives they were intended to achieve, the epistemic and political values they encoded. This book places the calculations at the heart of the story.

Chapter 1, "Finding the Money," describes the genesis of a new quantitative political culture in the aftermath of the Revolution of 1688. The narrative opens with a numerical mystery: the "Secret Service" payments made by former Secretary to the Treasury William Jephson, who had distributed hundreds of thousands of pounds of public money with no public record—money that suspicious members of Parliament desperately wanted to track down. Jephson's enigmatic accounts encapsulated a broader crisis of knowledge that beset Britain's political representatives. In the wake of 1688, as Parliament began to decide how much revenue they wanted to grant their new monarchs, the collected MPs realized they knew almost nothing about the nation's fiscal condition. A new Parliamentary Commission of Public Accounts, started in 1691, set about trying to put together a complete account of the nation's finances, a task that proved far easier said than done. In the process, though, MPs found that numerical calculations could be a potent political tool for critiquing Crown agents, measuring governmental success, and filling in the gaps in incomplete state information. Shortly thereafter, economic writer Charles Davenant seized upon this forensic approach to calculation in his reformulation of "political arithmetick" as a mode of political thinking. The efforts of Davenant and the Accounts Commissioners to use numerical calculation to solve the problem of "finding the money" exemplified how relative political outsiders were the driving force behind the emergence of Britain's quantitative age. The resistance those outsiders encountered also revealed that, in the 1690s, not everyone agreed that calculation was a welcome or reliable way to produce political knowledge.

The political promise and challenges of calculation would be even more evident in Britain's next constitutional upheaval: the union between England and Scotland. Previously, those two nations shared a monarch, Queen Anne, but maintained separate Parliaments and judicial, administrative, and ecclesiastical institutions. During union negotiations, it was agreed that the English government would make a large monetary payment, termed the "Equivalent," to compensate the Scottish people for accepting higher taxes. In 1706, a committee of six calculators determined that the precise value of that payment ought to be £398,085.10s. Chapter 2 tells the story of that number. Though subsequent generations have often seen it as a corrupt payoff, the "Great Project of the Equivalent" was also an extremely ambitious work of calculation, and testimony to the increasing, though hardly uncontested, authority of numbers in public life. For the two men most responsible for designing and calculating the Equivalent, the banker William Paterson and the mathematician David Gregory, that project was a statement about how better numbers could make a better nation, though each saw different virtues in calculation. The fate of the Equivalent figures after they were calculated, though, revealed that putting "calculs" (as the Scottish called them) into political practice was fraught with challenges. Once in circulation, their £398,085.10s. figure was misprinted and critiqued, reinterpreted and recalculated. Many questioned what role elite calculators like Paterson and Gregory ought to have in deciding their nations' futures. Yet critics often answered with calculations of their own, reinforcing the sense that, although £398,085.10s. might not be the correct, honest, or fair number, numbers still held the answer.

The battle over the Equivalent showed that political-economic calculations could inspire great political controversy. Nowhere was this more evident than in 1713–1714, the subject of Chapter 3, when dueling Whig and Tory calculators spent months wrestling over one number: the *balance of trade* between Britain and France. Beginning in 1712, peace negotiations between those two countries looked to end nearly a quarter-century of war. Incumbent Tory ministers saw a chance to pursue freer trade with France, a project their Whig opponents fiercely opposed. The resulting debate yielded a public numerical debate unlike anything that had been seen before. Two-party politics proved an incredible boon for the political career of numbers. Energizing the balance-of-trade debate was the fact that, while members of both parties contended that the argument should be settled by the numbers, they celebrated numbers for different reasons and thus had different visions for what made a credible calculation. Tory polemicists demanded sophisticated calculations to protect Britons from misrepresentations advanced by special interests; their Whig opponents felt unadulterated, "common sense" numbers better expressed commercial wisdom. Both sides carried on for months without either striking a definitive computational blow. Though inconclusive, these arguments were not inconsequential. Most crucially, the relentless stream of numbers that resulted was the most potent public advertisement yet for the value of calculation as a form of political reasoning.

As partisan conflict generated ever greater demand for public calculations, individuals with keen computational skills found new opportunities in the numbers game. Chapter 4 examines the people behind the calculations. It focuses on the stories of two prolific calculators: John Crookshanks, whose 1718 ode to "Truth and Numbers" is one of the book's opening epigraphs, and his rival, Archibald Hutcheson. By reconstructing how they fashioned their respective calculating careers, this chapter sheds light on the social history of numeracy during their lifetimes. As their eclectic stories show, there was no one, obvious way to become a calculator at the turn of the eighteenth century. Their stories take us from the port of Livorno to the Edinburgh Customs House, from London's Inns of Court to the Leeward Islands. Both men found that the calculating life could be a precarious one, and that computational skill did not necessarily translate into personal credibility. Crookshanks fought off accusations that he was a pirate; Hutcheson, that he was a Jacobite. Their quantitative paths ultimately clashed head-on in the late 1710s, in an intense—and intensely personal—pamphlet battle over Britain's national debt. In the course of that highly personal conflict, though, each calculator helped fortify a broader notion that there existed an impersonal realm of numerical truth transcending their individual efforts.

Chapter 5 turns to the most dramatic event in the financial history, and the computational politics, of the period: the 1720 South Sea Bubble. As discussed, 1720 was not only a pivotal moment for Britain's financial development but for its civic epistemology as well. The main character in this chapter is familiar from the previous one: Archibald Hutcheson. Suspicious of the political motivations behind the South Sea scheme, Hutcheson crafted sophisticated quantitative pamphlets designed to dispute the soaring prices Britons were willing to pay for South Sea Company stock during that fateful year, and to demand further transparency from Company leadership. At the center of those pamphlets was a novel, quantitative conception of the "intrinsick value" of stocks. Hutcheson's distinct approach to valuation drew upon decidedly *early modern* attitudes about information, secrecy, and political virtue; but his mathematically intensive methods, like the "discounting" of expected future profits, would become foundational to *modern* financial practice. Once the Bubble burst, Hutcheson's calculations found a powerful new purpose, used to prove that the Company's leaders had knowingly deceived the public. Hutcheson wrote his own computations into official Parliamentary reports, laying the foundation of an enduring explanation for what went wrong in 1720. Perhaps paradoxically, the South Sea Bubble was probably the greatest political triumph for calculation in the entire eighteenth century.

The collapse of the Bubble generated an array of problems. The South Sea Company needed to be reorganized, misled investors demanded compensation, and the massive national debt still needed addressing. Governmentally, the task of sorting out this mess fell to Robert Walpole, former Chancellor of the Exchequer. Fortuitously out of office during the disaster of 1720, Walpole returned to the helm of British government shortly after the Bubble burst. As Chapter 6 examines, Walpole's response to the Bubble was highly calculated. Walpole and a team of computational advisers carefully read and annotated pamphlet calculations and built elaborate quantitative models of their own to project the impact of different government policies. The frantic salvage mission that followed the 1720 crisis ushered in Walpole's tenure as Britain's first "Prime Minister," which would last more than two decades. During his lengthy tenure, Walpole spearheaded a vital transformation in the cognitive practices of the British state. While numerical calculation had previously attracted the greatest attention from critics of the government and other political outsiders, Walpole internalized calculation within government and made it an essential part of state thinking. The greatest example of Walpole's own numerical ambitions was his highly touted "sinking fund" scheme for managing the national debt, an ambitious set of fiscal rules that promised to bring public credit under predictable control by harnessing the mathematical power of compound interest. Not just a sober fiscal expedient, Walpole's sinking fund promoted a radical reimagination of political time, bringing the political future far closer at hand.

Walpole's use of calculation to project the political future was one of the most striking testaments to just how much authority numbers had come to garner over the preceding half-century. At the same time, his deft use of calculation to shape public attitudes and build his own reputation revealed that Britons' collective faith in numbers might open them up to new political dangers. In an irony befitting the era, Walpole himself was one of the first to worry that his fellow citizens might be too quick to believe any statements "cloath'd in the Dress and Appearance of Calculations and Figures." Chapter 7 will explore the expanding popularity and authority of numerical calculation in British life in the mid-eighteenth century—as well as the new kinds of discontentment that quantitative fervor had begun to generate. It will do so with the help of two discerning but divergent philosophical critics from the period: David Hume and Richard Price. Both thinkers diagnosed serious deficiencies in how their fellow citizens went about thinking through political problems, but offered contrasting conclusions. For Hume, the problem was that Britons had become too captivated by calculation. In his mind, economic numbers were epistemologically unreliable and politically distracting. Astute calculators could make figures say almost anything, while taking advantage of the extra credibility the gullible public granted to numerical claims. The contemporary fixation on calculation also led Britons to embrace misleading and even downright dangerous principles of political economy, like the obsession with the balance of trade, and to put faith in quantitative gimmicks like Walpole's sinking fund. For Price, on the other hand, numerical information and mathematical reasoning offered the best defense against the uncertainties that frustrated human life, whether in natural philosophy, finance, or politics. In political affairs, Price argued that the nation needed more rigorous public accounts and public policies guided by mathematical logic, including a revived sinking fund. By making collective political thinking more rational, Price contended, more numbers and a greater reliance on calculation could actually make the nation freer and more virtuous.

Hume and Price's disagreement portended a lingering tension in Anglophone political culture. Political numbers are fraught with uncertainties and dangers, as Hume noted; but they are also the best tool we have for making politics rational. The Conclusion will examine the legacies of the post-1688 British experience for subsequent generations, both in Britain and the United States, and explore some possible lessons for our own, twenty-first-century quantitative age.

CHAPTER ONE

Finding the Money:
Public Accounting and
Political Arithmetic after 1688

WE BEGIN WITH a story of a man who knew too much, and a nation that didn't. William Jephson had a secret. He was not the most powerful member of Britain's growing state finance apparatus, but at the beginning of the 1690s, he had access to privileged information that it seemed no one else had—save perhaps the new monarchs William and Mary themselves. In his capacity as Secretary to the Treasury since April 1689, he was responsible for paying over £100,000 per year out of public revenues for "Secret Service," money paid directly on order of the monarchs without public record.[1] From the perspective of the Crown, maintaining absolute secrecy over these payments was essential to exercising royal authority and preserving state security. Secret Service was indeed put to plenty of legitimate uses, including diplomatic transfers and small rewards to royal employees. Yet many Englishmen suspected that Jephson's account-books mapped a covert network of payoffs to royal creditors, kickbacks to court favorites, and even bribes to members of Parliament. It did not help that Secret Service had become a favored practice under the recent Stuart kings, especially Charles II (and his chief minister Danby) in the 1670s.[2] Jephson's secret accounts reminded some that, whatever strides had been made in the recent Revolution, the threat of autocracy was not dead.

So when the House of Commons formed a new Commission of Public Accounts to investigate the movements of public money in early 1691, Jephson quickly became a person of great interest.[3] But the Commissioners soon hit a serious obstacle. On June 7 of that year, Jephson died. Over the next six months, the Commission tried desperately to recover Jephson's secret accounts by pestering his employees and pleading with his employers at the Treasury. In return they received partial answers, artful dodges, and appeals to executive privilege. They knew Jephson had paid out a total of about £243,000 in Secret Service money since November 1688, but no one knew where most of it had gone. In the end, the Commission could only account for payments to eighteen members of Parliament, totaling just £26,815, about 11 percent of the total. Rumors abounded about corrupt payoffs, like £20,000 reputedly given to the Earl of Nottingham for the purchase of a house in Kensington. Other members of Parliament found it hard to believe that the Commissioners could not recover such crucial information. MP Sir John Thompson captured the mood best when he said: "I stand amazed that, in the best times and Governments, things should be in such darkness. . . .'Tis our misfortune, the person is dead that should give you Account."[4]

Britain's quantitative age arose out of problems like Jephson's secret—problems about the political control of financial information, about finding the money. Jephson's story was emblematic of a broader crisis of knowledge that beset England's Parliamentary representatives in the wake of 1688. As MPs looked to wrest greater power over the nation's finances from the Crown, many realized that, as a collective body, Parliament knew almost nothing about the public's money. Parliament's ignorance was put into stark relief in 1689, when the legislators were faced with deciding how much revenue they ought to grant their new monarchs William and Mary to fund government operations. Many MPs began to clamor that, if Parliament were to exercise its power over the nation's purse effectively, it needed far more knowledge about the nation's money. By 1690, the House of Commons had tasked a new institution, the Commission of Public Accounts, with compiling a consolidated account of public revenues and expenditures. This seemed like a reasonable aspiration at first. Scrupulous English men and women of business were well aware of how to fashion such a balanced account of inputs and outputs according to the best practices of double-entry bookkeeping.[5] But *public* accounting—casting up such an account for the whole of the nation's government—turned out to be a different thing altogether.

As they set to work, the Commissioners of Public Accounts confronted a massive divide between the data that they expected to find concerning government revenues and expenditures and the actual data that they managed to acquire—

several executive places since the Restoration, culminating in the lucrative position of Auditor of the Receipt in the Exchequer in 1673. In that position, he served as the primary broker of information between Parliament and the Lower Exchequer, the main clearinghouse for Crown funds. If members of the House of Commons wanted financial numbers—and this was certainly not a regular occurrence before 1688 anyway—they usually had to ask him.[16]

In the form of the Exchequer, the English Crown had long since developed elaborate mechanisms for managing the day-to-day movements of its money. The Exchequer was not primarily concerned with producing data that was useful to a broader public, though. Its peculiar methods of accounting were developed to safeguard the king's money and were highly opaque to outsiders. That was even true when protocols were followed exactly, and by 1688, exactitude was rare. Procedural standards had deteriorated under Howard's watch, especially when it came to distilling information for use by legislators. One Exchequer official, the Clerk of the Pells, was traditionally expected to make a "General Declaration" to Parliament of all the receipts and payments to and from the Exchequer every half-year. By the early 1690s, the current clerk, William Wardour, had become at least three years delinquent in making these declarations.[17] For several reasons, then, financial numbers were not a prominent feature of Parliamentary politics in the years leading up to 1688: the Crown had stable, permanent sources of funds; the institutions that managed those funds were opaque, uncoordinated, and often territorial; Parliamentary sessions were rare; and many MPs took little interest in financial affairs.

The Revolution shook all of this up. In a short period of time, a body of political representatives with little shared knowledge or experience in financial matters found itself trying to account for all governmental revenues and expenditures, debating projected tax yields, and weighing newfangled public credit schemes. Parliament's numerical awakening began very shortly after the landing of William of Orange. In December 1688, a makeshift Convention (composed of peers and MPs who had served under Charles II) undertook to establish the constitutional terms under which William and his wife, Mary, would rule. As that Convention transformed into the first official Parliament under the new monarchs, their "financial settlement"—what revenues the monarchs would be granted and under what conditions—quickly emerged as a pivotal problem. And soon the MPs began to realize how little they knew.[18]

On the first day of that new Parliament (February 26, 1689), the debate on monarchical supply proceeded without much apparent concern for financial particulars. The next day, some MPs pointed out that the easiest solution was an early example of the "expectations gap" that accounting theorist Michael Power has shown plagues the practice of auditing.[6] The accounts kept by government agents were often outdated, incomplete, or incompetent.[7] Public money lay entangled in complex obligations, obscured by antiquated administrative procedures, and hidden in the hands of private contractors. Most frustratingly, many officials—like William Jephson—simply would not disclose the information they had. The Accounts Commissioners tried various things to combat this. They cajoled and intimidated Crown officers. They tried to educate themselves on the arcane methods of the Exchequer. And they began to calculate.

The Accounts Commissioners and like-minded MPs discovered that arithmetical calculations could be a potent tool for interrogating and critiquing the Crown—an instrument of dispute. This was especially true for "Country"-minded Commissioners like Robert Harley and Paul Foley, who were deeply concerned about the abuse of executive power on the part of the Crown. Such Country critics found that forensic calculations could compensate for the imbalances of power and information they suffered. They could use calculation to estimate numerical information that they could not get their hands on, to extrapolate bigger conclusions from the limited sources of data that Crown agents would provide, and to cross-check different sources of information against each other. Through the work of such critical calculators, arguing about numbers soon became a familiar activity in the post-1688 Parliament. The efforts of such Country calculators highlight a seminal fact about Britain's quantitative age: it was born of politics, but it was not born of "the state" per se. At the end of the seventeenth century, the demand for numbers in political life did not come from the most powerful executives of the English state—key ministers and agents of the Crown—seeking to gather information on the English nation. It came from relative political outsiders—Parliamentary auditors and Country critics—seeking to gather information about the state itself.

The informational challenges that frustrated the Accounts Commissioners reverberated beyond Parliament proper. Public accounting controversies became the subject of conversations in coffeehouses and pubs. They also became the subject of new printed texts, which proliferated after the end of pre-publication censorship in 1695. Deciphering public finance data became a matter of collective political concern. One thinker in particular, Charles Davenant, seized upon that problem as inspiration for developing new techniques of economic analysis and a new approach to political calculation. Davenant characterized his method as "political arithmetick," borrowing the label coined by William Petty, who had died in 1687. Davenant adopted Petty's terminology but substantially changed what political arithmetic meant, transforming Petty's grandiose "programme" in social engineering into a functional political "practice" suited for the post-1688 era.[8] Davenant shared much with Country-minded Public Accounts Commissioners like Harley: a deep fear of ministerial corruption; an intolerance for administrative ignorance; and, most critically, the challenge of trying to craft computational arguments on the basis of limited data. In a series of writings in the second half of the 1690s, he created a new political arithmetic around these shared frustrations. Davenant took the political project of the Accounts Commissioners and codified it into an analytical method.

Davenant's reform of political arithmetic required reimagining the political value of numerical knowledge, but also the very nature of numerical knowledge itself. By necessity, he found himself—like the Accounts Commissioners—advancing uncertain arguments on the basis of imperfect data. Davenant's opponents, usually defending the "Court" position, castigated him for making unsubstantiated claims without sufficient authority or evidence. Critics not only dismissed his specific arguments, but his quantitative method as well. Political arithmetic was not capable of creating certain knowledge, they claimed. Some contended that the numbers Davenant deployed did not meet the standards of proper mercantile accounts; others argued that arithmetical calculations were no substitute for reason. Those critiques make it strikingly clear that the quantitative age was still in its infancy in the 1690s.[9] Davenant could not take it for granted that his contemporaries would accept, let alone welcome, numerical arguments. Instead, he had to confront this skepticism by thinking in new ways about what made a numerical assertion good enough to believe. Political pressure drove him to wrestle with questions about probability.

In the antagonistic and suspicious political culture of the 1690s, epistemology became a pressing public concern. Recently, historians have shown how disputes over credibility and trust, "public knowledge and private interest," lay at the very heart of politics in the late seventeenth century.[10] Nowhere was this more true than when it came to the problem of finding the money—whether tracking down the recipients of William Jephson's Secret Service payments, determining the current state of the nation's revenues and expenditures, or assessing the best future tax strategy. In the fraught realm of financial politics, certain knowledge was nearly impossible to come by. Instead of certainty, Britons got conflict—endless fights over and about numbers. But it was out of these fights that the nation's quantitative age would emerge.

"To Enquire into the Revenue Is Your Best Method"

The Revolution of 1688 gave Parliament unprecedented power, particularly over the nation's finances. In doing so, it created a massive crisis of communal knowledge for Parliament's members. As a result of the constitutional settlement reached with the new monarchs, the nation's MPs assumed considerable control over an entrenched, opaque, and largely archaic fiscal system—a system which, as a collective body, Parliament had basically no capacity to understand. This first became clear shortly after the Revolution, when MPs tried to decide what revenue ought to be granted to their new monarchs. Parliamentarians were forced to confront how little they knew about taxation and government expenditures, and how hard that knowledge was to acquire. What information did exist lay not with elected representatives but with a small handful of privileged individuals, whose loyalty and integrity could not be taken for granted. MPs had to come up with their own mechanisms for thinking through public financial problems. In doing so, they initiated a substantial reordering of the nation's civic epistemology, which would ultimately give special authority to a new mode of political reasoning: numerical calculation.

Public finance was not, strictly speaking, a new concern in Parliamentary politics. For centuries, Parliament had technically claimed the exclusive right to raise revenues and determine how much money was granted to the monarchy. Often this right had amounted to little more than a rubber stamp, though. Rarely did Parliamentary representatives scrutinize the public revenues in much technical detail, and financial numbers were hardly a regular part of Parliamentary politics. This began to change during the tumultuous seventeenth century, as MPs made efforts to take more direct control of the nation's money. During the English Civil Wars, Parliamentary representatives moved to make public revenues more transparent in order to cultivate the "publike faith," and even began trying to analyze how much money the government actually needed on an annual basis. Those analyses had helped to provide a benchmark for the revenue granted to Charles II upon his restoration as king in 1660. In that next decade, MPs showed even further informational initiative. In 1667–1668, opponents of the king's leading minister, the Earl of Clarendon, established a novel Parliamentary body, the so-called Brooke House Commission, to audit the accounts of the public finances. Their goal was to identify administrative mistakes and malfeasance by Clarendon's ministry, which critics believed had led to costly mistakes in the Anglo-Dutch war.[11]

The mid-century efforts at improving fiscal accountability and making Parliament better informed offered important precedents for what would come after 1688. In the near term, though, they amounted to something of a false start. The Brooke House experiment in public auditing was quickly stymied, as Charles II came to his ministers' defense. Though MPs had begun to analyze public revenues and expenditures during the Civil Wars, such analyzes effectively stopped after 1660. As had long been customary, Parliament voted to grant both Charles II and his successor James II substantial tax revenues that lasted for their entire lifetime. This meant that MPs only had to address the public revenues comprehensively once, at the beginning of each reign, aside from emergency funds needed for war. The funds granted to both Charles II and James II were quite generous and only became more so over time. English trade was booming during much of that period, meaning that customs duties, a main source of tax revenue, were booming as well. The Crown's "ordinary" revenues (revenues not raised for war) climbed from £1.2 million in 1670 to £1.9 million when James II took the throne in 1685.[12] Consequently, the kings were relatively flush with funds and had little need to appeal to Parliament for more. Throughout 1660–1688, Parliamentary sessions were often brief and became increasingly rare, sitting only once between 1681 and 1689 (in 1685, when James II succeeded to the throne and was granted generous lifetime funds). All told, this meant that by the 1680s, the nation's Parliamentary representatives had only the most limited experience dealing with matters of public finance.

This lack of collective financial knowledge was compounded by the decentralized, proprietary way public finances were administered. Many fiscal functions, like tax collection, public borrowing, and military payment and supply, were managed by individual "undertakers" with limited central coordination. Much of the data that was collected about the public finances was thus held by these undertakers, who treated it as effectively proprietary.[13] Administrative procedures had improved in some areas of government in the preceding decades, most notably the customs and excise, which had previously been "farmed" out to private undertakers but were placed under centralized management in 1671 and 1683, respectively.[14] Such reforms helped expand the government's ability to raise money, but they did not immediately translate into greater transparency or a better-informed legislature; a more powerful "military-fiscal" state was not necessarily easier to account for.[15] Few MPs at the time were skilled, or interested, in the complexities of financial calculation. Instead, a small number of individuals functioned as stewards of financial information for their Parliamentary peers. Foremost among these was Sir Robert Howard, who had held simply to renew James II's ordinary revenues permanently. But this put the nation in "danger of falling into the misfortunes of the last two governments." More thoughtful solutions required knowing what existing revenues actually were. "I find there will be occasion of discourses, what the Revenue is?" suggested merchant MP William Love. "Therefore I move, that *Howard* may give you the Revenue. When all is before you, you may consent to such a Revenue as may make the king great to all the world." Spirited calls for financial information came from Edward Seymour, a key Tory supporter of William III: "If we settle the Revenue, I would enquire into it; if you know not the value of what is given, you cannot do it effectually. . . . To enquire into the Revenue is your best method."[19]

Seymour's cries did not go entirely unanswered. Robert Howard offered a preliminary "Computation of the several Branches of the Revenue" on March 1, 1689. He followed up with further detail on expenditures and the excise later in the month. The MPs remained largely at Howard's mercy at first, as he "delivered in such Accompts and Estimates, as he conceived were necessary for the Service of the House." In time, though, MPs began to make more frequent and pointed numerical demands. Some MPs may have used such requests as a political tactic, designed to stall proceedings. Others, like the Whig Henry Capel, earnestly championed the need for better information and criticized those who were willing to make such major decisions on informational faith. Capel cautioned that Parliament still knew little about major areas of the public finances, like the customs and excise. What is more, "there are complaints of the ill administration of them," Capel urged, and "now is the time to correct it. . . .'Tis not a time to name the Revenue now."[20]

This message of caution was heeded. In mid-March 1689, the House of Commons passed a stop-gap bill continuing current revenue collection until June. March 20 featured a spirited debate in the Commons about what level of "ordinary" revenue the Crown actually needed. Technical questions about financial data became contentious political issues. MPs found themselves fighting over where to draw the line between ordinary and military expenditures, and about how to discern a fair baseline level for revenues granted to the previous two kings using available historical data. The veteran Whig William Sacheverell argued that expansions in regular military spending and pensions made it necessary to go all the way to the late 1660s to establish a fair budgetary benchmark. The moderate William Garway suggested that it was perhaps better not even to try to project from old figures. "We cannot say the Revenue will stand to 1000, or 10,000*l.*," he said, "there being several variations of the Customs in several Years." These methodological musings were rudimentary at first, but it would not be the last time MPs had to think critically about how to make deductions from numerical data.[21]

The House of Commons ultimately fixed upon a peacetime revenue figure of £1.2 million, the amount granted to Charles II in 1660. Over the course of April 1689, the House turned to the question of government expenditure—but only after they had settled upon a figure for revenue. After sorting through different governmental needs, the Commons agreed that ordinary expenses amounted to more than £1.7 million, well beyond the £1.2 million level already set for income. This prompted the House to revisit the revenue question (a move that did not please King William). It was not until May 1690 that Parliament settled upon a reasonably stable plan, granting the monarchs certain excise taxes for life and existing customs revenues for a term of four years.[22] These anxious revenue debates initiated a new era of legislative concern with financial matters. As England became entangled in a new war with France (the Nine Years War, 1688–1697), such "supply" debates became even more common, as the Crown was in constant need of new money to fund the military effort. Parliament was not about to surrender the newfound command it was gaining over financial information. In the money debates of March 1690, Paul Foley offered a succinct formulation of what had become a powerful mantra among many MPs: "I would give this King Money, but not by a Rule, because we have given other Kings. I stand upon it, to have more reason from the accounts before we give Supply."[23] This principle soon took institutional form in a new Commission of Public Accounts. Foley was one of its first members.

Misadventures in Public Accounting

The debate over how much to pay William and Mary prompted new expectations about what Parliament ought to know, and deserved to know, about the government's finances. Some MPs demanded that the Commons have its own infrastructure for managing information. This ideal took form in a new Commission of Public Accounts, a body of MPs that sat from 1691 to 1697 and was renewed from 1702–1704 and 1711–1714. The Commission's goal was to create a comprehensive Parliamentary account of all public funds, built from the ground up. This ended up requiring more than just asking people to open their books and then doing unproblematic arithmetic. The "public accounts" were something they had to make, through physical effort, computational ingenuity, and political imagination.

It was a military matter that first inspired calls for Parliament to take the public accounts into its own collective hands. The new war with France had barely begun when concerns arose about impropriety in its management. In October 1689, MPs began to look angrily upon agencies like the Victualling Office, the Admiralty Board, and the Army provisioners for Ireland. Waste-conscious Tories like Edward Seymour and Christopher Musgrave found common cause with Whigs who feared such offices continued to be infested by James II's "placemen." The Commons appointed two select committees to investigate. This pilot project fueled further enthusiasm for investigating public funds, as MPs grew frustrated with the limited account information the new king had conceded. In late May 1690, William proactively proposed to establish a royally appointed accounts commission, though jealous Whigs in Parliament blocked its creation. The king promised to provide greater access to government accounts, but this failed to satisfy zealous MPs who wanted their own autonomous accounting agency. In late December, the Commons and the Lords ratified a plan to create a Commission of Public Accounts composed of nine members of the House of Commons. (The Lords maneuvered to add members of their own, but were stymied by the Commons.)[24] Driven by distrust and dissatisfaction with Crown agents, the Commission's primary objective was increasing Parliamentary oversight over government functions. But advocates also promised that the increased transparency fostered by the Commission would make it easier for Parliament to raise revenues. William III ultimately acquiesced, in part to hasten Parliament's granting of military funds.[25]

The Commission featured members of varied political leanings, and it was not immune from partisan tensions at first. Commissioner Robert Harley reported that "some hot words did pass between some" at the first meeting in early March 1691. But soon the Commission's Whigs, like leading vote-getter Sir Robert Rich, found common cause with their Tory colleagues, like the Oxford clergyman Thomas Clarges. As Harley reported less than two weeks in, "business now has taken off the edge of our passions and we are fallen to it with calmness and diligence."[26] Their first order of business was to request information. The Commissioners did so through a flurry of written "precepts" sent out to government agents, many of which contained dozens of specific requests. The Commissioners hired a considerable staff, including a dedicated messenger, a secretary (with his own assistant), and two bookkeepers. They also established a numerical system for tracking the data requests they made (over 300 by late 1691). Their targets ranged from the most elevated executives to minor clerks and contractors. Some of their first precepts, issued March 11, went to: three top excise officials; the Excise Commissioners; the Paymasters General to the Army in England and Ireland; and the Secretary of War. By May 8, the Commissioners were sending precepts numbered 158–162 to: a major general in the army; a Bristol merchant; a Covent Garden linen-draper who had a military coat contract; the "Avener" of the Crown stables, who was responsible for feeding the king's horses; and several naval victualers. The Commissioners often pursued a microscopic level of detail. Sir Peter Colleton, for example, took a particular interest in the minutiae of military spending, carefully examining the efficiency of coat tailoring, the pricing of footwear, and the supply of hay for horses.[27]

The Commission's early procedures betrayed a remarkable level of confidence, both in the Commission's authority and in the competence and willingness of myriad government agents. Precepts were assigned ambitious deadlines, and "accountants" were called to testify to their numbers on short notice. But this early confidence soon came to appear naive. Rarely did government agents reply to the Commissioners without incident. Some individuals simply refused to deal with the Commission. William Bridges, a Commissioner of Provisions for Ireland, provided the most dramatic reaction. On May 6, a messenger for the Commission named Arthur Nicholson delivered a precept to Bridges. The latter angrily claimed that there were others "more principally concern'd than himself" in the matter, then threw the paper in Nicholson's face. When the messenger remarked how he had never before been treated this way by a gentleman, Bridges "pluckt out a hand full of money" and dismissively asked Nicholson whether he owed any fees. Expert data-mongers like Robert Howard preferred subtler forms of misdirection. A favored tactic involved sending lower officials to testify before the Commission, so they might plausibly deny knowledge of the requested information. When the Commission made a request for information on fees taken by Exchequer staff, for example, Howard and his underlings adeptly dodged the question for weeks.[28]

This kind of obfuscation was not the only disappointment the Commissioners encountered. Much of the information they expected to find was not where they assumed it to be, was in an obscure or unusable form, or sometimes was not recorded at all. Some Crown agents simply lacked the competence or desire to keep the kind of records the Commission expected. Taking the prize for the most disorderly (and probably dishonest) accounts was Richard Jones, 1st Earl of Ranelagh, who had served as Paymaster of the Forces since 1685. Ranelagh's records were so bad that the Commissioners had to report his negligence to the House of Commons only a couple of months into their work. The earl tried various tactics to put off the Commissioners, including claiming that he needed extra time because he wanted to provide far more information than they had actually requested. Ranelagh's embarrassing accounts would continue to be a source of frustration for public accountants, and of fodder for political gossip, for over a decade. (The matter eventually caught up with Ranelagh in 1703, though, when he was expelled from Parliament and his position as paymaster. Some £900,000 remained unaccounted for.)[29]

In many cases, the Commissioners asked for data from the wrong people—or at least people who claimed to be the wrong people. On May 2, for example, the Commission hosted William Jephson to inquire about back taxes in Ireland. Jephson rerouted them to the Treasury clerk William Lowndes, who informed them that he knew only of one particular tax backlog in Ireland, and told them to consult Robert Howard for more complete documentation. This kind of runaround was common.[30] In other cases, the Commissioners found that significant sums of public money moved around through back channels with limited documentation, including a massive £20,000 payment to a "Mr. Vanderesch," muster-master of the Dutch forces, for which no authorization could be found beyond a supposed Dutch royal seal.[31] Most frustrating, perhaps, were the cases where data the Commissioners expected to find had never been recorded, or had been recorded but not preserved. The Commissioners were disheartened to discover that the Treasury did in fact make up weekly accounts for many government agencies to assist with managing Crown cash flows, but that, after that immediate purpose was fulfilled, the Treasury "looke upon those acc[ts]. as wast[e] papers, and keep's them not." The Treasury officials regularly threw away the very kind of numbers the Commissioners were working so hard to compile![32]

For the Commissioners and many of their skeptical colleagues in Parliament, the reason for all these data disappointments was clear: Crown agents were intentionally obstructing the Commission's efforts to hide their corrupt activities. But it was not always obvious when data was being deliberately withheld, and when it simply did not exist for less nefarious, structural reasons. The riddle of William Jephson's Secret Service accounts was indicative. On June 10—after Jephson's death—the Commission sent a request to Jephson's clerk, Robert Squib, to bring in copies of various records Jephson had purportedly gathered for the Commission prior to his death. Two days later, Squib delivered an account of

Secret Service payments made by Jephson, but it only contained data on payments to MPs. The Commissioners were hardly satisfied by that answer, or by a second report produced by Squib, who pleaded that the king had instructed Jephson only to disclose information on payments to MPs, and no other state secrets.[33]

Squib suggested further details about the Secret Service money could probably be found from the Lords of the Treasury. The Accounts Commissioners had developed a very strained relationship with the Treasury Lords, though, and on the matter of Secret Service, they proved even less helpful than Squib. The Lords explained that they only approved the allocation of funds from the Exchequer to Jephson for Secret Service use. Thereafter, the monarchs "privately entrusted him [Jephson] in that affair without our knowing or being any way privy to the particular uses to which or the particular persons to whom the same were applied." That is, once Jephson got the money, he was the only one, save the king and queen, who knew where it went. "It is not in our power to send you such an account as you desire," concluded the Treasury Lords. The Treasury Lords' claims of ignorance were probably somewhat exaggerated, as at least one member of that group, Chancellor of the Exchequer Richard Hampden, seems to have had some access to Jephson's ledgers. Whether because of lack of knowledge or intentional misdirection, though, the Treasury Lords offered no further information. In late November, the Commissioners received a formal letter from Squib, on the order of the Treasury Lords, admitting that "there were other payments made by Mr. Jephson to several Members of Parliament," but explaining that "the secrecy of them being so absolutely necessary, his Majesty does believe that the House will not desire the Particulars of them." That was as far as the Accounts Commissioners managed to get in deciphering Jephson's secrets.[34]

With such disappointments, though, came a measure of education. The Commissioners gradually learned how to navigate the idiosyncrasies of England's various governmental bodies, and to put those disparate pieces together into some kind of coherent whole. Some of their interlocutors proved quite willing to help the Commissioners learn, or at least to explain why the information they gathered did not look like the Commissioners imagined it would. For example, the First Lord of the Admiralty, Lord Falkland, offered the Commissioners lessons in the vagaries of naval finances, explaining about how payments could be made through any of three different mechanisms (by "bill," "book," or "ticket"). No one did more to tutor the Commissioners than the Treasury clerk William Lowndes, who gave the Commissioners a series of impromptu lessons on topics ranging from the basics of Exchequer procedure to the history of the "fictitious loan" as a fundraising mechanism.[35]

Yet such lessons only did so much to close the gap between the kind of clear and certain information the Commissioners expected to find, and the muddle they actually found. In late November 1691, the Commons requested that the Commission submit a first "State of the Income & Issues of ye Publick Revenue," which the Commission provided to both Houses shortly thereafter. That report contained a lengthy, two-part table, summarizing government receipts on the left and expenditures on the right. This loose debtor-creditor structure tried to bring order to the informational bedlam the Commission had encountered, but in doing so it framed the imperfections in the Commission's efforts. The Commissioners had to explain that "this Account is as exact as we have been able to make," and provided several "reasons why it is not more perfect." Notably, the account did not quite balance; the Commission found £18,018,586 in receipts but £18,020,403 in payments. The difference of just over £1,800 was modest, and the Commissioners explained that it was a minor issue, stemming from the vagaries of short-term borrowing and lending. But it was meaningful that a difference existed at all. The numerical gap between debits and credits mirrored the "expectations gap" between the informational ideals of the Accounts Commissioners and the informational realities of the nascent English state.[36]

The Commission's first report detailed the many frustrations the Commissioners had encountered in their day-to-day investigations. One such problem, of course, was poor bookkeeping by government agents. The report complained that the "miscasting of accounts . . . has given the Commissioners a great deal of trouble, many of the accounts brought before them being greatly faulty." The report also spoke to more systemic problems. It frankly admitted that the network of individuals responsible for public funds was "so voluminous, that it will require much more time to examine and state" them than the Commission had been allowed. As the Commissioners reported, the government still relied heavily on single individuals for managing the flow of public funds. At the close of the Commission's account, almost £120,000 of public money lay "in the hands" of individual agents. Thomas Fox, Receiver of the Customs, held over £75,000, more than lay in the Exchequer (about £63,000). The Commissioners worried, for example, about the possible "double-charging" of certain revenues because of loans Fox had advanced on customs receipts. Such borrowed money was especially hard for the Commissioners to follow. The growing use of more sophisticated financial devices for public borrowing, like the fictitious loans Lowndes had tried to explain, further hindered the Commissioners' project.[37]

The Commissioners had hoped to find a government where there was documentation for every expense, where every pound could be vouched for, and where the accounts ultimately balanced. They did not, and their incomplete first report reflected that. That report, though, was not "a document of amazing ignorance," as one early twentieth-century historian has described it.[38] Rather, it was an aspirational document. By attempting to reconcile England's disorderly fiscal system into a single account, the Commissioners created a numerical image of what they wanted their government to be.

"The Accounts Are Amazing Things"

Not everyone saw the value in the Commissioners' zealous, intrusive, and imperfect project in public accounting. When a bill was proposed in 1693 to renew the Commission of Public Accounts, the sentiments in Parliament were mixed. Some complaints were predictable: Robert Howard claimed of the Commission's reports that "there was nothing in them but what any clerk of their office should be bound to present you with, but there are several mistakes." Others were more telling. Edward Seymour, among the loudest proponents of Parliament's inquiry into the Crown's revenues in 1689, claimed of the Accounts Commission that "he did not see any good was done by it answerable to what it costs you, which is almost £10,000 per annum." Many later administrative historians have echoed Seymour's ambivalence, contending the Commission was primarily a political gambit that did little to make the English state more professional, honest, or accountable in the long run. Yet the Commission's early exercises in public accounting had a far greater impact on British political practice and culture than scholars have realized.[39]

For one, the Commission's tenacious efforts did in fact help reshape how the English state treated and handled information. The clearest testament to this comes from late 1691 or 1692, when the Treasury began to keep a permanent, consolidated set of annual revenue and expenditure accounts (the T 30 series in the National Archives at Kew). In other words, the Treasury began to compile its own version of the account that the Commissioners of Public Accounts had worked so hard to put together, a kind of account that does not seem to have existed prior to that point. The timing—the new accounts began in the wake of the Accounts Commission's first audit—suggests that Parliamentary pressure was likely a critical factor behind the Treasury's new practice. (Notably, even after the Treasury accounts began to be kept in this consolidated form, the Accounts Commission continued to calculate its own autonomous version.) P. G. M. Dickson argues the consolidated Treasury accounts constituted a level of numerical organization and formal accountability unprecedented not only in England, but anywhere at the time. Those accounts offer an excellent example of what Theodore Porter has identified as a central feature of modern quantitative culture: the drive to implement rigorous, quantitative procedures frequently does not come from within bureaucratic institutions, but rather arises as a defense against outsiders.[40]

Even more dramatic, though, was the impact that the Commission of Public Accounts had on the nation's civic epistemology. The agitation of the Accounts Commissioners, and the recurring debates about public accounts they fueled, began to make numbers a durable part of political life in England. The 1690s were not strictly the first time that English politicians had fought over financial numbers. From 1679 to 1681, for example, Robert Howard had picked a numerical fight with the Earl of Danby over the debts Danby had supposedly racked up as Lord Treasurer.[41] Thanks to the dogged efforts of the Accounts Commissioners, though, such numerical disputes went from being an occasional event into a regular and essential feature of the nation's political proceedings. By harassing state agents to reveal the secrets of their books, the Commissioners brought numbers into the open—sometimes into existence—that had never been subjected to political scrutiny, or even attracted much political notice of any kind. They also learned that arithmetical calculation could be a powerful tool, both for political analysis and political maneuvering. The armature of accounting organized disparate information and isolated what potentially valuable data were unavailable, shining light on those monarchical agents who could not, or would not, reveal information to the public.

Energizing this enthusiasm for numbers was a particular political ethos: the "Country" spirit, part of a long tradition of republican political thinking that had a profound influence on political culture in Britain and the Atlantic world.[42] Accounts Commissioners Robert Harley and Paul Foley were leading exponents of Country sentiments in the post-1688 Parliament. Country MPs contraposed themselves to the monarchical "Court": that multitude of ministers, MPs, and retainers who gave their loyalty, and owed their power, to the Crown. Chief objectives of the Country cause included purging the government of corruption and curbing the power of Crown agents. The Country interest became a vibrant force in post-1688 politics, energized by Parliament's expanded constitutional powers and by an acute fear of Court influence following the disastrous reign of James II. In fact, as J. A. Downie has shown, the vexations faced by the Accounts Commissioners, and the startling levels of Court malfeasance they revealed, were critical in galvanizing a discernible Country faction in Parliament.[43]

Country MPs in the 1690s, like Harley and Foley, drew upon a tradition of English republican thinking that had been building throughout the seventeenth century. The Commonwealth period (c. 1649–1660) gave rise to a flourishing of republican ideas, led by thinkers like John Milton, Marchamont Nedham, and James Harrington. After the Restoration, republicanism would be nurtured and extended in various directions by thinkers ranging from the radical Algernon Sidney to the 1st Earl of Shaftesbury, a leading voice of the early Whig party. Over the long eighteenth century, republican ideas were developed by influential Country voices like John Trenchard and Thomas Gordon, authors of the influential *Cato's Letters* (1720–1723), and Henry St. John, Lord Bolingbroke, a key ally of Robert Harley in the early 1700s and later a leading theorist of the Country opposition to Robert Walpole in the 1720s and 1730s. Such eighteenth-century Country writers would form a crucial source for political thinkers in early America. Republicanism was a varied and evolving way of thinkers, which drew on an eclectic set of sources (ancient texts, Renaissance humanism, the moral imperatives of radical Protestantism) and traveled under many names (commonwealth, Country, "Real Whig").[44]

Despite its variety and continued transformation, republican thinking was organized by certain persistent themes, which were clearly in evidence in the Country tradition that took hold after 1688. At its heart lay a deep suspicion of corrupted and "arbitrary" political power. Republican thinkers thus held that the common welfare of the commonwealth was best secured through a mixed, balanced, and representative form of government. They were also keenly aware of the moral dimensions of political life, seeing the stability of the commonwealth as dependent on preserving civic virtue—that is, a commitment on the part of citizens and especially political leaders to put public good above private interest.[45] Republicans believed that government itself had to be vigorously watched, lest any of its components become too powerful or corruption be allowed to take root. For English republicans, including the Country partisans of the 1690s, the greatest political danger was therefore what J. G. A. Pocock has termed the "hydra-headed monster called Court Influence or Ministerial Corruption," whose heads included the venal "placemen," tax collectors, military officials, and government undertakers with whom the Public Accounts Commissioners tangled.[46]

This republican sensibility was vital to the growth of calculative thinking in the earliest years of Britain's quantitative age; conversely, calculation was an important component of republican thinking in the long eighteenth century, which has gone overlooked by scholars. Members of the Country interest were usually, almost by definition, political outsiders without positions in executive government. There was consequently a consistent asymmetry of information between Country and Court. It was Country partisans, who did not have regular access to information about the government and its finances, who came to value that information most highly. Emboldened by the frustrations of the Commission of Public Accounts, Country politicians like Robert Harley turned to arithmetical calculation as a tool for combating these informational asymmetries. Debating numbers offered MPs a new way to exert authority on matters of public finance, empowering the Country interest to challenge the Courtly institutions that traditionally wielded control. This was evident in the first Commons debates over Secret Service money on December 3, 1691. Frustrated by the failure to unearth Jephson's secrets, the MP Sir John Thompson expressed an emerging Country attitude about government information. "The Accounts are amazing things," declared Thompson. "We were told in the last session, 'Country gentlemen understood not Accounts,' and now, it seems, the Commissioners of the Treasury do not. If they understand not Secret-Service, then they are not fit for their places."[47] For too long, the nation had been beholden to Court insiders like the Treasury Commissioners and Robert Howard for information about their own government. But now, armed with numerical accounts of the public money, and with new computational skills, even "Country gentlemen" like Thompson came to believe they could scrutinize the operations of Court power.

Parliamentarians mobilized their growing accounting acumen to seek greater input into governmental decision-making. Robert Harley contended that the diligent efforts of the Parliamentary Accounts Commissioners to bring order to the public finances were evidence for why Parliament was a better steward of the nation's resources than the Crown and its Court agents. In fact, Harley even suggested that the Commission of Public Accounts should have greater authority for managing, and not just auditing, public money. In early 1693, Harley wrote to his father that he felt the Commission ought to be given control over the process of raising a new public loan. Harley suggested that the Treasury, which had "us[e]d ye Com^{rs}. of accounts so unk[i]ndly," ought to allow the Commissioners to "shew their skil[l]s" in managing such public funding transactions. After all, it was the Commission "who have found out all ye mon[e]y hither to."[48] It is unclear exactly what Harley meant by "finding out the money" in that sparse letter. Was he applauding the Commissioners' analytical "skills" in calculating the nation's fiscal condition? Or their administrative success in improving revenues? Or their ethical achievements in rooting out corruption? Perhaps all of the above. In the 1690s, finding the money was at once an analytical, administrative, and ethical challenge.

Harley's vision of a new system of financial administration run by the Accounts Commissioners was overly ambitious, but Parliament did find various ways to assert greater influence over government finance and government information. One key tactic was to demand new kinds of data from Court agents, and then to interrogate that data computationally. For example, MPs began to request forward estimates of military expenditures. Previously, executive agencies had rarely projected future monetary needs. As William Lowndes would later note of military budgeting for the years 1688–1689, "estimates and appropriations for the years were few and imperfect." Parliamentary pressure, though, forced government agents to do so beginning in the early 1690s. The MPs soon began to question and critique the numerical projections put before them, and to fashion their own calculations assessing military expenditures.[49] In November 1693, for example, the Commission of Public Accounts produced an account of Parliamentary grants for the Navy, with an analysis of what funds had ultimately gone to naval use. They computed that, of just under £6.6 million granted for naval support since 1689, only £4.6 million could be shown to have been spent on naval use—implying there was some £2 million of naval appropriations unaccounted for. This kind of comparative audit quickly became one of the most common styles of numerical argument in British politics: some ideal numerical benchmark (in this case, naval funds appropriated) was lined up against data on actual government performance (naval funds spent) in order to demonstrate a governmental shortcoming in numerical terms. Such calculations were especially appealing to Country partisans fixated on fraud and mismanagement.[50]

The agents of the Court rarely let such numerical critiques go unanswered, though. In the debate over naval funding that began in 1693, the Treasury Lords responded with their own computations to defend against the implication of mismanagement made by the Accounts Commissioners. As the two sides squared off, they came up against tricky, technical questions. In the midst of the naval appropriations debate, the Accounts Commissioners and the Treasury had an intensive debate about how to appropriately account for £98,000 in payments that the Treasury had claimed to make in 1690–1691 for the purchase of saltpeter, a key ingredient in the making of gunpowder. The debate came down to a question of when appropriated funds could officially be said to have been spent. While the Treasury Lords claimed that they technically spent the funds when they bought the saltpeter, the Accounts Commissioners contended that the Treasury Lords could not lawfully claim to have spent those funds until the saltpeter had actually been converted into gunpowder and put into naval use. What was effectively an arcane, methodological question about inventory accounting became a matter of political controversy.[51]

This became one of the defining features of the era's new computational politics: abstract questions about calculating procedures, even about the epistemology of calculation, became heated topics of political conversation, including within Parliament.[52] The spotty data and obscure administrative procedures that frustrated the Accounts Commissioners also created space for computational creativity—and computational controversy. Very often, neither the Court nor its critics could offer a definitive account on a particular question, and so both sides had to resort to arithmetical speculation. This was evident in another 1693 dispute between the Accounts Commission and Robert Howard. Early in 1693, the Commission had offered an analysis of the funds available to the government to pay for war needs over the next three to four years. They found the government had roughly £1.3 million that could be expected "to come in in time toward defraying the Charge of the Warr," net of loans—the implication being that the government had considerable resources at hand and ought not to ask for quite so much from Parliament. On February 16, Robert Howard pushed back against the notion that the government was quite so flush with cash, arguing that the Commission had made over £1.1 million of "Errors in their Acco^{t}." In fact, the government only had £172,000 at its disposal. Howard particularly claimed that the Commissioners had grossly overestimated the likely revenues from certain special taxes, such as customs on sugar. The Commissioners had calculated their estimates on the basis of the most recent year's returns, but an irregularity in the timing of voyages had meant that an exceptionally large number of ships had returned to England from the West Indies that year. Howard proposed a creative alternative to smooth out these irregularities: to project future sugar customs based on a four-year historical average instead. By that model, expected sugar duties would be £282,000 below the Commission's prediction. The pressures of Court-Country conflict encouraged both the Parliamentary public accountants and Court agents like Howard to craft inventive new calculations to analyze public money.[53]

These early numerical showdowns in Parliament marked the beginning of an energetic culture of computational combat in English politics. Arguing about accounts became an increasingly frequent occurrence in the 1690s. More frequent Parliaments provided one venue; the pamphlet press provided another, especially after the end of pre-publication censorship regulation in 1695. The conflicts surrounding the Revolutionary financial settlement had educated a number of MPs in the rudiments of public finance and the political power of calculation, while the powerful republican sentiments of the era drove representatives to audit Court power. Over the 1690s, Country MPs turned the tools of public accounting aggressively against insiders whom they feared had betrayed the public trust for personal gain. In 1694–1695, for example, MPs carried out a series of investigations into possible embezzlement of subsistence money for military troops, corruption in licensing of hackney-coaches, and bribery regarding legislation for the East India Company (EIC). The investigations of the latter issue, which led to the sacking of Speaker of the House John Trevor and the elevation of Accounts Commissioner Paul Foley in his place, involved detailed forensic investigations into the account books of key figures like EIC governor Thomas Cooke.[54]

Public accounting also began to capture political attention beyond Westminster. Details of the 1694–1695 investigations into "briberies and corrupt practices" were publicized in two pamphlets, containing extensive accounting data showing numerical evidence of the transgressions. A few years later, it was the chronically terrible accounts of Lord Ranelagh that became a prominent topic of political conversation. The attention given to the fiasco of Ranelagh's accounts is vividly documented in the remarkable journal of James Brydges, future Duke of Chandos (1674–1744), who used public accounting as his own entry point into power politics. In March 1698, Brydges worked with Harley, Foley, and others on drafting a new bill to renew the Commission of Public Accounts. He was elected to Parliament for Hereford later that year, and, in early 1700, he spearheaded a Parliamentary investigation into Ranelagh's management of military funds. As reported in his journal, Brydges spent a great deal of time talking and strategizing about that accounting affair. On January 13, 1700, Brydges and three others met at "ye goat in Bloomsbury Square,"—a public house, presumably—"where we staid till toward 11 concerting about L^d. Ranelagh's accounts." On January 26, his friend Godfrey Copley and his cousin William

Brydges "staid talking with me about Ld. Ran: accounts &c. till towards 12." Brydges no doubt had an unusual level of interest in public accounting, not least because it helped him build his own political resume; one of the first major advances in his career was appointment to the Commission of Public Accounts in 1702. Yet the fact remains: at the turn of the eighteenth century, irregularities in government accounting was a topic that kept enterprising English politicians up until midnight talking in pubs.[55]

From Public Accounting to Political Arithmetic

As disputes about public accounts moved beyond the halls of Westminster, they also became entangled with more expansive public conversations about the nation's political economy.[56] For at least one pioneering economic thinker, Charles Davenant, the problem of finding the money became the inspiration for a novel approach to political-economic analysis—and a new vision of the place of calculation in political life. Driven by the same Country political values that had motivated Harley and the Accounts Commissioners, Davenant refashioned William Petty's ambitious, but outdated, project of political arithmetic into something far better suited to the contested politics of his era. Though Davenant was less renowned than his predecessor Petty, it was Davenant who would create the more enduring style of political arithmetic—what he called "the Art of Reasoning, by Figures, upon Things relating to Government."[57]

Charles Davenant was born in London in November 1656, the son of William D'Avenant, a renowned playwright. He studied at Oxford and Cambridge, receiving an LL.D. from the latter in 1675; hence he would regularly be styled as "Dr." Davenant. He took an early interest in his father's theater business, and briefly tried his own hand at drama with his *Circe: A Tragedy* in 1677. His next venture proved more lasting. In 1678, he took a post as Commissioner of the Excise, the office that assessed taxes on beer and ale (and which was rapidly becoming one of the most sophisticated branches of England's nascent fiscal state). As Miles Ogborn has reconstructed, Davenant was deeply immersed in the reform of the excise administration after the end of excise "farming" in 1683. He made regular circuits through towns in southern and western England, visiting local excise officers to ensure they were acting ethically and following appropriate technical protocols, such as the "gauging" of beer barrels. His excise experience would significantly shape his later thinking on finance, administration, and ethics. In 1685, Davenant first became an MP, elected for St. Ives in Cornwall.[58] After 1688, though, Davenant found himself out of Parliament and out of work. Having served under James II, he was tarred as a Stuart "placeman" and had to fight rumors of disloyalty. He spent much of the decade searching for a stable job in public life. In 1692, the House of Lords, seeking to exert greater influence on the Commission of Public Accounts, proposed Davenant as a potential non-MP addition to the Commission, though their efforts were blocked by the Commons. Such professional disappointments became a recurring theme: between 1694 and 1696, he came close to, but ultimately failed to obtain, positions as surveyor-general of the salt duty, surveyor-general of the excise, and as a member of the Board of Trade (losing out to John Locke).[59] So he turned to writing on political and commercial matters to revive his career. From 1695 to 1698, Davenant wrote prolifically on taxation, trade, the coinage, national credit, and "public virtue," first making his name with his 1695 *Essay upon Ways and Means of Supplying the War*.

Davenant's political-economic writings were motivated by his personal, professional, and political interests. For much of the decade, for example, he was either in or trying to be in the pay of the East India Company, and thus crafted diligent arguments in print and manuscript defending the Company's special trading privileges in Asia.[60] His thinking was not entirely mercenary, though. His excise work gave him a firsthand look at the maladministration and corruption plaguing English government. That experience, alongside his personal animosity toward the Court Whigs who dominated government in the 1690s (and shut him out of office), made him a natural ally of the Country cause. In the late 1690s, he regularly associated with a circle of Country politicians that included James Brydges and Godfrey Copley. Brydges's journal for the year 1697 recalls frequent meetings with Davenant; much of their time was spent talking about how to get one or the other a job.[61] Davenant found that his writing skills were useful tools for advancing both political values he believed in and his own career. He quickly became one of the most influential spokespeople for the Country cause and a key contributor to the English republican tradition. By 1700, Davenant would be called upon by Robert Harley, whose political star was also rapidly rising, to serve as one of his chief publicists.[62]

At the heart of Davenant's political writings in the 1690s was a belief that governmental information was a shared public good and that knowledgeable administration was essential to good government. Like the Commissioners of Public Accounts and allied Country MPs, he was aghast at how little the nation's government seemed to know about its military and fiscal condition. In a 1696 manuscript *Essay on Publick Virtue*, Davenant argued that the short-sightedness and ignorance of the "young and giddy statesmen" at the nation's helm—the so-called "Junto" of Court Whigs who had assumed key ministerial positions over the 1690s—had left the nation's foreign trade, currency, and public debt in a miserable condition. A key part of the problem was that the nation's ill-informed leaders had never accurately assessed the nation's real economic resources—"the annuall Income ariseing from Land Manufactures and Trade"—and thus pursued expensive military strategies the nation could not afford. "They should have duely weighted the strength,—& Wealth, of this Kingdom in order to know how long, & upon what foot, We were able to carry on ye Warr," Davenant explained. He surmised that the ignorance of such courtly statesmen was partly a reflection of a deeper, unethical agenda. Courtly schemers did not want well-informed, transparent, accountable government; they "better like infinite expences, which produce Infinite Reckonings and at last no accompt."[63]

Davenant recognized that agents in England's decentralized administration did not always have incentives to keep good records. He admitted that one potential solution was to radically consolidate the management of state information under a powerful minister. This was what England's leading enemy, France, had appeared to do. In the introduction to his *Essay upon Ways and Means,* Davenant lauded French ministers like Cardinal Richelieu and Jean-Baptiste Colbert who "brought . . . perfection in the Public Revenue." Davenant did not want England to adopt the autocracy of France's "great machine of government," though.[64] England's government needed to become much better-informed while maintaining its mixed and balanced constitution. The solution lay in capitalizing on the remarkable informational and analytical capacities of Parliament itself, for example by establishing a new, thoroughly Parliamentary council of trade, a proposal Davenant laid out in another 1696 manuscript.[65]

As Davenant undertook his own analyses of the nation's finances in the mid-1690s, he witnessed just how far England's government fell short of the ideal he envisioned, coming face-to-face with the fragmented and territorial administrative system that had so frustrated the Commissioners of Public Accounts. Just like the Commissioners, he struggled mightily to acquire the data he felt was necessary to carry out his analyses. "The Writer of these Papers has met with extream Difficulty and Opposition," he wrote in his 1698 *Discourses on the Publick Revenues,* "in procuring the sight of the Accompts relating to the Revenue. . . . The Books of the Principal Offices have been in a manner shut up against any Inquiry he desir'd to make; and this has render'd his Work more imperfect."[66]

And just like the Commissioners again, Davenant found that the best way to work around these informational challenges was calculation—political arithmetic, as William Petty had labeled it.

In combining Petty's quantitative techniques with Country values, and bringing them to bear on the problem of finding the money, Davenant fundamentally transformed how political arithmetic was practiced and what it signified in English culture. In his superb biography, Ted McCormick has shown that Petty had originally imagined political arithmetic to be an instrument of social engineering focused on "the manipulation of populations," the ultimate goal being the "transmutation" of Ireland's population into an English one. This grandiose, alchemical vision went largely unrealized, having little impact on governmental practice in his own day. In order to succeed, Petty's calculated social engineering project required the backing of a powerful sovereign, something he never managed to attract. After Petty died, a subsequent generation of calculators led by Davenant would rework Petty's grand, ideological "programme" into a usable "practice." The reformation of political arithmetic in the 1690s reflected a changing political environment, "centred less and less on the shadowy comings and goings of a handful of courtiers and more and more on Parliamentary party politics."[67]

Petty's political arithmetic had been a courtly project designed to appeal to monarchical power; Davenant's would be a Country tool designed for Parliamentary debate and public dispute. Inspired by the values and experiences he shared with the Country public accountants in Parliament, Davenant gave political arithmetic a new forensic function. It was to be a tool for combating governmental secrecy and disorder, and for making useful knowledge in the absence of perfect information. In other words, it was a tool for finding the money. Davenant's earliest publications revealed a close affinity between his approach to political arithmetic and the public accounting exercises carried out by Country MPs. The first numerical arguments in his *Ways and Means* involved reorganizing government "Accompts" about tax revenues to reveal trends over time. Davenant expanded upon those basic accounts through creative arithmetic, combining public account information with data and estimates on other relevant economic quantities, like population and trade volumes, culled from various sources. An early display of such creativity came in his critique of "poll money," a wartime tax on individuals based on status and occupation, which he suspected was not producing the revenues that ought to be expected of it. Adopting demographic estimation techniques pioneered by Petty and John Graunt, Davenant developed a bottom-up calculation of what that quarterly poll tax should have yielded based on the size and make-up of the English population. The poll tax legislation specified different imposts for different social groups—nobility, gentry, tradesmen—and so Davenant found artful numerical proxies to estimate the size of each class. He used hearth tax returns to estimate the number of "Richer Families" eligible for one provision of the tax (800,000 families yielding £373,333), while he used the total number of land tax commissioners, a position reserved for local elites, to estimate the number of wealthy gentlemen subject to a different tax bracket (80,000 people yielding £320,000). He ultimately concluded the poll tax should have yielded £800,000 per year. The actual returns from the previous year, around £597,000, missed this projection by over 25 percent; current-year returns suggested a shortfall approaching 50 percent. Davenant concluded that such a dramatic deficit could only come from a dramatically flawed policy. "When a Tax yields no more than half what in reason might be expected from it," he argued, "we may plainly see it grates upon all sorts of People."[68]

Davenant's *Ways and Means* made clever use of limited data to frame the difference between an ideal benchmark—what the poll tax should produce—and the disappointing reality of government at the time. It was a similar style of calculation to the one ventured by the Accounts Commissioners in November 1693, when they asked naval administrators why their actual expenditures (the reality) fell so far short of the money appropriated (the ideal). Davenant's 1698 *Discourses on the Publick Revenues* took this method of calculative criticism to a new height of sophistication. In Discourse III, "On the Management of the King's Revenues," Davenant systematically calculated the difference between the expected and actual yield for some eight different taxes (Figure 1.1). He concluded that the nation was losing £736,075 annually as a result of mismanagement—a shockingly large figure at that time. For Davenant, this figure not only revealed the distressing state of the nation's financial management, but also demonstrated the power of calculation. "Political Arithmetick may peradventure be an uncertain Guide in all these Matters," Davenant admitted in a bit of satisfied self-deprecation, "But if his Computations should happen to be right, Seven hundred thousand Pound Annual Income, is a Sum not to be slighted."[69]

Davenant's analysis of the state of excise revenues, where he was formerly employed, was especially intricate. He recalled how, beginning in the early 1680s (when he began his own service), major improvements had been made in the management of the excise, marked by an increased number of well-trained gaugers, upright new Commissioners, and better Treasury oversight. Subse-

In the Excise on Beer and Ale, of	£318,000
In the Duty on Malt, of	200,000
In the Duty on Salt, of	38,075
In the Duty upon Leather, of	30,000
In the Duties on Parchment and Paper, of	15,000
In the Duties on Marriages, Births, and Burials, of	26,000
In the Duty on Windows, of	89,000
In the Duties on Glassware, Earthenware, and Tobacco-Pipes, of	20,000
Total, yearly	**£736,075**

Figure 1.1   Summary of Charles Davenant's calculations on "Improvements to be made Annually" in various underproducing parts of the public revenue. From Charles Davenant, *Discourses on the Publick Revenues, and on the Trade of England*, vol. 1 (London: Printed for J. Knapton, 1698), 119.

quently, the produce of the excise taxes had risen consistently from 1683 to 1689, such that annual returns grew by £150,000 over that period.[70] He admitted, though, "that the Improvements made in the Six Years of the former Management, did not all arise from the Conduct of the Managers, but in part from the natural Increase of Wealth, and Numbers of People in the Kingdom." He proceeded, "Computing by Political Arithmetick," to assess what proportion of the £150,000 aggregate annual increase in the excises resulted from each of these three forces: the nation's increased "Stock of Wealth" (£9,000), its increasing population (£39,000), and improvements in excise management (£102,000).[71]

Davenant was vague on how he determined these proportions, and it seems likely those figures were little more than round guesses. (Elsewhere, the foundation of his figures was more solid; in his analysis of the malt tax, for instance, he drew upon recent data gathered by Gregory King regarding England's social structure and arable land.) Though they may fail to meet modern standards of empirical rigor, Davenant's arithmetical estimates were nonetheless highly innovative. His goal was to analyze chronological trends in excise data and to explain those trends in terms of the separate influence of a range of different independent variables. In a rudimentary way, these calculations anticipated the objectives of modern regression analysis.

He followed his analysis of the 1683–1689 excise returns with an even more elaborate exercise concerning the excise receipts of 1690–1696, during which time annual excise revenues dropped from £694,000 in 1689 to £473,000 at the low point in 1695 (Figure 1.2). His goal, once again, was to determine how much of this change in revenue resulted from poor management. In order to isolate this, he first tried to account for eight other mitigating factors that might have impacted excise revenues: a change in the standards of measure used for beer and ale effected in 1690; a specific change in excise assessment procedure ("the Tenth Allowance"); the effect of brewers producing ale to higher alcoholic strengths to reduce their tax burdens; the influence of wartime "Additional Duties" on beer and ale, which might have driven some brewers out of business; an increase of private home-brewing; the quartering of soldiers; the general decline in trade due to war; and, finally, the high price of malt and hops.

Correcting for the estimated impact of all of these different forces, and making a series of other technical corrections, he then compared the revenue yield in each year since 1690 to the benchmark of £694,000 set in 1689. (In the year 1693, for instance, the government received £488,000 in actual revenue, £206,000 short of the 1689 level. Davenant calculated that £102,000 of that amount was the result of those eight mitigating factors; this left a shortfall of £104,000 attributable simply to poor management.) Davenant then contended that, because excise rates had effectively been doubled in 1691, any revenues squandered by maladministration after that point were actually doubly punitive to the government, on the thought that the excises actually ought to be producing twice what they had been producing in 1689. He thus multiplied those lost funds by two in determining their ultimate effect. All told, he concluded that mismanagement had cost the public over £1.1 million in total over those seven years.

Calculations like this one were full of unsupported estimates, like the various "allowances" Davenant calculated, and dubious computational maneuvers, like his bold decision to double the size of the effective losses due to doubling of excise rates in 1691. But they were also unquestionably inventive. Davenant's creative attempts at estimation and extrapolation, his use of proxy data, his analysis of quantitative trends over time, and his efforts to control for multiple economic variables all deserve mention in a deeper history of "econometric" thinking.[72] Yet these methodological innovations were driven not by the pursuit of purely epistemological virtues, like rigor or objectivity, but by his political objectives and circumstances. His calculative strategy was built to overcome informational obstacles and develop incisive political arguments from inadequate data. As he learned, though, calculating with incomplete information could be a dangerous game to play, no matter how clever his figures.

A SCHEME, shewing in what Proportion the Eight fore-mentioned Heads may every Year have affected the Revenue of Excise.

	1690.	1691.	1692.	1693.	1694.	1695.	1696.	Totals.
The Alteration of Measure	20,000	21,000	21,000	20,000	19,000	18,000	17,000	136,000.
The Tenth Allowance	10,000	11,000	12,000	11,000	10,000	9,000	8,000	71,000.
The Additional Duties,	8,000	9,000	30,000	27,000	26,000	26,000	26,000	152,000.
The Drink brew'd of extraordinary Strength,	1,000	2,000	3,000	4,000	5,000	6,000	7,000	28,000.
The private Brewing,	3,000	5,000	7,000	9,000	11,000	13,000	16,000	64,000.
The Quartering of Soldiers,	3,000	3,000	4,000	5,000	6,000	7,000	8,000	36,000.
The Decrease of Trade, Wealth and People,	3,000	5,000	8,000	11,000	14,000	17,000	20,000	78,000.
The high Price of Malt and Hops,		5,000	5,000	15,000	7,000	5,000	5,000	37,000.
Total of the Allowances,	48,000	56,000	90,000	102,000	98,000	101,000	107,000	602,000.
The Revenue actually receiv'd, in round Numbers,	634,000	555,000	515,000	488,000	475,000	473,000	512,000	3,652,000.
The Allowances and Revenue together,	682,000	611,000	605,000	590,000	573,000	574,000	619,000	4,254,000.
Revenue answer'd, Anno 1689.	694,000	694,000	694,000	694,000	694,000	694,000	694,000	4,858,000.
Revenue, and Allowances together, since 1689.	682,000	611,000	605,000	590,000	573,000	574,000	619,000	4,254,000.
There appears then to be lost, over and above all reasonable Allowances,	12,000	83,000	89,000	104,000	121,000	120,000	75,000	604,000.
From 1691. inclusive, to 1696. inclusive, the Duties upon Strong Drink were doubled within 3 Pence, and 3 Pence more than doubled upon Small Beer: So that the Loss would come double to the King, thus		166,000	178,000	208,000	242,000	240,000	150,000	1,184,000.
But deduct out of each Years Loss, One 25th part for the 3 Pence it wants of being doubled in the Strong, which is sufficient, there being an Over-plus of 3 Pence in the Small,		6,640	7,120	8,320	9,680	9,600	6,000	47,360.
And the Neat Loss seems to be		159,360	170,880	199,680	232,320	230,400	144,000	1,136,640.

Place this Scheme, p. 102.

Figure 1.2 Charles Davenant's analysis of the causes of recent shortfalls in the excise revenues, 1690–1696. From Davenant, *Discourses on the Publick Revenues*, vol. 1, after p. 102. Courtesy of the Kress Collection of Business and Economics, Baker Library, Harvard Business School.

The Politics of Probability

Charles Davenant was well aware that his style of "Reasoning, by figures, upon Things relating to Government" would be unfamiliar and potentially untrustworthy to contemporaries. He made so much clear in his essay "Of the Use of Political Arithmetick," the opening chapter of his *Discourses on the Publick Revenues,* a text which could reasonably be called an early manifesto for a dawning quantitative age. He began by admitting that he was undertaking a substantially new way of thinking—"advanc[ing] a new Matter," as he put it. He acknowledged that he was not the first to practice political arithmetic; it had been Petty who coined the name and "brought it into Rules and Method." But he was also aware that invoking Petty's quixotic project would not have inspired confidence from many readers, and so he was quick to distance himself, laying out a series of sharp critiques of Petty's earlier arithmetical work. Davenant would have to provide readers with his own, new justifications for why they ought to give any credit to his calculations.[73]

By the time Davenant wrote this new defense of political arithmetic in 1698, his calculations had already come under harsh criticism for being imprecise, unaccountable, and even irresponsible. A telling example came in 1696, when Davenant drafted two pamphlets (one published, the other remaining in manuscript) on the East India trade. These writings were prompted by pending legislation to ban the import of silk and calico textiles from India, measures strongly supported by the dominant Court Whigs and staunchly opposed by the East India Company.[74] Writing in support of the Company, Davenant estimated that the East India trade brought an average profit of £600,000 per year to England, even though it appeared that England sacrificed substantial amounts of precious metal in exchange for East Indian goods. His argument was that England ended up a net gainer because of the extensive profits it made re-exporting East Indian imports to other countries for high prices. He offered a hypothetical calculation to that effect. In general, England "Exported for this traffick, either in Bullion, or our Manufactures, about *per annum* 400,000 *l.*," he explained. "Suppose we consume at home the Returns of 200,000 *l.*" That left £200,000 in Indian goods that could be re-exported for a four-fold return of £800,000, Davenant argued. This meant England brought in a total value of £1,000,000 per year from the East India trade (£200,000 in Indian goods consumed at home, plus £800,000 in re-export profits), for a cost of only £400,000. Davenant concluded "that the East-India Trade did annually add to the gross Stock of England at least 600,000 *l.* per Annum."[75]

Davenant's Whig opponents were quick to censure his venturesome style of argumentation. One critic wrote of Davenant's East India pamphlet that "this Book is filled up with Flourish, and Rhetorick, and Conclusions, from false Hypothesis, that it is not easie at once Reading to grasp it." A more pointed criticism came from John Pollexfen, recently appointed to the new Board of Trade, who specifically attacked Davenant's calculations about the profitability of the East India trade. "Instead of offering Vouchers to prove this Accompt, it is said, That it must be clear Gains . . . because no one versed in Merchandize will deny it."[76] Pollexfen's implication was that Davenant fabricated his figures, citing little more than the hearsay of merchants, in order to advance his own interests and the interests of the East India Company. Pollexfen's critique revealed something more than just partisan distrust, though. He rejected the validity of Davenant's calculations as a mode of political reasoning, and he did so by invoking what, at the time, was the most familiar standard for judging numerical statements: mercantile accounting. "Vouchers" referred to the written records that were necessary to verify accounts; because Davenant could not vouch for all of his data, his numbers did not constitute a credible account.

Yet Davenant's political arithmetic aspired to do something different from just rendering a numerical account. It tried to extend the limited data that was available to create new political knowledge. In fact, Davenant was not "accounting" in a traditional sense. He was not an insider testifying to numbers that were within his individual control, but rather an outsider making claims about distant quantities that lay far beyond it. Admittedly, Davenant's calculation of East India trading profits did not rest on especially solid empirical ground. But Pollexfen's criticism might have been leveled against almost any of Davenant's calculations, because all of his political arithmetic involved some kind of quantitative leap where he had to estimate, extrapolate, or hypothesize without proper "vouchers." The disagreement between Pollexfen and Davenant was implicitly a clash about epistemology: Pollexfen thought that the ability of numerical calculations to constitute valid knowledge depended on those numbers qualifying as valid *accounts;* Davenant thought calculations could be valid irrespective of such considerations. This demonstrates an essential feature of the emerging quantitative age. The calculations that would constitute Britain's new public culture of quantification, like Davenant's, were not simply mercantile accounting cast upon a grander, political scale. They constituted an epistemological departure from the strictures of that accounting tradition.

Davenant faced a fundamental problem. Pollexfen's critiques suggested that rough numerical estimates and speculative calculations might not pass muster with many in the English public, who—if they had any familiarity with numbers at all—might expect to find numbers that met the exacting standards of (idealized) mercantile bookkeeping. But if political arithmetic were designed to make sense out of uncertain situations, how could it ever achieve such standards of certainty? Davenant's solution lay in *probability,* in showing that the conclusions of political arithmetic could justify very high levels of belief even if they were not completely certain, and even if their underlying data were incomplete and lacking the requisite vouchers. One pamphlet rebuttal to Pollexfen's criticisms of the *Essay on the East-India Trade,* possibly by Davenant himself, expressed the essence of this position. Shooting back at Pollexfen's unrealistic demands on the calculations in Davenant's *Essay,* the author wrote: "it does not concern the Truth of the Proposition, That the East-India Trade is Beneficial to the Nation . . . whether the Computation in the Essay be exact or no." As to Davenant's claims about the monetary benefits of the East India trade, "the certainty of the Sum does not so much concern the Point in question, as the Proportion."[77] Calculations did not need to be exact or certain to offer crucial political insights.

Davenant expanded on this thought in his 1698 manifesto "Of the Use of Political Arithmetick." "In the Art of Decyphring," Davenant wrote, "'tis said where three or four Words, perhaps Letters, can be found out, the whole Cypher may be discover'd; in a great measure, the same holds, in the Computations we are treating of: And very probable Conjectures may be form'd, where any certain Footing can be found, to fix our Reasonings upon."[78] Fellow political arithmetician Gregory King, whose work on population was a key source for Davenant's calculations, had made a similar point in a 1696 manuscript: "To be well apprized of the true State and Condition of a Nation, Especially in the Two main Articles of it's People and Wealth, be a Piece of Politicall Knowledge . . . the most usefull, and Necessary . . . But since the attaining thereof (how necessary & disireable soever) is next to impossible, We must content our selves with such near approaches to it as the Grounds We have to go upon will enable."[79]

This argument that numerical analyses could be valuable because they produced probable, rather than certain, knowledge was a bold departure from traditional attitudes about the nature of mathematical knowledge, which had long been prized precisely for its certainty. (Compare how John Dee had described the essence of mathematical knowledge in his 1570 *Preface* to the English translation of Euclid's *Elements:* "In Mathematicall reasoninges, a probable Argument, is nothing regarded. . . . But onely a perfect demonstration, of truthes certaine, necessary, and inuincible.")[80] Davenant and King were not, of course, the first to argue that useful knowledge should not be cast aside for failing to live up to the strictest standards of certainty. A turn toward probable knowledge rather than certain demonstration was a prominent feature of intellectual activity in seventeenth-century England across many spheres, from law to natural science to political economy, as scholars like Ian Hacking, Barbara Shapiro, and Carl Wennerlind have shown.[81] Petty's political arithmetic had also been grounded in the belief that numerical knowledge did not have to be perfect to be valuable. As McCormick observes, the defense of "rough, relative proportions," rather than accurate numbers or demonstrative proofs, was a hallmark of Petty's project. But Petty was interested in using numbers to discern underlying laws governing the social world, not in the nature of numerical evidence *qua* evidence. Questions of methodology and epistemology were never a major preoccupation.[82] In the 1690s, the imprecise calculations Petty celebrated would not have withstood the assault of skilled opponents like Pollexfen, at least not without a more vigorous defense. The task of justifying the validity of political arithmetic as a tool for making political knowledge therefore fell to Davenant.

Davenant began to ponder: To what degree could the claims of political arithmetic be justified by the numbers themselves? What made a calculation credible, even if based on irregular evidence? In one instance, he proposed a rudimentary standard for what qualified as a sufficiently good projection of the produce of a future tax: "That Art is therefore to be prais'd, the Rules of which, if rightly follow'd, will show within a Seventh or an Eighth, what any Branch of Revenue shall produce."[83] In a very preliminary way, Davenant was gesturing at the question of statistical significance. Davenant definitely did not get very far on this question of significance—or probability more broadly—as a mathematical matter. But he did frame the epistemological challenges facing political arithmetic in a way few before him had. He would continue to invoke the language of probability and to reflect on the nature of probable numerical knowledge throughout his calculating career. It is probably no coincidence, for example, that his 1699 treatise on commercial policy was entitled *An Essay upon the Probable Methods of Making a People Gainers in the Ballance of Trade.*

Not everyone accepted Davenant's probabilistic argument for trusting in arithmetic, of course. One of the sternest rebukes came in an anonymous 1698 pamphlet—*Remarks upon Some Wrong Computations and Conclusions, Contained in a Late Tract, Entitled,* Discourses on the Publick Revenues—which attacked Davenant's methods even more directly than Pollexfen had. An irreverent epigraph on the title page set the pamphlet's tone: "Miserable is that Country, where the Men of Business do not Reckon right"; the source was Davenant's own *Discourses,* page fourteen.[84] The author began by reflecting on his experience reading the *Discourses* for the first time. "I read those Discourses, particularly that on *the Management of the Revenue of the excise . . .* with an Expectation of finding an *Exact Account* of the *Rise* and *Fall* of that Revenue, and of the true Causes thereof," the pamphleteer began. It seems his expectations were unfounded, though. "Instead of this, I met with such a Series of Wrong Computations and loose Reasoning, as do sufficiently expose the Vanity of the Author, in pretending to a Knowledg in *Political Arithmetick,* and a *Faculty of Computing* and *Reasoning upon things by Figures,* not common to other Men." Like the Commissioners of Public Accounts, and like Charles Davenant, the critical pamphleteer had gone looking for certain answers—"*Exact Accounts*"—and found them lacking. But while Davenant and the Commissioners had taken this as an occasion to try to calculate what they could not find, the pamphleteer took it as an occasion to find fault with calculation itself.[85]

The calculations that troubled the pamphleteer most immediately were Davenant's analyses of excise revenues in Discourse III, especially his claims about the massive losses (over £1.1 million) the public had suffered due to bad management since 1688 (see Figure 1.2). The pamphleteer contended that, rather than rendering a true account, Davenant's figures had been designed instead to "*Magnifie* the *good Conduct* of the first Commissioners (of which number himself was one)" and to "Censure the Conduct" of the new Commissioners who took over in the 1690s. The pamphleteer employed a variety of tactics to undermine Davenant's numerical assertions. He pointed to evidence of bias in how Davenant discussed the new Commissioners and identified several instances of seemingly "wrong computations" within the *Discourses.* For example, he rebuked Davenant for failing to recognize how a prohibition on French brandy had hindered the revenues from beverage excises.[86]

Woven throughout these particular contentions was a running critique against political arithmetic as a mode of political reasoning. The pamphlet repeatedly disparaged the notion that "great things may be done by the help of Political Arithmetick" and chided Davenant for his "Air of Assurance." At one point, noting an inconsistency in Davenant's data on beer and ale excises in London, the pamphleteer quipped: "But perhaps *Political Arithmetick* gives a Man a Priviledge to contradict himself." Later, he waxed sarcastically about how "one cannot imagine what Strange Feats may be done by *Reasoning by Figures upon Things.*" What especially bothered the pamphleteer was Davenant's hubristic assumption that his command of calculation gave him the authority to make assertions about numbers he did not actually know. "He complains (*p. 34*) That *the Commissioners of excise have refused him any Inspection into their Books,*" the pamphleteer noted, quoting the text of the *Discourses,* "and tells us (*p. 266*) He has met extream Difficulty and Opposition in procuring the Sight of the Accompts relating to the Revenue.*" Yet, the pamphleteer said, this did not excuse the arrogant and irresponsible use he made of the numbers he could access. The Excise Commissioners probably had good reason to refuse Davenant access to their books, "and such Refusal would much better have Excused his not writing on this Subject, than the false Account he has given."[87] To the author of *Some Wrong Computations,* Davenant's method of making numerical estimates and extrapolations without proper vouchers was deceptive, even dishonorable. In fact, Davenant's political arithmetic—his new *"way of arguing upon Things by Figures"*—was an arrogant and unnecessary innovation, which obscured well-trodden paths to the political truth, like what the pamphleteer called "the Old way of Arguing upon Things by *Reason.*" In the pamphleteer's mind, calculation was not necessarily a manifestation of reason, but a different and potentially antithetical way of thinking altogether.[88]

The challenges leveled against Davenant's political arithmetic by Pollexfen and the author of *Some Wrong Computations* highlight the novel, and precarious, place of calculation in public life in the 1690s. Not everyone at the time was naturally inclined to trust in numbers. In Davenant's age, some—probably most—people saw numerical calculation as a new and strange way of thinking through political problems, somehow different from traditional reason. Even many of those who were acquainted with numerical thinking were likely inclined to be suspicious of Davenant's political arithmetic, which did not produce the certainty associated with mathematical demonstration and did not follow the exacting standards expected of mercantile accounting. The authority of calculation as a mode of public reasoning was something that would have to be made.

The people who did the most to advance that project, to make numerical calculation an established and eventually esteemed part of political practice, were relative outsiders in the world of politics. That is the central argument of this chapter—and a central argument of this book as a whole. Those who were most enthusiastic about calculation as a way of political thinking, from the public accountants in Parliament to the political arithmetician Davenant, were those who did not have access to original governmental data and who needed calculation to overcome such asymmetries of information. This created an essential predicament for those calculating outsiders: they could not provide certain answers to every question and could not provide a voucher for every number they produced; if they could have, they would not have needed to calculate in the first place. This meant that, from the very beginning, numbers began their career in British political life under a cloud of suspicion—an imperfect, presumptuous, partisan way of thinking. How, then, could numerical calculations come to garner such pervasive authority in British civic epistemology only a few decades later? As the remaining chapters will show, it turned out that calculations did not necessarily need to provide certain answers to be politically powerful. They just needed to make good arguments.

CHAPTER TWO

The Great Project of the Equivalent: A Story of the Number 398,085½

IN THE MIDDLE of May 1706, six men—three Scottish and three English—convened in London to undertake one of the most demanding economic calculations perhaps ever contemplated at that time. Included among the group were two reputable English accountants, an Oxford mathematician, and a key founder of the Bank of England. Broadly, the question at hand was how to balance the terms of a potential constitutional union between Scotland and England.[1] For Scotland, agreeing to a union meant relinquishing fiscal autonomy and accepting the far higher tax burdens levied by its southern counterpart, much of which went to servicing England's mounting national debt. By early summer 1706, negotiators from the two nations had determined that the best way for Scotland to be compensated for those future burdens was through an upfront monetary payment, an "Equivalent." The six-person calculating committee was tasked with putting a precise figure on that payment. After about six weeks of work, they came up with a number: £398,085 and 10 shillings. In a sense, this was the price of Scottish nationhood. It was also an unprecedented exercise in using a calculation to solve a major political problem.

That peculiarly precise number would eventually be ratified by both English and Scottish Parliaments and become encoded within the Union Treaty that created Great Britain—but not without becoming one of the most discussed, debated, and derided numbers in British history. Officially, the Equivalent payment was intended to solve a range of interconnected political-economic problems that stood in the way of a thriving, unified Britain. In addition to helping balance out the terms of the deal and allay Scottish fears about higher taxes, the money advanced was supposed to help improve ailing Scottish trade and industry. It was also designed to help Scottish investors recover from the recent collapse of the Company of Scotland and its "Darien Scheme," a calamitous project to establish a Scottish commercial colony in Panama. As the Union Treaty was still under debate, the Equivalent—or "Equivalents," plural, as contemporaries sometimes said, denoting the various streams of money involved—elicited considerable confusion, not to mention suspicion. In late 1706, Scottish Treaty Commissioner Sir John Clerk of Penicuik reported that "amongst all the *Articles of the Treaty of Union,* there has been none more talked of, and less understood, than the . . . *Rise, Nature, and Management of the Equivalents.*"[2]

In the years that followed, sentiments about the Equivalent would become increasingly embittered, particularly among Scottish observers like George Lockhart of Carnwath. "The Equivalent was the mighty Bate," he explained in 1714. "Here was the Sum of 398085 Pound Sterling to be remitted in Cash to *Scotland.* . . . This was a Fund say they sufficient to put *Scotland* in a Capacity for prosecuting Trade, erecting Manufactories and improving the Country: But in Reality here was a swinging Bribe." Subsequent observers echoed Lockhart's cynicism, casting the Equivalent as part of a massively corrupt bargain by which Scottish independence was—as poet Robert Burns famously put it in 1791—"bought and sold for English gold."[3] Most modern historians have agreed that the Equivalent was essentially a pile of "English gold" intended to sweeten the union deal, though they have disagreed about how important it was to the outcome of the union and what it reveals about that process. Some have followed Lockhart's line, seeing it as a devious "confidence trick" enriching corrupt Scottish politicians. Others have taken it as a more benign consolation prize by which pro-union politicians in Scotland could "secure prestige at home," or as simply "testimony to inept Scottish negotiation" because of its paltry size.[4]

In his 2007 account of the Company of Scotland and its connections to the union, *The Price of Scotland,* historian Douglas Watt contends that, overall, "historians have underplayed the significance of the Equivalent." He argues that the financial device of the Equivalent was critical to the entire union design, particularly the plan to salvage the Darien disaster—what he calls a "shareholder bail-out with cash provided by a foreign government." Watt's study has brought much-needed attention to the fact that the Equivalent was not just a "swinging Bribe" or a diplomatic concession, but a complex financial technology. Yet with the exception of one excellent article by accounting scholar Crawford Spence, almost no serious attention has been paid to another key feature of the Equivalent: *it was a calculation,* and an incredibly complicated one at that. For the most part, historians have assumed that the Equivalent calculation was essentially for optical effect, designed to make the English payoff appear rational and fair. For example, Watt writes that the Equivalent was ultimately a "lump sum dressed in political clothing," for which "financial mathematics were utilized to assure politicians and the 'public' that the sum was equitable."[5]

In supposing that the numbers were just for show—all too common throughout the political historiography of the long eighteenth century—historians fall into a subtle anachronism. They assume that numbers *already* merited widespread esteem in public reasoning, such that calculations like that of the Equivalent could be used to exploit the public's pre-existing faith in numbers. In 1706, though, the position of numbers in English and Scottish civic epistemology was not nearly so secure. That was what made the Equivalent so remarkable: the implicit assertion that a massively important political, economic, and constitutional question like the union could be solved, at least in part, through mathematical calculation. Indeed, what Scottish union commissioner John Clerk of Penicuik described as "the great project of the Equivalent" was not just *a* calculation.[6] It was an argument *for* calculation as a mode of political reasoning.

So how did anyone come to believe that complex mathematics was the best way to resolve the fiscal complications of a potential union? How did opaque calculations become a credible way to decide what was fair?[7] As it happened, the numerical vision behind the Equivalent project was largely the work of two individuals. The first was the Scottish financier William Paterson, one of the age's greatest financial projectors and the visionary behind both the triumphant Bank of England and the fateful Darien Scheme. The second was Dr. David Gregory, Savilian Professor of Astronomy at Oxford, ardent Newtonian, and described by Newton himself as Scotland's greatest mathematician. Both halves of this odd couple were instrumental in crafting the idea that calculation could balance the terms of the union, and both then took leading roles in calculating the Equivalent and justifying that figure to the English and Scottish publics. Paterson and Gregory's calculated project illuminates several key lessons of this book as a whole. First, it shows that excitement about numbers in this period was not uniform but eclectic, stemming from diverse intellectual and cultural sources. Both Paterson and Gregory extolled the benefits of numbers, but their zeal for numbers came from different places. Paterson's inspiration came from commerce. He believed that merchants and bankers possessed specialized practical knowledge, exemplified by financial mathematics, that ought to play a much greater role in government. Gregory's inspiration was scientific. He was a leader among a group of Scottish Tory philosophers and physicians who believed that "mixed" mathematics in Isaac Newton's style held the answers to myriad problems, from medicine to history to government. Paterson and Gregory spoke two different quantitative idioms, but they converged on a single political objective in 1706.

Both Paterson and Gregory fostered ambitious visions for how numerical thinking could transform their nation. Political realities did not necessarily match those mathematical dreams, though. This was a second key lesson of the Equivalent episode. Putting calculations—or "calculs," in Scottish dialect—into political action was riddled with frustrations. Paterson, Gregory, and others put great effort into explaining the £398,085.10s. figure to the broader public, particularly in Scotland. Yet they could not control what happened to that number once it entered public circulation. Several obstacles arose. For one, the numbers proved remarkably unstable; they did not always "travel well." Printers regularly misprinted the intricate Equivalent calculations, including the "398,085" itself. Once printed, the Equivalent calculations proved highly malleable. They were subjected to various creative readings, reconstructions, and criticisms, from both proponents and critics of union. The meanings that members of the public attached to the Equivalent figures were often quite different from what Paterson and Gregory intended. While the Equivalent calculation aimed to provide a conclusive solution to challenges facing the union, it opened up new realms of controversy at the same time.

In turn, the disputes about the Equivalent brought profound, unsettled questions of civic epistemology to the fore. Perhaps the biggest problems concerned the future. The Equivalent was fundamentally a forward-looking projection, which promised to compensate the Scots immediately for a series of higher tax payments to be made in the coming years. By accepting that payment, the Scots were agreeing to accept a present benefit in exchange for future costs. This brought with it deep constitutional confusions. Could present generations bind the future in this way? It also prompted difficult questions of value, in particular the problem of how to assess the present value of money to be transferred in the future. The financier Paterson and the mathematician Gregory calculated the present value of future Scottish taxes using the mathematical technique of exponential discounting, cutting-edge for its time. For many of their contemporaries, that esoteric calculation had disconcerting consequences. In particular, it placed almost no value on anything that happened beyond one human lifetime. This peculiar claim clashed violently with many Britons' intuitions about what the future was worth to them.

This debate over the future was, in part, a debate about what it meant to be reasonable in a changing political and economic world. It was also a debate about who got to determine those definitions for the polity. Gregory and especially Paterson felt the failure of most people to understand certain key financial and mathematical principles was indicative of the collective ignorance that was holding their country back. Others, though, did not want to believe self-appointed mathematical experts over their own traditional sense of value and fairness. Ultimately, political representatives from both England and Scotland approved the Equivalent figures as an official policy; nonetheless, few Britons assented completely to the expert view of mathematical governance Paterson and Gregory promoted. Here lay a third key lesson about Britain's emerging quantitative age: accepting the political authority of calculations did not mean submitting to the personal authority of calculators. The number itself—£398,085.10s.—became far more powerful than any of the people who calculated it.[8]

The Anatomy of the Equivalent

The Equivalent was an odd twist in a long-standing constitutional and diplomatic saga between England and Scotland. The two nations had gained a single monarch with the accession of King James (VI of Scotland, I of England) in 1603, though they continued to maintain separate Parliaments, legal systems, and established churches. Many English leaders wished to bring their northern neighbor under greater control, particularly because Scotland represented a strategic weak point in England's defenses against France. Throughout the seventeenth century, various unification plans had been floated, though it was not until Queen Anne's accession in 1702 that the idea began to gather serious momentum. Anne was a staunch advocate for union, largely for dynastic reasons. Because Scotland had not assented to the 1701 Act of Settlement that specified the future monarchical succession, it was possible that the Crowns of England and Scotland might split after Anne's death. The fact that many Scots harbored loyalty to the Stuart (Stewart) dynasty that had been ousted with King James II / VII in 1688 exacerbated this fear.

Economic circumstances also fueled union interest on both sides. English imperialists looked to tap into the "burgeoning entrepreneurship" of Scottish commercial networks across the globe.[9] In Scotland, a series of major setbacks caused Scots to look south for economic support. One trigger was a substantial famine in 1695–1699. Another was the languid condition of Scottish trade, which some blamed on England's harshly protective commercial policies. The most acute cause was the Darien disaster. In 1695, the Scottish Parliament had chartered a new "Company of Scotland Trading to Africa and the Indies," which raised £400,000—roughly a quarter of the circulating money in Scotland—from Scottish investors with the promise of building a trading colony connecting the Atlantic and Pacific Oceans. By 1700, two expeditions to the Panamanian isthmus had failed catastrophically due to famine, disease, and military siege by Spanish forces. The failure cost many Scottish lives, vast reserves of Scottish capital, and Scotland's ambitions as an imperial power.[10]

Queen Anne initiated a first attempt at a union deal in 1702. Those early negotiations helped set the terms for subsequent union talks, highlighting key obstacles that would have to be overcome for a union to succeed. It became clear that finding equitable economic terms would be a major hurdle. The acting Scottish commissioners outlined several essential demands in December 1702: "free trade betwixt the two kingdoms, without distinction"; equal and integrated taxation; and equal commercial access to the "Plantation Trade." Many of these points were generally agreeable, though others were far less straightforward. One problematic question involved compensation for shareholders of the ill-fated Company of Scotland. Many Scots felt that responsibility for the fiasco lay with King William III, who failed to support the Darien venture out of preferential concern for England's imperial agenda and for English commercial interests like the East India Company. A union provided an opportunity for the English to pay back this debt to the Scots. Another sticking point was the Scots' demand that "neither of the Kingdoms be burthened with the Debts Contracted by the other before the Union." The Scottish commissioners in 1702 proposed two potential paths forward: either the Scots be exempted from taxes levied to support the English debts; "or, if an Equality of Imposition on Trade be thought necessary, that there be allowed to *Scotland* an Equivalent." A figure of £10,000 a year may have been mooted.[11]

By the late seventeenth century, "equivalent" was commonly used as a term-of-art to designate targeted political or diplomatic compensation. Equivalents of various kinds were a common part of international diplomacy, for example. During peace negotiations, like those leading up to the Treaty of Ryswick in 1697, warring parties contemplated offering "equivalents" in exchange for ceding certain disputed territories.[12] Equivalents were also used to describe the monetary subsidies that more powerful nations sometimes paid to less powerful allies in exchange for their loyalty.[13] In domestic politics, MPs spoke of equivalents in describing possible shifts in tax policy. When the Commons determined in the late 1690s that it was best to remove certain existing taxes on glassware, earthenware, and tobacco pipes, they agreed that "an Equivalent be granted to his Majesty" in the form of new taxes on whalebone and Scottish linen.[14] While many of these equivalents were monetary, the most familiar equivalent in the late seventeenth century was not. In 1687, King James II had pushed to repeal various laws that limited the rights of non-Anglicans, particularly the Test Acts of 1673 and 1678. Opponents feared James's goal was to secure greater toleration for Catholics and undermine the supremacy of the Church of England. In order to allay such fears, the King offered to replace the Tests by some "equivalent" (or "equipollent") as assurance that the Church was secure. James's vague promise was sharply rebuked by George Savile, Marquess of Halifax, in a pamphlet entitled *The Anatomy of an Equivalent,* which helped make "equivalent" a familiar entry in the English political lexicon (and which would be republished in revised form in 1706 in relation to a very different kind of equivalent).[15]

So the idea of paying Scotland some kind of equivalent as part of a diplomatic negotiation was not especially strange in 1702. At first it did not get very far, though. The English negotiators resisted the Scots' demands for economic concessions. They stressed that it was essential that England and Scotland pay equal taxes under any union plan, and argued that it was reasonable to ask Scotland to assume some responsibility for English debts because those funds had been raised to pay for a war that greatly benefited Scotland "by the Opposition made to the Growth and Power of *France.*" They demurred on the question of "what Equivalent should be allow'd the Scots," contending instead that "Scotland would be abundantly compensated" by the broader economic benefits any union would bring. That point would be reiterated in coming years by English commentators, including Charles Davenant, who had served as secretary to the English commissioners for the abortive 1702–1703 talks. In 1705, Davenant drafted a manuscript treatise extolling the benefits Scotland would receive from "a free Communication of our Trade" and arguing against the payment of any direct monetary equivalent to the Scots.[16]

Around that same time, though, others were beginning to ponder whether in fact a monetary equivalent was precisely the way to solve the riddle of the union. What seems to be the first detailed plan for such an equivalent appears in a two-page manuscript, "Heads proposed for an Union between England & Scotland," dated 1705, saved among the Stowe Manuscripts in the British Library.[17]

The paper laid out a twelve-point plan for a proposed union. "Heads" 1–7 called for a unification of the two nations as "Britain," under one monarch, with shared Houses of Commons and Lords, but with no change to legal or administrative structures. Heads 8–9 specified that there be a "free Communication and Intercourse of Trade, between all parts of this United Kingdom," and that all parts of the kingdom be subject to equal taxes. The final three points addressed how to make that commercial union equitable. Head 10 proposed Scotland be exempted from the English land tax. Head 11 offered a more radical balancing measure:

> As an equivalent for the present debts of the Kingdom of England, as likewise for reparation of the late losses of the Indian & Affrican Company of Scotland in their attempts to the Indies, that Her Majesty will be pleased to allow a Rent-charge of 30,000£. per Annum by quarterly payments to be made upon Her Royal Revenue untill a Sum of 600,000£. proposed to be raised thereupon and allowed to the Kingdom of Scotland for the reasons and purposes aforesaid shall come to be repaid.

The manuscript gave only a brief explanation for where the £600,000 figure came from. A small marginal note stated that "This Sum is proportionale to the debts of England and revenues of the two Kingdoms." The final Head, number 12, specified three targets where this £600,000 "equivalent" was to be spent: £230,000 would refund losses from Darien (including accrued interest), £100,000 would be dedicated to repaying Scottish public debts, and the remaining £260,000 would serve as a "fund for imployment and maintenance of the poor, supporting, promoting, and encouraging the manufactures, fisheries, products and improvements of that end of this Island." (It is unclear why the three targets added up to only £590,000, instead of the £600,000 total specified.)[18] The 1705 manuscript proposal seems to have been the first to take the vague diplomatic concept of an "equivalent" and give it a precise value. The 1705 manuscript was unsigned, but its author was almost certainly William Paterson.[19]

William Paterson and the Costs of Ignorance

Born in Dumfriesshire in southern Scotland in 1658, Paterson probably relocated to England relatively early in his life, perhaps to Bristol, before beginning a commercial career in London. He joined the Merchant Taylors' Company in 1681 and was promoted to the company's "livery" in 1689. In the intervening years he began building an ambitious mercantile operation, primarily in the West

Indian trade. He also spent considerable time in continental Europe, particularly the Netherlands. From around 1686, he took an increasing interest in more abstract and technical questions of commerce and public finance, and, by 1691, he had gained sufficient command over such matters to testify before a Parliamentary committee about a proposal to create a new credit-based bank. That project was initially rejected, but Paterson helped design a subsequent version that was accepted. The latter plan, approved in 1694 and promoted by leading merchant Michael Godfrey and Treasury Commissioner Charles Montagu, envisioned a new bank that would provide loan financing to the government and circulate paper credit notes backed partially by a silver reserve. This became the Bank of England. Paterson served briefly as a director before professional disagreements and personal ambitions led him to leave for another visionary project: the establishment of a Scottish port colony on the Isthmus of Panama. Paterson had apparently been pushing that idea for about a decade by the time the Scottish Parliament approved a new "Company Tradeing to Affrica and the Indies" in June 1695. Paterson became a director and major investor in the company. He was integral in raising domestic and foreign investment and in planning the company's Darien venture, even traveling to Central America on the first voyage in July 1698. Needless to say, Paterson's two great projects of the 1690s—the Bank and the Darien Scheme—came to very different ends.[20]

The failure of Darien was a turning point for Paterson, and for Scotland as a whole. David Armitage writes that, in reaction to the Darien failure, Paterson and other like-minded Scots "came to calculate that their vision of empire could only be realized by Union, and within a British state." This calculative metaphor was especially apt in Paterson's case. After Darien, he set to work on two goals. One was narrow: trying to compel the English to pay compensation for Darien, including to Paterson personally. The other was broader: trying to implement a new "oeconomy" in Scotland, a centrally directed program of commercial modernization grounded in calculation. Paterson was a skilled student of commerce and a creative political thinker; he was also a shameless self-promoter who, according to one observer (David Nairne) in 1696, "talks too much."[21] Between 1701 and 1706, Paterson outlined a series of provocative political-economic reforms that aimed to strengthen both England and Scotland—and to salvage something from the Darien wreckage and advance his own career at the same time. Behind these entangled agendas was a distinctive attitude about how political knowledge ought to be made. Paterson wanted his fellow citizens, and especially their political leaders, to calculate like shrewd businessmen.

Paterson felt that both the English and Scottish governments suffered from severe ignorance, which produced tremendous financial waste and compromised the nations' credit at home and authority abroad. He laid out this charge as it applied to England in a 1701 manuscript, "Letter to King William III" (a text which, among other things, sought to resuscitate his plan for a trading colony in Panama as a jointly British project).[22] Nowhere was England's collective political ignorance more evident, or more damaging, than in the management of public credit, which Paterson believed was of supreme importance to national welfare. He lamented how a "very Considerable part of Your Majesty's Revenue [is] usually Lost and given Away, in dear, and bad Bargains, High Interests of Money, Discompts, Premiums, and other Extortions."[23] This was symptomatic of a broader malady afflicting the nation's political thinking. Not only were political leaders themselves ignorant, but members of the political nation had become so preoccupied with partisan jealousies and so paranoid about corruption that they had completely ignored the far more dangerous specter of mismanagement. ("Where a Peny shall be in danger by Wilful Fraud or Bankrupsie of Councellors of Trade," he warned in another pamphlet from 1701, "there will be at least Ten if not Twenty, so by Ignorance, Presumption and Neglect.")[24] In his mind, the solution was to put more power into the hands of people who knew what they were doing. "Mankind are commonly divided into three parts," Paterson wrote. "The first thereof, are those, who understand of themselves; The second tho' not of themselves, yet when it's Explained; The Third, neither understanding by themselves nor by an Explaination. . . . [O]f the first sort or rate, there are att best but few in any Country."[25] It was clear who Paterson thought ought to have control.

The case was even more acute in Scotland than in England, as Paterson explored in an ambitious pamphlet entitled *Proposals & Reasons for Constituting a Council of Trade,* published in Edinburgh in 1701. The pamphlet was targeted explicitly to a Scottish audience and began by lamenting the "the Injustice and Inequality of our Treatment" by the English. Yet the pamphlet repurposed many themes from his "Letter to King William," most notably his call for smarter, more commercially minded government. Paterson's central idea was to revive the "Trade, Industry and Improvements" of Scotland by combining a political innovation—a new, expert "Council of Trade"—with a financial one, a "National Fund of Money." In his plan, the new Council would administer the collection of £1,000,000 in new tax revenues, including a 2.5 percent tax on land transactions, new taxes on manufactures and lawsuits, and a radical 10 percent tax on grain sales. A fixed amount would be paid to the Crown, and all revenues above that would be contributed to the National Fund, plus an upfront contribution from the Crown in compensation for Darien. The Council of Trade would be charged with using this new Fund to repay Darien shareholders and to institute a range of economic improvements, including employing the poor in government workhouses, erecting national granaries to insure against famines, and increasing domestic salt production to aid the fishing industry.[26] Paterson's pamphlet depicted a new, calculated vision for the governance of Scotland. It also rehearsed the kind of calculated thinking he championed. The pamphlet's opening pages offered Paterson's own exercises in political arithmetic, citing quantitative estimates of Scotland's population and national wealth to illuminate the deteriorating condition of the nation since 1603. Later in the pamphlet, more intricate calculations used historical data on the cost of poor relief in Edinburgh and Leith to argue that, through the new National Fund, the Scottish people would be absolved of a poor-relief burden of £135,000 per year.[27]

After 1701, Paterson continued to press his position that Scotland and England needed expert management of their national trade and revenue. In 1703, for instance, he promoted the creation of a national library of works on "trade, revenue, and navigation."[28] Soon thereafter, he penned another pamphlet on Scotland's "National Trade," reminding readers that "Poverty is the disease of this Nation" and once again contending that the nation needed "a *Council of Trade,* composed of Gentlemen of the best Sense, and Merchants of the greatest Experience." In the latter pamphlet, Paterson suggested that such a Council might allow Scotland to maintain its economic autonomy from England.[29] Paterson would soon reverse his position on the importance of Scottish independence—but not on the need for intelligent government. He laid out this new vision in his April 1706 *Inquiry into the Reasonableness and Consequences of an Union With Scotland.* The text reported the deliberations of a fictitious English discussion club, the "Wednesdays Club in Friday Street," a format which allowed Paterson to dramatize the practices of political thinking he found so essential.[30]

The Preface to his *Inquiry into the Reasonableness and Consequences of an Union* began by lamenting how an incorporating union between England and Scotland was yet another "reasonable, and plain" plan to improve national welfare that, like so many others, had been obscured and undermined by the "prejudices, humours and secret designs of a few." In order to clear this up, the members of the "Wednesdays Club" had empowered "Lewis Medway" (Paterson's pseudonym) to report their recent discussions on the benefits of union.[31] In a loose dialogue form, the first half of the 150-plus-page pamphlet described the Club's preliminary discussions about the key issues facing any union plan and about the history of Anglo-Scottish relations since 1603. Thereafter, Medway reported that a group of Club members had been "appointed to draw up a Scheme for an *Union*" of their own. The core of the "Scheme" they presented was, almost to the letter, the twelve-point plan laid out in the 1705 manuscript now in the Stowe Manuscripts.[32] Effectively all of the remaining seventy-five pages of the pamphlet were spent elaborating technicalities related to that scheme, like calculating the allocation of Parliamentary representatives between the two nations, integrating the tax systems, and regulating Scottish fisheries.

One technical detail attracted particular attention: calculating an appropriate Equivalent to compensate Scotland for various economic concessions. Whereas the 1705 manuscript had offered £600,000 as a figure with little justification, the *Inquiry* offered a much more detailed calculation. Paterson began by estimating average English revenues, £5,610,000 annually, followed by a similar "estimate of what Scotland may produce when upon the foot of the present Taxes of England," which he reckoned at £216,000. He then presented a table detailing the current state of England's national debt, roughly £21 million. Deducting £30,000 per year from Scotland's anticipated revenues for the ordinary costs of government, he projected that Scotland would annually contribute £186,000 toward the English debt, or one-thirtieth of the £5.61 million annually paid by English taxpayers. Taking one-thirtieth of the total English national debt of £21 million, Paterson concluded that the Scots were effectively taking on a burden of £700,000. Lastly, he took a rough discount of £100,000 to reflect that the Scots would be largely exempted from some land taxes. Paterson came to a final figure of £600,000—the same figure presented in the 1705 manuscript, though with a much more extensive justification.[33] This first Equivalent calculation would come to seem rudimentary compared to the intricate analysis behind the final figure of £398,085.10*s.*, but as a first take on the question, Paterson's initial calculation was remarkably inventive.

The very calculation itself made a bold argument—that the terms of union were a problem that could be calculated. Yet Paterson also acknowledged that the calculation was conceptually complex and sensitive to key assumptions about technical matters, like interest rates. Paterson noted, for instance, that the fair value of the Equivalent would increase by as much as £210,000 if the interest rates in Scotland, around 6½ percent, fell to the 5 percent levels current in England. Behind these claims lay Paterson's assumptions about a crucial financial-mathematical question: how to calculate the "present value" of future property, particularly a future stream of income that paid out regularly for a long period of time, like the annual payments on an annuity, the rents on a tract of land, or the future tax revenues of a newly annexed nation. According to one contemporary theory, the present value of a future income stream was inversely proportional to prevailing interest rates. Consequently, if Scottish interest rates went down from 6½ to 5 percent, the value of the Equivalent would increase in inverse proportion, from £700,000 to £910,000.[34]

This cryptic commentary on interest rates was no minor technical point, but spoke to deeper questions about the financial future. Paterson's argument about the relationship between the Equivalent and interest rates was representative of his belief in a specific, mathematically intensive method for thinking about the future. The central premise was that any property to be acquired in the future could be assigned a value in the present, using the logic of *compound interest.* That "present value" was however much had to be saved *today* in order to have the desired amount in the future, assuming whatever you saved today earned compound interest in the meantime. For example, assuming 5 percent interest (and using decimal notation for clarity): £100 one year in the future has a "discounted" present value of £95.23 today; £100 ten years in the future is worth £61.39 today; £100 one hundred years in the future is worth £0.76 today; and so on. This technique can be used to value both single payments in the future—like £100 in one hundred years—and streams of future payments, for example, an annuity that pays £100 every year for one hundred years.

This method of compound-interest discounting had been formulated at least as early as 1202 by the Italian merchant-mathematician Leonardo of Pisa, better known as Fibonacci. Practical tables for solving compound-interest discounting problems emerged in the last half of the sixteenth century, and the techniques had been publicized in English vernacular texts at least since 1613, in Richard Witt's *Arithmeticall Questions: Touching the Buying or Exchange of Annuities.* Over the following century, practical mathematicians crafted more user-friendly texts and tables to facilitate compound-interest reckoning.[35] Though not exactly new, compound-interest discounting was hardly universal among Paterson's contemporaries. It remained a relatively arcane way of thinking, employed by a small cadre of skilled merchants, financiers, and mathematical practitioners. If they needed a way to think about the financial future, most Britons relied on a much simpler heuristic called "years purchase," which had long been used for working out the purchase price for landed property, and which did not involve any complex calculations.[36] For Paterson, such obsolete modes of reckoning were indicative of the general ignorance holding both England and Scotland back. For a pamphlet ostensibly about a constitutional union between England and Scotland, Paterson's Wednesdays Club pamphlet spent a remarkable number of pages offering didactic lessons about the technicalities of compound interest. At one point in the dialogue, Paterson likened Scotland to a financially ignorant family who, instead of saving money and benefiting from the power of compound interest, instead squandered its funds and thus had to pay compounding interest on its exponentially mounting debts.[37] Economic ignorance and economic poverty were intertwined; for Paterson, the laws of compound interest governed both.

In one of the oddest—but also most indicative—passages in the pamphlet, Paterson actually dramatized the debate between the traditional years-purchase view of the future (played by a "Mr. North") and Paterson's preferred, compound-interest view (played by a "Mr. Brooks").[38] The details of their debate related to an obscure question about the present value of long-term annuities, a technical issue that arose in the calculation of the Equivalent for several reasons.[39] But the crux of the matter concerned which approach to the future was more reasonable. A key point of contention concerned the way that Paterson's compound-interest approach, voiced by Mr. Brooks, seemed to put a very low price on the distant future. It suggested that the "present value" of any property received beyond about a hundred years in the future had negligible value for the present. (Recall that the present value of £100 one hundred years in the future is only £0.76 today, according to compound-interest discounting at 5 percent interest.) Paterson, by way of Mr. North, acknowledged that this idea flew in the face of early modern intuitions. Most people at the time were simply unwilling to say that the distant future was so unimportant, particularly given that the ownership of perpetual property in land was so central to contemporary conceptions of economic well-being, social prestige, and political power. But Paterson also felt that such customary ideas about time and value were mere "fancies," long in need of updating.

That was the essence of the Equivalent project for Paterson. It was not just about paying money to compensate Scotland for higher taxes. It was about changing how people, in both England and Scotland, thought—about government and commerce, about value and time. He wanted his fellow citizens to dispense with their traditional values, prejudices, and jealousies and put their trust in calculation. Whether they would agree to do so was, of course, another matter.

The Mathematical Principles of Doctor David Gregory

Paterson was not the only person who believed that mathematical calculation could make a better polity. If Paterson was responsible for laying the conceptual groundwork for the Equivalent, it was likely his calculating colleague David Gregory who gave the £398,085.10s. figure its distinct precision and mathematical intricacy. Gregory first entered the story of the Equivalent in early 1706, when he would be called upon by the Scottish Union Commissioners to advise them on financial data under discussion. He would quickly assume a leading role in calculating the Equivalent, alongside Paterson. Like Paterson, Gregory was truly a Briton, born in Scotland but finding professional success in England, embedded in personal and political networks that bridged the two.[40] Both men saw mathematical thinking as an instrument of national advancement. But Gregory's mathematical values stemmed from a different source. Whereas Paterson's commitment to calculation derived from commerce and finance, Gregory's derived from academic mathematics, especially the new mathematical natural philosophy of Isaac Newton.

Born in 1659 in Aberdeen, David Gregory was the son of a laird and the nephew of James Gregory (1638–1675), a similarly gifted mathematician whose research on topics like the "quadrature" (area) of trigonometric curves and the properties of infinite series made key contributions to the development of calculus.[41] Like his uncle, David was educated at Marischal College in Aberdeen and the University of Edinburgh, where, in 1683, before even taking his M.A., he was offered the chair of mathematics left vacant by his uncle's early death. Shortly thereafter he published his first text, *Exercitatio Geometrica de Dimensione Figurarum,* on the geometric properties of algebraic curves. Over the ensuing decade, his relationship with the Edinburgh faculty became strained, in part because of his tendency to teach new and controversial material, but more proximately because of his religion and politics. A devout Anglican and a committed Tory, Gregory refused to assent to the Presbyterian religious tests required of university faculty after 1688–1689. He soon left Edinburgh and relocated to Oxford in 1691, a haven of Toryism and Anglicanism, where the previous holder of the Savilian Chair, Edward Bernard, had recently retired. Gregory earned the lofty chair thanks largely to a favorable recommendation from Isaac Newton himself, who wrote that Gregory was "very well skilled in Analysis & Geometry both new & old. . . . & is respected the greatest Mathematician in Scotland."

During his time in Oxford, Gregory drafted one of the earliest manuscript commentaries on Newton's new fluxional calculus (1694) and also published a text on optics (1695), *A Treatise of Practical Geometry* (1695), and a substantial tome entitled *Astronomiae Physicae & Geometricae Elementa* (1702) that sought to integrate Newton's theory of gravitation into traditional astronomical instruction.[42]

In the late 1680s, Gregory emerged as a leader among a tight-knit coterie of "Tory Newtonians." His key collaborator was fellow Scot Archibald Pitcairne, a renowned physician whose fervor for the Tory-Anglican cause surpassed Gregory's. Mostly composed of Scottish Episcopalians, including many of Gregory and Pitcairne's students, their circle stretched from Edinburgh to Oxford to Leiden, where Pitcairne was briefly a professor. It encompassed many leading thinkers, including the physiologist George Cheyne, the probability theorist and physician John Arbuthnot, and the mathematician John Keill, who would publish the first popular exposition of Newton's *Principia Mathematica*. This group was committed to applying Newton's style of mathematical inquiry to natural, social, even theological problems. In medicine, for example, Pitcairne, Cheyne, and James Keill (John's brother) sought to compose a *Principia Medicinae Theoreticae Mathematicae,* which used Newtonian models of attraction to explain physiological processes like glandular secretion. Similarly ambitious, and even more inflammatory, was John Craig's 1696 *Theologicae Christianae Principia Mathematica,* which sought to develop a Newtonian style of theology grounded in the mathematical analysis of sacred history.[43]

Historian of science Simon Schaffer argues that, for Anglican monarchists like Gregory and Pitcairne, this reverence for mathematics had political motivations. They felt that the events of the Glorious Revolution marked an "unleashing of fanatic power in civil society," exemplified by the Presbyterian enthusiasm that had disrupted the Scottish universities and forced Gregory's departure to Oxford. Pitcairne and company saw the "prescription of 'mathematical' authority as a cure for enthusiast mob-rule." By contrast, their religious and political rivals—Presbyterians and low-Church Anglican Whigs—saw the "dogmatic mathematical elitism" espoused by Pitcairne and Craig as indicative of the dangerous "patriarchalism" favored by Episcopalians and Tories (not to mention Catholics and Jacobites). Whig physicians like Thomas Sydenham rejected Pitcairne's "'tyranny' of mathematics" in favor of a clinical approach to medicine based on experience and observation. An especially scathing rebuttal came from Edward Eizat, whose 1695 satire *Apollo Mathematicus: or the Art of Curing Diseases by the Mathematicks, According to the Principles of Dr Pitcairn* harangued the Tory Newtonians for believing that "Mathematical Spectacles" did anything but obscure the doctor's trained eye.[44]

Not every mathematical project promoted by the Tory Newtonians was quite so controversial, though. One of the era's most influential odes to mathematical thinking came from a member of the Pitcairne-Gregory circle: John Arbuthnot's popular *Essay on the Usefulness of Mathematical Learning,* published in 1701 and republished in 1721 and 1745. Among other things, Arbuthnot contended that mathematical learning was essential for all "those that would judge or reason about the state of any nation." Throughout his career, David Gregory actively promoted a similar, wide-ranging vision of the value of mathematics for both academic learning and national welfare. One place he expressed this was in his pedagogy. While a professor at Edinburgh, he actively sought to make mathematical learning accessible to all students, tailoring his lessons to various levels of ability. For general undergraduates, he taught well-trodden techniques that could be applied to practical pursuits like accounting, military engineering, and navigation. For more advanced students, he eagerly introduced cutting-edge topics, including Newtonian mathematical philosophy.[45]

Beyond his own classroom, Gregory also campaigned to make mathematics a more central part of Scottish education. In 1687, he drafted a manuscript memorandum, "To the Committee of Parliament for Visiting Schools and Coledges," arguing for a widespread increase in mathematics instruction. Since "Parliament had laid down such measures for encouraging trade and Manufactures," Gregory explained, "it is highly reasonable that such parts of learning be encouraged in our universitys as chiefly tend to advance those, which with the common Consent of Mankind are Mathematiques." He urged that students learn geometry, arithmetic, and even the mathematics of infinite series (which happened to be essential to financial applications like annuity valuation). Gregory also argued that the university professors of mathematics, and not the university regents, needed to play a much larger role in examining degree candidates. After all, "of the sober arts sever[al] are Mathematical, of which the Regents know nothing."[46]

After his move from Edinburgh, Gregory drifted away from the hot-tempered agitation of Pitcairne and settled into a more moderate Toryism. Yet he remained frustrated with the state of Scottish politics and committed to reform. (Rather than fomenting Jacobite revolution, Gregory and friends preferred to protest Presbyterian predominance by drinking toasts with illicit French claret.) Gregory therefore brought a distinctive political perspective to his work on the Equivalent. He was hardly a natural champion of Anglo-Scottish union, seen in both Scotland and England as primarily a Whig project.[47] Yet Gregory was deeply interested in the problem of Scotland's improvement, and he ultimately accepted that union with England was the best expedient. As with many of the leading calculators in this book, Gregory was always a kind of political outsider—an Episcopalian in Presbyterian Scotland, a Scottish Tory in Whiggish England—who saw mathematical calculation as a tool for changing the political status quo. Given his formidable scientific reputation, his reformist political spirit, and his deep commitment to mathematics as a tool of national improvement, it is little surprise that the Scottish Union Commissioners called him to be their mathematical guide in spring 1706.

Calculating the Equivalent

On April 22, 1706, appointed Commissioners from England and Scotland once again set about negotiating terms for a possible union between the two countries. Expectedly, questions of trade and taxation quickly came to the fore. One of the Commissioners' early orders of business was to request accounts showing the debts and revenues of the respective nations. The Scottish Commissioners were taken aback by the technical complexity of the task at hand. They could not produce especially detailed data on Scotland's finances, while they struggled to decipher the intricate accounts they received from their English counterparts. "We met at my Lord Glasgow's & continued upon [the accounts] a great part of that night but cou'd not make much out of them," recounted one of the Commissioners, John Clerk of Penicuik, in early May. In need of assistance, the Commissioners sought out "the Help of Doctor Gregory." Armed with Gregory's mathematical insights, the Scottish Commissioners returned to the talks and submitted a new set of demands to their English colleagues on May 9. Critically, they demanded that "an Equivalent be allowed" to Scotland in exchange for those taxes "already appropriated for the Payment of Debts properly belonging to *England*." Being "so desirous" of "an entire Union," the English Commissioners agreed to such a payment in principle the following day.[48]

The two sides agreed on the basic principle of an equivalent rather quickly. Figuring out the details was much harder. On May 11, John Clerk recalled that he was worried the negotiations had hit an impasse because the two sides could not agree on an appropriate equivalent to balance out the nations' tax burdens.[49] Instead of trying to hash out an agreement themselves, the negotiators agreed to approach the problem a different way: through calculation. They assembled a six-person committee who were instructed to carry out such a calculation "with the greatest exactnes, so as to be answerable for what they did." The committee included: the Commissioners' secretaries, a Mr. Dorington on the English side and David Nairne on the Scottish; two English representatives named Townsend and Taylor; plus, as the Scottish appointees, William Paterson and David Gregory—"all th[e]se Except the Secretar[ie]s being the most expert men in England, in matter of calculation," as Clerk put it.[50] It should be stressed: the decision to determine the Equivalent through this kind of expert calculation was not the only, or even the most obvious, choice the Commissioners could have made. They easily could have settled on a round figure through bargaining or decided to negotiate some alternative, nonmonetary compensation. Instead, the Commissioners, likely prompted by Paterson and Gregory, opted for a novel method.

The crack committee worked for a month-and-a-half on the calculation. It was often a frustrating task. Most labor-intensive was the meticulous analysis of English public accounting data. Using historical data, the committee estimated the future produce of thirty-eight distinct taxes, ranging from the excises and the customs, the traditional backbones of the English revenues (estimated at £286,000 and £257,000 per year, respectively), to the massive new land tax (nearly £2 million a year), all the way to a tiny duty on French shipping expected to yield £81. The calculators produced a similarly detailed "State of the Debts of the Kingdom of England." These revenue and debt figures were testimony to the rapid proliferation of new taxes and new debts raised during the Anglo-French wars. The fact that such projections were even possible—that data existed with which to make them—was also a sign of the substantial strides the English government had made in public accounting since the toils of the Commission of Public Accounts in the early 1690s.[51]

The Scottish side of the equation was a different matter. The majority of the Scottish public taxes were "farmed," or sold to private undertakers for a fixed fee: £33,500 (in English pounds sterling) for the excises on beer and ale and £28,500 for the customs. Those two numbers constituted the brunt of the data available about Scotland. The calculating committee bemoaned these limitations in an interim report to the Union Commissioners, requesting further detail on what goods were taxed in Scotland, what level of duties were assessed, and what those duties produced annually "upon a Medium of some years." They also admitted that "there seems to us to be no prospect of obtaining what we have thus humbly laid before your Lordships." Under the system of tax farming that prevailed in Scotland, the private undertakers who collected Scottish taxes were not required to account for the revenues they actually received. The six calculators did the best they could "under these difficulties."[52]

Most of the day-to-day work that went into calculating the Equivalent remains hidden from historical view, but we can get a few glimpses, from an interim report sent to the Union Commissioners and from David Gregory's papers at the University of Edinburgh. The calculators' efforts were intense and thoughtful. We can see, for example, that in June Gregory drafted a detailed series of queries to Sir Humphrey Mackworth, a Welsh-English MP, industrial entrepreneur, and fellow Anglican Tory, about constitutional and administrative questions relating to the earlier union of England and Wales (presumably 1536).[53] Another manuscript in Gregory's papers, likely produced slightly after the Equivalent calculation was finalized, offers an even more vivid picture of the committee's computational labors. Covered in disorganized decimal numbers and arithmetical scrap-work, the manuscript is a valuable relic of computation-in-practice (see Figures 2.1a and b).[54] The manuscript concerned one specific technical aspect of the Equivalent calculation. Many of the taxes that Scotland would be taking on after the union were temporary, levied by the English Parliament to serve wartime needs. All of these taxes were projected to expire in different months in the year 1710. Gregory's manuscript calculations wrestled with how this motley collection of short-term taxes would affect Scotland over the near future. For example, Parliament had levied a duty on coffee that was dedicated toward a specific loan. Those taxes were set to expire June 24, 1710. The Equivalent calculators collected historical data on those English taxes and projected how much Scotland would pay under them after the union. The English were estimated to pay £116,475 yearly for the coffee tax, implying that Scotland would pay £2,605, based on the historical ratio of English to Scottish customs taxes.[55]

The challenging part was determining the "present worth" of those new taxes for Scotland. If Scottish taxpayers agreed to accept a duty on coffee and pay an expected £2,605 per year through June 1710, how much money did that represent in 1706 as one lump sum? What difference did it make if the tax terminated in June 1710 or August 1710, or if it continued indefinitely? (These technical questions intersected with political ones. Just because these taxes were scheduled to expire in 1710, was it safe to assume that they would actually cease then, rather than be renewed by Parliament?)[56] As we saw, William Paterson had strong opinions about present value, contending that the strict mathematics of compound interest were superior to the conventional years-purchase approach. In his manuscript calculations, Gregory gave Paterson's ideas on present

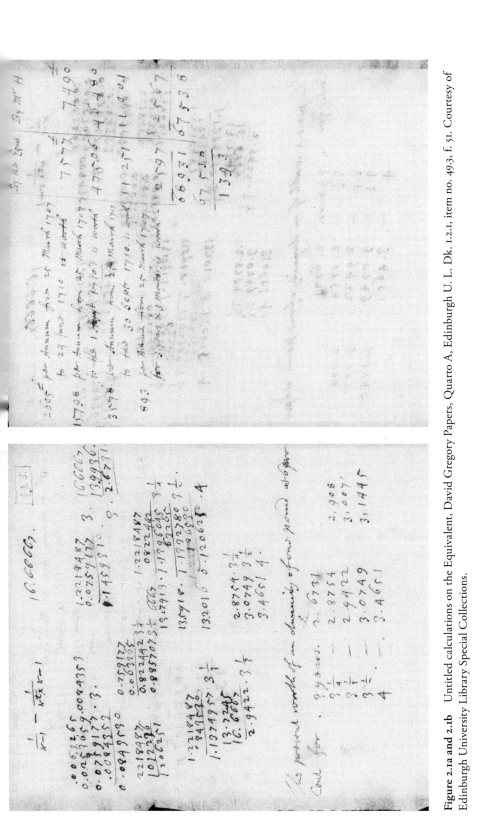

Figure 2.1a and 2.1b Untitled calculations on the Equivalent, David Gregory Papers, Quarto A, Edinburgh U. L. Dk. 1.2.1, item no. 49.3, f. 51. Courtesy of Edinburgh University Library Special Collections.

value a heightened level of mathematical formality, as evidenced by an algebraic formula scrawled at the top left corner of the manuscript page: $1 / (r-1) - 1 / (r^t \times r - 1)$. Mathematically, this represented the closed-form sum of a geometric series; financially, this was an algebraic formula for calculating the present value of an annuity using compound-interest discounting, with an interest rate r and time t.[57] For at least one person, fraught questions about the constitutional future were best worked out algebraically.

Gregory's manuscript calculations testified to the diligence and meticulousness of the calculating committee. So too did their final analysis, reported to the Union Commissioners on June 25.[58] It was an extraordinary work of quantitative skill and ingenuity (see Figure 2.2). The first part of the calculation combined their analysis of past customs and excise revenues in England and Scotland to project the respective proportions of taxes each country would pay under a unified tax system. For example, Scotland was projected to deliver £30,000 in customs revenue, versus over £1.3 million for England, implying Scotland would shoulder roughly 2.2 percent of future customs burdens. (Excises came to 2.4 percent.) The calculators broke down the English revenues tax-by-tax, distinguishing those that were earmarked for preexisting English debts. (79 percent of customs and 62 percent of excises were dedicated to debt.) Using the proportional breakdown between English and Scottish revenues, the calculation projected how much of the future Scottish tax revenues would be directed toward servicing preexisting English debts. Finally, the calculators applied compound-interest discounting to assess the "present value" of those expected Scottish taxes, differentiating among taxes with different expected durations and using a "discount rate" of approximately 6½ percent.[59] Adding up all of these present values for nine different relevant taxes, they came to a final number for the Equivalent: £398,085 and ten shillings.

The final Equivalent calculation reflected many subtle analytical choices on the part of the calculators. Some of these assumptions increased the size of the final figure, to the benefit of Scotland. For example, by choosing to distinguish between those taxes that were devoted to the national debt and those that were not, rather than simply comparing all Scottish taxes to all English taxes, the calculators may have augmented the final Equivalent by almost £60,000.[60] Other choices diminished the Equivalent's value, like the relatively high discount rate, the assumption that many of the taxes Scotland would assume would expire by 1710, and the choice to use current Scottish tax levels (namely £30,000 for customs and £33,500 for excise) rather than the expected levels after the union went into effect (£50,000 for both customs and excise). Minor adjustments to any of

	England Annual	Scotland Annual	Present Value
Customs:			
Total per Annum	£1,341,559	£30,000	
For Civil Government	253,514	5,669	
Unappropriated for Debts	25,480	570	
Customs Appropriated for Debt:			
Coffee Duty: to June 1710	116,475	2,605	7,577
Impositions: to August 1710	706,471	15,798	47,506
⅔ Subsidy: to September 1710	160,000	3,578	11,251
⅓ Subsidy: for 98 Years	79,619	1,780	27,145
Total Customs Appropriated for Debt	**£1,062,565**	**£23,761**	**£93,479**
Total Customs	£1,341,559	£30,000	
Excise:			
Total per Annum	£947,602	£33,500	
For Civil Government	269,837	9,539	
Unappropriated for Debts	85,581	3,026	
Excises Appropriated for Debts:			
For 99-year and Perpetual Annuities	192,400	6,802	103,731
From 1692: For 99 Years	132,433	4,682	71,004
For the Bank of England	137,460	4,860	70,866
For Lottery Tickets	104,624	3,699	56,410
Excises on Low-Wines	25,267	893	2,597
Total Excises Appropriated for Debts	**£592,184**	**£20,936**	**£304,606.10s.**
Total Excise	£947,602	£33,500	
Total Equivalent			**£398,085.10s.**

Figure 2.2 Summary of the official Equivalent calculation. Adapted from *The Minutes of the Proceedings of the Lords Commissioners for the Union* (London: Printed for Charles Bill, 1706), 61–68. All figures rounded to the nearest £, except **Totals**.

these assumptions might have justified a higher final figure—up to the £600,000 Paterson originally suggested, and possibly more.

Some historians have argued that the Equivalent was unfairly low, a sign that the Scots had been duped by the English and that the figure was a political construction, "negotiated" rather than truly calculated in an objective way.[61] There is no question that the Equivalent calculation was *political,* and it involved many negotiations on the part of the calculators. But it would be wrong to suggest that the £398,085.10s. figure was somehow *only* negotiated or *merely* political. The archival record shows no evidence that the calculators were trying to back into some predetermined figure, or that all their computational labor was just an elaborate act.[62] To assume that politicians and negotiators were using calculations to give legitimacy to a "negotiated" figure is to misunderstand the civic epistemology that prevailed in the early eighteenth century. In 1706, few in the English or Scottish public would have expected that the union negotiations should be balanced out by a precise mathematical calculation. If anything, they would have been more familiar with the idea that such an equivalent was arrived at by negotiation, rather than the obscure, closed-door work of mathematicians. It was calculation itself that needed justifying.

Reading the Equivalent

The Equivalent calculation marked the intersection of separate strands of quantitative enthusiasm that were emerging in the early eighteenth century. On the one hand was Paterson's reformist plan to improve the nation through calculated commercial thinking; on the other was Gregory's Tory-Newtonian appreciation of the mathematical order that governed heavens, bodies, and hopefully nations. These converging mathematical values show that Britons had many reasons to get excited about numbers. But they also had many reasons to doubt them. Many in the public, especially in Scotland, proved reluctant to accept the £398,085.10s. figure for the Equivalent. Many factors fed into public skepticism about that number: the forbidding technical complexity of the calculation, rampant errors in published versions of it, and especially resistance to the view of political knowledge for which that number stood.

Between the conclusion of Treaty negotiations in late July 1706 and the Treaty's final approval by the Scottish Parliament in mid-January 1707, the terms of the union were a subject of extensive discussion in various forms—from Parliamentary debate to pamphlet publishing, from petitioning in the outlying constituencies to full-fledged rioting in Edinburgh. Debates about the Union Treaty marked the maturation of Scotland's own public sphere.[63] As John Clerk reported, among all aspects of the union deal "none [were] more talked of, and less understood" than the Equivalent. Both supporters and critics of union had a hard time making sense of it. Union supporters worried that the calculations were too complex for their own good, casting doubt on the entire union deal. Anti-union commentators voiced an array of criticisms. It was hard to argue that a massive payment to some number of Scottish people was inherently unappealing. But many key questions remained unanswered: Was the number large enough? Were the calculations accurate and trustworthy? Who was really going to get paid, and in what order? Would the Equivalent really benefit Scotland in an enduring way?[64]

Many things made the Equivalent unclear to public audiences, especially in Scotland. One problem appears almost comically trivial in retrospect, but posed serious difficulties at the time: typographical errors. As union supporters like David Nairne lamented, published versions of the Union Treaty were riddled with "errors in writting or printing the equivalents."[65] Obvious and embarrassing problems arose, like the failure of a column of numbers to add up to its stated total. Union opponents looked upon these manifest mistakes with derision. The anonymous author of one *Essay upon the Equivalent* observed that, in the version of the Equivalent calculation he was looking at, "in the Summing of the Annual Incomes and Revenues of *England,* there is an Error of 125200 lib." That error actually benefited the Scottish, but such an irregularity bespoke a larger worry: "Either the Overseers of the Press, or the Accomptants have been at no great pains in the said Accompts; and if they have fallen into such a great Mistake in the addition of a few Articles; what Error may they not commit, in calculating of Proportions, valuing Annuities, &c."[66]

Union advocates like John Clerk tried to reassure readers that the calculation was correct and that all the wrinkles would be ironed out. "No Body can be answerable for faults in the Printing or Transcribing the Minutes of the *Treaty,*" Clerk explained, "but when the *principal Calcul* that was adjusted at *London,* comes to be produced, all Difficulties of that sort will be cleared." But it was not just in official accounts of the Equivalent that errors crept in. Numerous pamphlets on the union question printed various numbers incorrectly, even the most high-profile number of them all: the £398,085 total value for the Equivalent. John Clerk himself had a torrid time of it. His *Letter to a Friend* contained a misprint—"398088"—in the very sentence following his explanation of why typographic errors should not be a cause for concern. (See Figure 2.3.) Another pamphlet of his had the figure misprinted at different points as "398093" and "398084." One anonymous pamphlet, entitled *A Letter to a Member of Parliament, Anent the Application of the 309885 Lib: 10 Shil: Sterl: Equivalent,* got the number wrong in the title! (See Figure 2.4.)[67]

This proliferation of printing errors revealed broader challenges facing the use of calculation as a public form of political argument. For one, complicated numbers simply did not travel well in existing public media. Printers did not seem to care about getting the numbers right. This suggests that the eventual success of numbers as a mode of public knowledge cannot be explained in terms of their intrinsic mobility, resilience, or fidelity as a mode of communication.

Q. 9no. What is the reason that some People gave out a few weeks ago, that there were gross mistakes in the Calculation of the Equivalent?

R. No Body can be answerable for faults in the Printing or Tranfcribing the Minutes of the *Treaty*, but when the *principalCalcul* that was adjufted at *London*, comes to be produced, all Difficulties of that fort will be cleared:

Q. 10mo. What is the Reason, that the Scots *Commissioners for the* Union *did not foresee, That the Equivalent of* 398088 l. 10 fh: Sterlin *will be exhaufted in Payments for the* Englifh *Debts in the space of eleven Years or thereby, as seems to be agreed by two late Pamphlets.*

Figure 2.3 Page from [John Clerk], *A Letter to a Friend, Giving an Account How the Treaty of Union Has Been Received Here* ([Edinburgh]: s.n., 1706), 32. Seligman Collection 1706E C59, Rare Book & Manuscript Library, Columbia University. Note the erroneous figure "398088 l. 10 sh.," shortly after the discussion of "gross mistakes in the Calculation of the Equivalent." Seligman Collection, Rare Book & Manuscript Library, Columbia University. (1706E C59).

Numbers can easily be mistaken, miscopied, and misunderstood—in some cases even more easily than words. Calculations are not automatically "immutable mobiles" that traverse political distances and transmit facts faithfully.[68] They must be made to do so through human care and effort; numbers are only as faithful as people want them to be. In 1706, while some people, like Paterson and Gregory, cared deeply about the precision of the Equivalent calculations, others, like the Scottish printers, were not even concerned enough about those numerical values to avoid the most blatant errors.

These problematic printings were indicative of an even deeper complication. The different communities involved in calculating, interpreting, and approving the Equivalent were not playing with a common set of skills, rules, or expectations. Bringing their own knowledge and assumptions to bear on the figures, public readers and respondents often interpreted the Equivalent in their own ways, differently from what the six-person calculating committee intended.[69] One pro-union pamphlet, fittingly called *The Equivalent Explain'd*, offered an "explanation" for the complex number that bore almost no resemblance to the actual calculation Gregory, Paterson, and company had carried out. The pamphlet began by repeating the official line on why the Equivalent was necessary: "*England* did agree to repay to *Scotland*, the Equivalent of what *Scots* Money should be employed for paying of *English* Debt." The annual amount Scotland would pay toward preexisting English debts, the author explained, was projected to be "26536 lib. *Sterl.* yearly." Valuing this annual cost

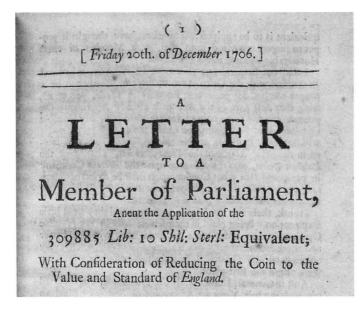

(1)

[*Friday* 20th. of *December* 1706.]

A

LETTER

TO A

Member of Parliament,

Anent the Application of the

30988 5 *Lib:* 10 *Shil: Sterl:* Equivalent;

With Confideration of Reducing the Coin to the Value and Standard of *England.*

Figure 2.4 Title page from [Anon.], *A Letter to a Member of Parliament, Anent the Application of the 30988$ Lib: 10 Shil: Sterl: Equivalent; With Consideration of Reducing the Coin to the Value and Standard of England* ([Edinburgh]: *s.n.,* 1706). Reproduced with the permission of Senate House Library, University of London. (ESTC: No20160).

"at the highest Value of Publick Funds, tho in Perpetuity, *viz.* at 15 Years Purchase," the author determined that the Equivalent "could at most amount to 398085 lib."[70] With the exception of the 398,085 final total, all of these numbers were fabricated. The explainer assumed that the calculation followed a basic years-purchase approach—it did not—at a rate of fifteen years purchase—wrong again. The pamphleteer then back-solved to determine the annual amount that Scotland was projected to contribute to the English debts, supposedly £26,536 annually—a number that never appeared in the official calculation. (Even that number was apparently a typographic error for £26,539, the value of £398,085 divided by fifteen!) The *Equivalent Explain'd* was almost a complete computational fiction. Unacquainted with the advanced financial mathematics employed by the calculating committee, the pamphlet made sense of the Equivalent in the only way its author knew how.

Given that union supporters often found aspects of the Equivalent calculations unintelligible, it is little surprise that opponents found much to dispute. Some of the arguments against the Equivalent had little to do with the calculations themselves. Andrew Fletcher worried about whether the Scottish would actually be able "to force the English to pay their Equivalent," given the limited representation they would receive in a unified Parliament (forty-five MPs for Scotland versus 513 for England and Wales). Fletcher and other critics like Roderick Mackenzie also argued that the Equivalent was not nearly large enough to repay all that had been lost with the failure at Darien. Others contended that, in fact, the Equivalent would only benefit a narrow segment of the Scottish nation, namely Company of Scotland shareholders. As one pamphleteer, likely William Black, put it: "a Bargain that concerns that particular Society . . . can never be understood or reckoned as an Equivalent to the Nation." The most sweeping criticism was exemplified by Robert Wylie, who wondered, "How can the *English* ever give unto the *Scots* an Equivalent for their Resign'd Sovereignty?"[71]

Several anti-union critics chose to take on the technicalities of the Equivalent directly. The Scottish Parliament itself played host to a flurry of critical computations beginning in October 1706. Opponents felt the Union Treaty placed "intolerable and unjust burdens" on the Scottish and railed against the Treaty's advocates for "mock[ing] them with the notion of an equivalent." Critics began to create their own numbers to show that the Equivalent was unjust. "They pretended to make calculations upon the proportions of the said equivalent," recalled one contemporary observer, Daniel Defoe. "And some, indeed, brought in their own rough draughts of disproportions." Amid this numerical chaos, Scottish MPs demanded to "enter upon more exact calculations," and consequently appointed a new Parliamentary committee for examining the Equivalent calculations. More outside help arrived in the form of "two very able accomptants, or arithmeticians," Dr. James Gregory, Professor of Mathematics at Edinburgh—David's brother—and Dr. Thomas Bower, Professor of Mathematics at Aberdeen. In early December, with the assistance of the two professorauditors, this new "Committee for Examining the Calculation of the Equivalent" reported to the Scottish Parliament that "they find that the Computation of the Equivalent mentioned in the Article is just, and that the Calcul is exact."[72]

This kind of political peer review hardly settled the matter. Many opponents chose to dispute the Equivalent in a more public way, through printed countercalculations of their own. One influential argument held that the Equivalent was nothing more than a short-term loan that the Scottish would have to repay quickly, doing little to improve the nation over the long term. One short but astute pamphlet, *An Essay upon the Equivalent,* presented a precise, step-by-step table showing that the Equivalent would be repaid in thirteen years according to the rules of compound interest. The author assumed 6 percent interest, and that the Scots would be effectively paying £44,697 annually in customs (£23,761) and excise (£20,936) toward the English national debt, per the *Minutes* of the Treaty Commissioners (see Figure 2.2). The author walked year-by-year through the repayment, deducting £44,697 from the outstanding balance each year, and then adding the 6 percent interest to the interim total to determine a new year-end balance.[73] The clever argument in the *Essay upon the Equivalent* demonstrated how the complex calculation carried out by Paterson and Gregory's team could be interpreted in very different ways by outside viewers. The *Essay* diverged from the Equivalent calculators on some substantial assumptions. It presumed a 6 percent interest rate, rather than the roughly 6½ percent used by the Equivalent calculators. It also did not assume, as the Equivalent calculators did, that many of the new Scottish taxes would actually end as planned in 1710. The most significant interpretive choice the author of the *Essay* made was simply to think about the Equivalent as a loan. In a way, the *Essay* employed the same kind of compound-interest calculation the Equivalent calculators had, only in reverse—turning the present lump sum of the Equivalent into a stream of future payments. The £398,085.10s. took on a very different political character when reimagined as a loan rather than a grant.

Another pamphlet entitled *Some Considerations in Relation to Trade,* often attributed to William Black but possibly the work of Andrew Fletcher, reimagined the Equivalent calculation in an even more radical way. Like the *Essay upon the Equivalent,* the author of *Some Considerations* reconceived the Equivalent as a loan. The pamphlet concluded with a table entitled "A Stated Accompt, by which it appears, that the Sum of 398085 lib. 10 sh. *Sterling,* given by *England* to *Scotland* as an Equivalent for engaging to pay their Debts, is repaid by *Scotland* with Interest at 6 per cent. per annum, in less as Eleven Years." (See Figure 2.5.) The author took a different computational approach from the author of the *Essay,* setting up a double-entry or "debitor-creditor" account. Strikingly, the author assumed not compound interest, but simple interest. On the "Debitor" side of the account were the funds Scotland would receive from the Equivalent: the £398,085.10s. the Scottish would be paid upfront, plus eleven years of 6 percent interest on that sum (£262,736 total), amounting to £660,821 in total. The "Creditor" side was the value of what Scotland had to pay for the Equivalent. This amounted to £44,697 in yearly taxes for eleven years, or £491,667, plus the interest that the English could earn over that period on the yearly payments the Scottish made. So the value of the first year's payment of Scottish taxes was not just £44,697, but that amount plus eleven years of 6 percent simple interest, or an additional £29,500. The second year's payment was worth £44,697 plus ten years of 6 percent interest, and so on. After eleven years, the total "Debitor" side added up to £668,667, more than repaying the total upfront value of the

Scotland is Debitor				Scotland is Creditor				
	Lib. St.	*sh.*	*d.*	"So much payable Yearly by *Scotland*":				*Lib. St*
Paid so much by *England*	398,085	10		Out of the Customs	23,761			
11 Years Interest at 6 *per Cent.*	262,736	8	5	Out of the Excise	20,936			
	660,821	18	5	"In all"	44,697			
				"For 11 Years . . . the principle"				491,66
England is overpaid	7,845	1	7					
				"11 Years Interest of our first Years Payment, at 6 *per Cent.*"	29,500	0	0	
				"10 Years Interest of 2d. . . . Payment"	26,818	3	9	
				"9 Years of the 3d. Payment"	24,136	7	4	
				"8 Years of the 4th. Payment"	21,454	11	0	
				"7 Years of the 5th. Payment"	18,772	14	7	
				"6 Years of the 6th. Payment"	16,090	18	2	177,000
				"5 Years of the 7th. Payment"	13,409	1	10	
				"4 Years of the 8th. Payment"	10,727	5	4	
				"3 Years of the 9th. Payment"	8,045	9	1	
				"2 Years of the 10th. Payment"	5,363	12	9	
				"1 Years of the 11th. Payment"	2,681	16	4	
	668,667	0	0	**Total**				668,667

Figure 2.5 "Stated Accompt, by which it appears, that . . . [the Equivalent] is repaid by Scotland with Interest at 6 per cent. per annum, in less as Eleven Years," in [William Black?], *Some Considerations in Relation to Trade, Humbly Offered to His Grace Her Majesty's High Commissioner and the Estates of Parliament* ([Edinburgh?]: *s.n.,* 1706), 15. The calculation uses a simple interest technique to project the future repayment of the Equivalent, compared to the compound interest technique used in the original Equivalent calculation.

Equivalent (£660,821). As the table states, England was effectively "overpaid" £7,845.[74]

The mismatch between the compound-interest calculations behind the official Equivalent figures and the simple-interest ones in *Some Considerations in Relation to Trade* is striking. But it should not be taken simply as evidence that critics could not follow the complex mathematics involved. The divergences were indicative of much deeper ambiguities surrounding value, time, and calculation in the midst of Britain's financial revolution. Was compound interest really a better way to interpret the Equivalent "repayment" than simple interest or the well-held years purchase technique? Was the complex Equivalent project really, in effect, a loan? These questions did not have settled answers.

Paterson, Gregory, and the Equivalent calculators had their own perspectives, which became encoded in the £398,085.10s. figure itself. But once their numbers were circulating in public, there was nothing to stop others from reprinting, recalculating, and reimagining their figures in their own ways. The reception of the Equivalent showed that calculation could be a rather unruly mode of public political reasoning. Critics of quantification in the modern era argue that efforts to address political questions through quantitative means often have the effect of foreclosing political controversy, by translating substantive political questions into "technical" ones.[75] This was not the case in 1706; if anything, the opposite was true. The Equivalent calculation brought new (technical) questions into public view and revealed their political stakes. It did not foreclose debate, but enabled it.

"It Is Not Arithmetick Only That Can Make a Man Understand"

The official Equivalent numbers only begat more numbers. As critics of the union produced counter-calculations to discredit the Equivalent, union advocates produced even more calculations to counter the counter-calculations. In doing so, the Scottish public came upon a broader set of questions *about* calculation itself, and the place it ought to hold within the nation's civic epistemology. How much should calculations be relied upon in answering tough public questions? What made a specific calculation valuable or trustworthy? And did trusting in calculations mean trusting in calculators?

These questions lay at the heart of a heated exchange of anonymous pamphlets pitting William Black against a sharp-witted opponent, probably James Donaldson. As we have seen, Black (possibly alongside Andrew Fletcher) penned multiple computational texts disputing the Equivalent and projecting the union's negative effects on Scottish trade and revenue. Donaldson, in turn, castigated Black's arithmetical criticisms as "Misrepresentations, and ill grounded Jealousies." Donaldson offered crafty computations of his own to point out holes in Black's reasoning and make his adversary look incompetent and deceitful. Donaldson's primary conclusion was that "What ever skill our Author [Black] may have in matters of his own vocation, he seems to be no great Mathnmatitian (*sic*)." Black shot back, calling his opponent to account for his computational trickery. Black pointed out that, in giving "his Judgment upon these Papers," Donaldson had "pass[ed] by the several Heads contain'd in them, which I think is the Substance," instead going straight for "the Calculs." For Black, the numbers were only part of his argument; for his opponent Donaldson to fixate on minute computational details was a dirty move. Black admitted that he did not have the most elite mathematical credentials, but he rejected the notion that one had to be a "great Mathnmatitian" to speak political truth with numbers. Instead, he challenged his opponent to take seriously the real meaning of his figures, and not just fixate upon their technical shortcomings: "If he takes a second View . . . he will find that all these are as real as his Imaginary Mathematical Calculs." Even if Donaldson could claim to have the more sophisticated calculations, it was Black's that were more real.[76]

The squabble between Black and Donaldson revealed deeper disagreements about the place of calculation in political life. Both clearly acknowledged a place for numbers in advancing their political values. But they had different attitudes about how numbers ought to be judged and who had the authority to produce them. For the anti-unionist Black, calculation was a tool that all citizens could use to challenge supposedly authoritative claims made by those in power. The calculations did not have to be unassailably certain, nor did the calculators have to be impeccably skilled. For the pro-unionist Donaldson, that kind of computational criticism was deceptive and dangerous. The question of the Equivalent demanded the most meticulous calculations, carried out by the foremost mathematical minds. Black's position somewhat echoed the "Country" calculations advanced by the Parliamentary Accounts Commissioners in the 1690s; Donaldson's recalled the elitist approach of William Paterson.

Many people, including both critics and supporters of union, shared Black's uneasiness with the Equivalent and the elitist circumstances of its production. This revealed itself in the negative feelings many contemporaries expressed about the project's chief engineer, William Paterson. Daniel Defoe, man-on-the-ground in Edinburgh for Robert Harley—who had since risen to the position of Secretary of State—observed this mounting resentment toward Paterson. In late November, Defoe reported that he had been called before the Scottish Parliamentary committee examining the Equivalent, "where I understand Mr Paterson is to be also." While admitting he had not seen much of Paterson in action yet, Defoe relayed that he had not heard good things about him: "that Gentleman is full of Calculates, Figures and Unperforming Numbers, but I see nothing he has done here nor does any body Elce speak of him but in Terms I Care not to Repeat."[77]

A more poignant, if less colorful, reflection on the problem was offered by David Nairne, Secretary to the Scottish Union Commissioners and, with Paterson and David Gregory, a member of the Equivalent calculating committee. In a November I letter from London to the Earl of Mar in Edinburgh, where there was public rioting against the pending Union Treaty, Nairne lamented how many Scots seemed unable to understand the tremendous commercial benefits the union would bring. "For my part I am convinced," Nairne asserted, "if that very particular were understood it would turn the mob on the other side." In particular he worried that the complexity of the Equivalent had so confounded people that it hindered support for the union. Nairne was frustrated that no one had been successful in explaining the Equivalent to public audiences. He suggested the best person for that job was not the officious Paterson but the more reticent David Gregory. "This leads me to regrait that Dr. Gregorie was not there about the Equivalent," Nairne lamented. His correspondent Mar, one of the Scottish Commissioners for the union, agreed.[78]

Yet Nairne also acknowledged that the negative Scottish reaction to the Equivalent was not just a matter of the public's ignorance. Even he was not sold on the idea that so much responsibility for Scotland's future prosperity ought to be left to the cryptic calculations of pedantic specialists. "These two may be men of letters and figurs," he noted, "but if they are not acquented with . . . the practicall part of business and even the custome of this place soe that they know the grounds upon which things were done, its impossible they can understand what they are about." The commitment of both Paterson and Gregory to the governance of mathematics left them out of touch with commercial and political realities. Nairne cited a telling example, concerning—of all things—the valuation of annuities. Paterson and Gregory seem to have continued pressing the patronizing point from Paterson's Wednesdays Club pamphlet, reproaching members of the public for not following the rules of compound interest and thus offering seemingly illogical prices for certain annuities in public markets. Nairne admitted he had little patience for their presumption that their calculations trumped practical experience and local knowledge. Whatever "th[eir] notion or th[eir] reason or the practise of other countries," Nairne explained, that was not how things were done in Britain.[79]

Nairne then offered a most telling comment. "So your Lordship may easyly judge from this," he wrote to Mar, "it is not arithmetick only that can make a man understand the equivalent." The problems that made the Equivalent so confounding and so controversial went far deeper than the intricacies of the calculation. To accept that £398,085.10s. was a fair price to compensate Scotland for its economic sacrifices in the union was not just about "understanding" the mathematics. It was a question of believing in what the mathematics said—or, perhaps more precisely, believing in the people who said what the mathematics said. It meant choosing to grant authority to numbers over custom and experience. In 1706, that was a disconcerting idea for all but a few in Britain. The disputes over the Equivalent calculations exemplify a phenomenon Sheila Jasanoff has observed in the case of modern political disputes over controversial sciences and technologies: namely, that public attitudes about such complicated issues can never be fully explained in terms of the "public understanding" of science and technology (or lack thereof). Rather, the way that publics respond to expert knowledge claims reflects much deeper political and epistemological values.[80]

In the end, the Great Project of the Equivalent achieved a kind of partial success. Most immediately, the plan for an Equivalent to Scotland was ratified and the number £398,085.10s. was woven into the constitutional fabric of the new Great Britain. The Act of Union, including the Equivalent, was approved by the Scottish Parliament on January 16, 1707, and in England shortly thereafter. This was not until the Treaty had undergone substantial debate, including the contentious Article XV on the Equivalent, whose complicated clauses took weeks of Parliamentary sessions to work through.[81] Disputes about the technical calculation of the Equivalent did not ultimately affect the final number, £398,085.10s., but public protests and written critiques were not without influence. Such public pressure led to at least one crucial adjustment to the economic terms of the Treaty, namely, a revision of the tax rates at which Scottish "twopenny" ale was assessed under the new excise system.[82]

Yet, beyond the number itself, few things about the Equivalent project went as planned. After the union went into effect, yet another commission was appointed to administer the Equivalent funds. The first payment was made with great fanfare on August 5, 1707, delivered by twelve wagons guarded by 120 Scottish dragoons. But that payment itself was incomplete—only £100,000 came in silver, with the remaining £289,085.10s. delivered in paper credit notes. Some of the recipients of Equivalent money were paid relatively quickly, notably the former shareholders of the Company of Scotland, who received £232,884 of the funds (about 59 percent of the total, or almost exactly the amount Paterson had suggested in 1705). But in the ensuing years, endless practical and interpretive problems arose regarding how to institute the Equivalent, how to raise the requisite funds, and how to distribute them to the designated creditors. Those who were not paid immediately, like many existing creditors of the Scottish government, waited a long time for payment, and many ended up selling their paper Equivalent notes on the secondary market at steep discounts.[83] An enterprising set of Equivalent creditors—or, probably more accurately, speculators who had bought up discounted Equivalent claims—petitioned for the right to incorporate as the "Equivalent Company" in 1724. Three years later, that Company cut a Parliamentary deal to begin a new banking business in Scotland. That enterprise became the Royal Bank of Scotland (RBS).

William Paterson would no doubt have been pleased with some aspects of the outcome of his great Equivalent project. The calculating committee's intricate analyses were officially affirmed. His tireless efforts to extract compensation for Darien were finally answered. In a more circuitous way, his efforts advanced the Scottish financial industry. In almost every other way, though, the Equivalent failed to deliver on Paterson's calculated vision for Scotland's future. The Equivalent money never became the basis for a "National Fund" for improving Scottish commerce. Various different commissions and committees would be assembled to deal with the disorderly legacy of the Equivalent. (One Parliamentary committee composed in 1715 to look into the chronic mismanagement of the Equivalent would contain many key characters in this book, including William Lowndes, Archibald Hutcheson, and Robert Walpole.)[84] But none of those bodies amounted to anything like the Council of Trade Paterson had envisioned. Beyond these practical failures, the political reality of the Equivalent fell short of Paterson's ambitions in a more fundamental way. For Paterson, the Equivalent symbolized a polity where political leaders looked to business experts for guidance, where citizens accepted the rational dictates of financial mathematics, where people feared ignorance more than corruption. Instead, the publication of the Equivalent calculations revealed a polity full of careless printers, skeptical citizens, and quantitative critics, all unwilling to entrust Scotland's future to Paterson's elitist project.

In many ways, the actual *numbers* associated with the Equivalent thrived in that disorderly political environment. Those numbers got reprinted and misprinted, explained and excoriated, reconstructed and deconstructed. And those numbers generated more and more public numbers. Even if many people doubted that the actual Equivalent figures were correct, each numerical argument that arose lent publicity and credence to Paterson's innovative notion that calculation could solve the riddle of union. Citizens across Britain found many different reasons to put political trust in numbers—whether it was William Paterson's reverence for commercial expertise or David Gregory's practical Newtonianism or William Black's Country-style computational criticism. Envisioning a calculated future for the nation did not mean accepting any one specific set of numbers or one set of quantitative values.

It also did not mean accepting the authority of any one group of calculators. In the tale of the £398,085.10s., *calculation,* as a mode of political thinking, met with far greater success than the *calculators* themselves. David Gregory made out reasonably well, at least momentarily. His work in advancing the union was rewarded with command of reorganizing the Scottish Mint, no small honor given that the parallel position in England had been held by Isaac Newton. (But he died little over a year later, in October 1708.) Paterson's fate was much more fraught. He was widely distrusted, ignored, even mocked, all while his computational project was reshaping the nation. Instead of being praised for his expertise and foresight, he was dismissed as a man "full of Calculates, Figures and Unperforming Numbers." Adding injury to insult, Paterson had to spend years trying to get his own cut of the Equivalent money. He did not get paid until 1715.[85] Paterson hoped for a new quantitative age in Britain, in which experts like himself were trusted to calculate the nation's future. He helped bring about a quantitative age, but not the one he wanted—and not one that necessarily wanted him.

CHAPTER THREE

The Balance-of-Trade Battle and the Party Politics of Calculation in 1713–1714

ONE OF THE GREATEST things to happen to the political career of numbers was the rise of two-party politics. In the quarter-century that followed the 1688 Revolution, the division of Tory versus Whig increasingly dictated British political life. Partisan enmity hit its peak in the early 1710s, when the Tory Party took control of Parliament, led by former Commissioner of Public Accounts Robert Harley. Not coincidentally, that period also gave rise to the most intense computational conflict yet seen in political history. Beginning in 1713, Tory and Whig calculators spent month after month trading computational blows over the value of one number: England's *balance of trade* with France, the value of England's exports to France minus its corresponding imports.[1] The metric translated the mechanics of merchant accounting to the transactions between nations. Many economic thinkers in Britain and elsewhere had come to believe that, like a trader who bought more than he sold, a nation that imported more than it exported had to make up the difference in hard currency, decreasing its national wealth and enriching its trading partner. The Tories were convinced this number was positive, indicating that French trade benefited the nation; their Whig opponents believed it was negative, and thus trading with France was harmful. The two sides never managed to settle on the correct value, or even whether it was greater or less than zero. But through their heated disagreements, the Tories and Whigs did forge an implicit agreement about the value of calculation as an instrument of political reasoning.

The debate stemmed from a pending Treaty of Commerce between Britain and France. Harley (recently ennobled as Earl of Oxford) and his ministry engineered the Commerce Treaty as an economic complement to the Peace of Utrecht, which promised to end the decade-long War of Spanish Succession. The Tory Commerce Treaty proposed to eliminate tariff barriers that had restricted Anglo-French trade for decades. The Whigs, aggressively anti-French in foreign policy and supported by many established mercantile interests, fervently opposed it. In an effort to generate political support for their Commerce Treaty, Tory leaders like Harley and Secretary of State Henry St. John (made Viscount Bolingbroke in 1712) went to the press, sponsoring a dedicated trade newspaper called the *Mercator: or, Commerce Retrieved,* led by Charles Davenant (now the Inspector-General of Imports and Exports), with assistance from others like ace polemicist Daniel Defoe.[2] In its first issue in May 1713, the *Mercator* explained that its agenda was to overturn the widespread misconception that trading with France made Britain poorer—and to disprove it using numbers. The paper opened with a numerical dare: "Take a Medium of any three Years for above forty Years past," the *Mercator* instructed optimistically, "and calculate the Exports and Imports to and from *France,* and it shall appear, the Balance of trade was always on the *English* side."[3]

The *Mercator*'s bold computational challenge echoed the numerical enthusiasm recently voiced by quantitative boosters like John Arbuthnot, William Paterson, and Davenant himself. But the calculators behind the *Mercator* soon found that, in the day's bitterly partisan climate, settling such a heated controversy with numbers was easier said than done. The *Mercator*'s early numerical agitation encouraged people to question their long-held prejudices against the French trade, but those quantitative efforts did not yield a swift and comprehensive victory in the Tories' favor. Instead, they triggered a furious computational battle between the two parties—in the words of one leading economic historian, "the first big argument about a major piece of English economic policy in which was deployed that familiar modern weapon: statistics."[4] In terms of its intensity, longevity, and the political significance accorded to the calculations, the balance-of-trade battle far surpassed earlier quantitative controversies, including the Equivalent debates. The amount of ink spilled during the controversy was unprecedented, some 10,000 copies of commercial texts per week at the highest point.[5] The Tory *Mercator* published a new issue three times a week without interruption from May 1713 to July 1714, a total of 181 issues. Articles ranged across a variety of topics—the state of Britain's distilleries, buccaneering in Jamaica—but the authors regularly circled back to the calculation of trade balances and constantly published new numerical data. Whig propagandists pled their case in a slew of articles, pamphlets, and, after August 1713, their own twice-weekly newspaper, the *British Merchant.* The compiled issues of that paper would be republished along with other Whig writings in three weighty volumes in 1721, totaling some 1,300 pages. That compilation "long served as a text-book of mercantilist economics."[6]

Like the calculation of the Equivalent, historians have often dismissed the "battle of figures" as a cynical and peripheral affair, irrelevant to the real story of the Commerce Treaty—a story more about international diplomacy, party intrigue, or economic ideology than trade statistics. As economic historian D. C. Coleman puts it: "the figures were meaningless when they were not fraudulent, and in either event were almost certainly wrong."[7] Coleman is right that the numbers marshaled by the Tories and Whigs would fall short of modern standards of statistical accuracy. But they were certainly not meaningless. The sheer amount of time and effort that went into producing the calculations, the vocal interest that Parliamentary representatives showed in them, the fact that the Whig *British Merchant* would be republished four times—all testify that the numbers mattered significantly to the practice of politics surrounding the divisive Commerce Treaty. So why did Tory and Whig calculators produce such vast quantities of numbers for so long? What functions did these numbers serve, and what did the calculators and their party patrons think they could do? And what does this most spectacular calculative conflict reveal about the development of Britain's quantitative age?

One explanation that does not suffice to explain the cacophony of calculations in 1713–1714—just like it cannot explain the calculation of the Equivalent— is that the numbers were just rhetorical decoration, designed to give cynical partisan projects a luster of objectivity. Of course, Tory and Whig polemicists were trying to make a political case. But to contend that the parties were trying to dupe the public by couching their arguments in the authority of numbers begs the question of this entire book. Another possible explanation is that the numerical turmoil was simply a consequence of more fundamental, preexisting conceptual commitments, particularly about the balance of trade. In other words: Britons in 1713 set out to calculate trade balances because they had long believed trade balances were indispensable to understanding international commerce. As an idea, the balance of trade had indeed played a significant role in political-economic thinking for at least a century. But that did not make it automatic that Britons would become so fixated on technical *calculations* about trade balances in 1713. For one thing, the balance-of-trade model was hardly hegemonic at the time. For decades, critics had pointed out problems with it and proposed other ways to think about international commerce.[8] Even more to the point, despite the fact that the balance-of-trade concept had been around for a long time, there had been at best sporadic interest in actually trying to calculate it.

So what made Britons so obsessively concerned with calculating this number in 1713? The answer lies in the distinctive political culture that had emerged in Britain since 1688. The decisive factor was two-party politics. The contentious events of the previous quarter-century had shown Britons that calculations were an excellent tool of political contention—for attacking policies, debunking arguments, and calling out opponents. As politics in Britain became more polarized, numbers became even more prized. For both parties, the initial impetus to calculate the Anglo-French balance of trade was adversarial, driven more by a desire to attack their opponent's position than to make a positive argument for their own policy. The Tory *Mercator* took to calculating trade balances to discredit what it saw as a decades-old misconception about the dangers of French trade, long fostered by self-serving commercial interests; Whig polemicists, in turn, took to calculating trade balances to discredit the imperious *Mercator*. The balance-of-trade battle exemplified how, in post-1688 politics, numerical calculation thrived because of its usefulness as an instrument of dispute.

In the heat of political combat, calculation also proved highly adaptable. Numbers could speak for diverse partisan perspectives. In Chapter 2, we saw how numerical enthusiasts like William Paterson, David Gregory, and William Black embraced numbers for different reasons. This was representative of an essential feature of the history of Britain's quantitative age. The public authority of numbers did not arise from one, unitary vision of what made numerical calculation a valuable way of knowing. Rather, it arose from the interaction of multiple, sometimes conflicting, perspectives on "the values of quantification." Lorraine Daston has observed that different scientific communities have identified different *epistemic virtues* in quantitative reasoning, including accuracy, precision, communicability, and impartiality.[9] This was evident in 1713 as well: voices for both parties contended that the Commerce Treaty question ought to be solved by the numbers, but they recognized different virtues in numerical thinking. Tory polemicists argued that an expert calculation of the nation's trade balance was necessary to protect Britons from misrepresentations advanced by special interests. Their Whig opponents countered that plain-spoken numbers justified the good sense of experienced merchants. For the Tories, the critical virtue of numbers was their incisiveness, their ability to "undeceive"; for the Whigs, the chief virtue of numbers was their "common sense" clarity.[10]

These differences were not limited to what the two sides said about numbers, either. They were evident in their different computational styles—in the very numbers they produced. Tory calculators developed more intricate calculations, foregrounding theoretical and methodological issues, while Whig calculators tried to stick close to the "raw data." These different calculating styles, combined with the highly irregular import-export data the calculators had to work with, created the conditions for a nearly interminable numerical struggle. The parties agreed on *what* number could settle the trade question— the Anglo-French trade balance—but never came close to settling what that number was. But that inconclusiveness was anything but inconsequential.

The Balance of Trade, as Concept and Calculation

From a certain perspective, it may seem hardly surprising at all that Britons became so obsessed with calculating trade balances in 1713. According to one account of early modern economic thinking, an account that dates to comments by David Hume and Adam Smith from later in the eighteenth century, a fixation on the balance of trade had long been the organizing fallacy of "mercantilist" political economy, dating back at least to the early seventeenth century. ("Nothing, however, can be more absurd than this whole doctrine of the balance of trade," Smith chided in his *Wealth of Nations*.)[11] Smith and company were right that the balance-of-trade concept had deep historical roots. But the attention it garnered in 1713, and particularly the quantitative attention, dwarfed anything that went before. That surge in prominence cannot be chalked up to intellectual momentum or the hegemony of "mercantilism"—if such a thing existed.[12] Looking at the earlier history of the balance-of-trade idea reveals two crucial points, which help bring into focus what was distinct about the 1713–1714 debates. First, the balance of trade as a *concept* and the balance of trade as a *calculation* followed distinct historical trajectories, and the prominence of one did not necessarily imply the prominence of the other. Many early modern thinkers had called upon the balance of trade as a model of international trade but showed little interest in calculating it precisely. The trade-balance concept could be a very useful heuristic tool—for explaining past changes in exchange rates, for example, or projecting the likely effect of a new tariff—without any real numerical data at all. Second, the balance of trade was not a completely hegemonic doctrine before 1713. By the early eighteenth century, many observers had identified limitations with the idea, from conceptual problems with a "zero-sum" picture of global trade to practical challenges in actually calculating balances from available data.

The kernel idea behind the balance of trade—reckoning the difference between imports and exports to judge the health of the nation's commerce—was floated at least as early as the mid-fourteenth century. The English monarchy had expressed explicit interest in recording import-export data in the 1560s, and economic treatise writers, like the author of the 1581 *Discourse of the Commonweal of This Realm of England,* began to give the idea more systematic consideration shortly thereafter. Interest in the differences between imports and exports swelled in the 1610s, when concerns about the nation's sagging woolen exports prompted Crown officials, like Surveyor-General of the Customs Lionel Cranfield, to compute summary import-export balances.[13] The term "balance of trade" was possibly coined in 1615–1616 during debates in the English Privy Council, perhaps by Francis Bacon.

England's economic hardships worsened in the 1620s, bringing a sharp decline in woolen exports, a dire scarcity of money, and social unrest. It was during that decade that balance-of-trade thinking began to attract sustained political attention. In 1622, King James I arranged official inquiries into the ongoing crisis. Debate hinged on England's critically low levels of specie money. Gerard de Malynes advised that the Crown ought to directly enforce exchange rates to prevent English money from fleeing to the Continent. In an effort to thwart Malynes's argument for currency control, merchant polemicists Edward Misselden and Thomas Mun rallied around the concept of the balance of trade. They argued that England's commercial distress stemmed from the movement of goods and not precious metals. If England's stock of gold and silver declined, they argued, it was because the nation imported more goods than it exported and had to pay the difference in cash—something that could not be changed by monarchical meddling with exchange rates.[14] These early seventeenth-century merchant-theorists did much to popularize the balance-of-trade idea, but they were ambiguous about whether it was important, or possible, to calculate trade balances accurately. Malynes was expectedly hostile to the idea of reducing the complexities of commerce to a static number; in his mind, any such attempt was a fanciful exercise in conjecture. Misselden was enthusiastic about developing a rigorous, quantitative study of trade based on import-export data, though he did not get very far in doing so. Mun fell in between the two, agreeing that the balance was a valuable heuristic, but skeptical about whether a truly accurate "account can possibly be drawn up to a just balance" given available data.[15]

In ensuing decades, the notion of calculating trade balances to guide commercial policy would continue to attract a mix of enthusiasm and ambivalence. Within government administration, sporadic trade councils were appointed to calculate official trade balances in 1622, 1650, and 1668. Each time, they made no more than limited progress. Economic writers would periodically make their own efforts in that direction. In 1663, for example, Privy Councilor Samuel Fortrey advised the newly restored King Charles II that, in order to revive the nation's "store and trade," it was necessary to curtail trade with France. To underscore this point, Fortrey produced a rough balance-of-trade calculation based on sixteen different groups of French imports, estimating that "our trade with France, is at least sixteen hundred thousand pounds a year, clear lost to this kingdom."[16]

Fortrey's computation marked the beginning of a period of intense concern about England's woolen exports and the destructive effects of recent anti-English trade policies put in place by France. In the mid-1670s, merchant lobbyists pressed Charles II's government to redress the inequalities in Anglo-French trade. In late November 1674, fourteen eminent London merchants produced a *Scheme of Trade* purporting to show that England had a negative trade balance with France amounting to £965,128, based on evidence collected by the Customs administration. This key negative figure became a rallying cry for anti-French economic sentiment, encouraging the establishment of stringent restrictive tariffs against French goods in 1678.[17] The 1674 *Scheme* was a remarkable early example of a public calculation used to influential political ends—an early sign of the quantitative age to come. But like other examples of political calculations in pre-1688 politics, the *Scheme* remained a somewhat isolated occurrence. It produced relatively little contemporary debate and did not yet produce durable interest in calculating more accurate trade balances, among either government administrators or public commentators. Agents for Charles II made a passing attempt at an updated trade balance in 1679, though that effort, and subsequent ones in the 1680s, made only modest headway.[18]

At the same time, observers in the late seventeenth century began to voice an array of doubts about the theoretical coherence and practical usefulness of the balance of trade as a concept. Thinkers like William Petty and Josiah Child began to question whether a simple reckoning of imports and exports could say anything meaningful about the complexities of global trade, while government administrators debated whether maintaining a positive trade balance ought to outweigh other imperatives in shaping fiscal policy. An especially disruptive new line of thinking, exemplified by John Houghton and Nicholas Barbon, contended that material consumption—even imported luxuries—might in fact be a powerful driver of economic growth, an evident contradiction to the anti-import premises of balance-of-trade theory.[19]

In the 1690s, interest in import-export statistics flared up once again, this time motivated in large part by the broader enthusiasm for public accounting taking hold in post-1688 politics. In 1695–1696, during vigorous debates about the nation's troubled silver coinage, commercial thinkers like Abraham Hill and John Pollexfen, both appointed to William III's new Board of Trade in 1696, contended, as did Charles Davenant, that the ability to compute up-to-date trade balances was a reasonable expectation of an accountable government. To feed those demands, a new office of Inspector-General of Imports and Exports was created in 1697, first occupied by experienced Customs clerk William Culliford, followed by Davenant himself in 1703.[20]

The foundation of the Inspector-General's office created an apparatus for managing national trade data unprecedented among European contemporaries, and that office did help make official data on the nation's trade more widely available for reference and reflection.[21] Yet, the Inspector-General's office also brought analysts face-to-face with the tremendous difficulty of actually calculating trade balances. The main source of data available were the record-books kept by Customs tax officials at English ports, most prominently London. These were relatively reliable for determining the quantities of goods shipped in and out of England, but they were deeply problematic in other ways. Customs records were much less reliable when it came to assessing the *value* of goods moving through the ports. Merchants' tactics also distorted the information in the Customs books. Davenant complained, for example, that some merchants would deliberately misstate the volume of woolens they exported to trick their domestic competitors into thinking certain foreign markets were oversupplied.

In 1711, Davenant drafted a lengthy report to the Commission of Public Accounts detailing the nation's trade balances with seven different trading partners for 1699–1704, and explaining the considerable obstacles he had encountered in trying to determine where those balances had stood historically. He specifically questioned the conclusions of the 1674 *Scheme of Trade*, hinting that there were serious analytical reasons to doubt the Anglo-French trade balance was negative at that point. Elsewhere in his 1711 report, Davenant ventured an even more penetrating critique of balance-of-trade reasoning. His concern was with two historical years that were sometimes cited as proving England's negative balance with France, 1663 and 1669. Davenant noted that, according to a strict reckoning of England's imports and exports with *all* nations based on extant Customs data, England had a *total* negative trade balance with the rest of the world during those two years. This implied that during those years England's international trade had made the nation poorer overall, an absurd conclusion given that the 1660s were widely regarded as thriving times for English commerce. "Here will arise a Question," he reflected, "how far the Excess between the Exports and Imports may be deem'd a certain Rule whereby to judge, whether a Country gets or loses by its Trade." Perhaps Davenant was trying to cast doubt on the balance of trade, and thus fears about the negative Anglo-French balance, in order to prepare the ground for the Tories' push to re-open French trade.[22] But whatever the motivation, it would seem that, shortly before the outbreak of the Commerce Treaty controversy, the very person tasked with tracking the nation's trade balances was unsure how useful or legible that metric even was. The balance of trade would soon come back with a vengeance in 1713—led by Davenant himself. But that surge of enthusiasm was not just the inevitable result of some unquestioned economic dogma. Some other force was needed to give that concept a new immediacy.

Tories, Whigs, and the Specter of Partisanship

That force was party politics. Beginning in the late seventeenth century, the boundary dividing "Tory" and "Whig" became the defining axis of political life in Britain, structuring contests over elections, legislation, and administrative control. The two great parties first coalesced in the late 1670s around the "Exclusion Crisis," when a group of politicians attempted to bar King Charles II's Catholic brother James—future King James II—from succeeding Charles as king. Those who supported this Exclusion Bill were labeled "Whigs" by their opponents, who in turn earned their own abusive epithet, "Tory." (The source of the names remains unclear.) By 1688, both Whigs and Tories largely agreed on the need to oust James II. But otherwise they envisioned the nation's future course—constitutional, economic, diplomatic, religious—very differently. Tories sought a strong and independent monarchy, economic policies that favored agriculture and landowners, a more insular foreign policy focused on strong naval defenses, and the unquestioned moral authority of the Anglican Church. Whigs wished for expanded authority for Parliament, aggressive support for domestic manufacturing, a muscular foreign policy targeted at limiting the French threat, and religious policies generous to dissenting Protestants. Those competing visions shaped political conflict for decades. The remarkable frequency of elections—eleven between 1688 and Queen Anne's death in 1714—meant party candidates faced off regularly across the country, giving the parties ample opportunity to define themselves against one another.[23]

The dynamics of partisan conflict did not remain constant over Britain's long eighteenth century. The relative significance of Tory-Whig affiliations fluctuated over time, as did the individual makeup of the parties. At times, competing systems of political identification arose. In the 1690s, the animosity between the ministerial "Court" and "Country" outsiders cut across and sometimes superseded the Tory-Whig axis. At other times, intraparty fissures proved just as significant, like during the "Patriot" Whig revolt against Walpole in the late 1720s. The composition of the two great parties often fluctuated, and some key politicians—including major leaders like Harley—and writers—like the prolific Defoe—shifted allegiances at various times. However much partisan geographies shifted, though, the specter of partisanship in general remained a constant in post-1688 political life. Britons imagined themselves living in an age in which political decisions were under the control of large-scale factions. Behind party animosities lay conspiratorial fears. Many Britons imagined that the opposing party designed not only to implement bad policies, but to thoroughly transform the nation's constitutional, religious, even moral character. Militant Whigs, styling themselves the rightful defenders of the 1688 Revolution, warned that their Tory opponents were covert "Jacobites" seeking to restore James II's line and implement a Gallic autocracy. Hard-line Tories feared the Whigs meant to destroy the Anglican Church and that Whig policies concealed schemes by "moneyed" interests to co-opt government for pecuniary gain. Political events regularly seemed to confirm these fears. Investigations frequently uncovered corrupt efforts to exploit the government from within and identified plots, both real and sham, to overthrow it from without. These distressing discoveries reflected and exacerbated Britons' "paranoid style" of politics.[24]

Partisan rancor hit its apex between 1710 and 1714. The trigger came on November 5, 1709, when the fiery "High Church" clergyman Henry Sacheverell preached a notorious sermon at St. Paul's Cathedral questioning the legitimacy of the 1688 Revolution, impugning Whig leadership, and warning that the Anglican Church was under threat from Dissenters, whom Sacheverell called "false brethren." When the offended Whig ministry impeached Sacheverell for sedition, the resulting trial in the House of Commons turned into a public sensation. The Whig prosecution and the Tory defense each pled their case in fiercely antagonistic terms, demonizing the other side and presenting what Mark Knights has described as "essentially two rival conspiracy theories." Sacheverell's impeachment also provoked widespread popular outrage among his Tory-Anglican supporters. Pro-Sacheverell mobs attacked the homes and meeting-places of Dissenters in a wave of violent riots that began in London in March 1710.[25]

That summer, Queen Anne dismantled Lord Godolphin's increasingly unpopular Whig ministry and replaced it with a Tory one led by Robert Harley (who had collaborated with the Whig ministry as Secretary of State until 1708). The Tories soon won a landslide electoral victory, but the slighted Whigs were emboldened by defeat. Harley took over a government facing a series of complex and consequential problems: how to manage, fund, and hopefully end the protracted war with France; whether to accept moderate religious diversity or make new efforts to protect the established Church; and how to secure the monarchical succession beyond Anne. With so much on the line, Britons worried that any political loss by their favored party might tip the balance in favor of the opposition, with catastrophic consequences. During the period, Tory-Whig conflict was highly public, and party leaders like Harley relied heavily on the press to generate support and discredit enemies. Harley employed a diverse legion of skilled writers, including Davenant, Defoe, John Arbuthnot, and Jonathan Swift. The Whigs countered with a considerable literary force of their own, including the *Spectator*'s Joseph Addison and Richard Steele.[26] Few argumentative devices went unused on "Grub Street": fictional "letters to friends," artful parody, fearmongering, bawdy rhyme, innuendo, and outright lies.

It also turned out that few modes of argumentation were better suited to this boisterous brand of politics than numerical calculation. By the early 1710s, both Tories and Whigs would come to find the value in public calculation, for reasons that will be discussed later. But it is worth noting that perhaps the first true master of the partisan numbers game was Robert Harley. As a member of the Parliamentary Commission of Public Accounts in the 1690s, Harley had been one of the most enthusiastic adopters of calculation, particularly as a tool for critiquing Court adversaries. As he transitioned from Country outsider to leading Tory minister, Harley repurposed those same calculative tools to demonstrate his party's reforming spirit. After assuming Parliamentary control in 1710, the Tories zealously audited the account-books of outgoing Whig administrators, looking for evidence of corruption (a project Defoe called the "Search After the Great Plunderers of the Nation").[27] The most spectacular result came in April 1711, when the Auditor of the Imprest Edmund Harley, Robert's brother, presented a detailed numerical report to the Commons showing that £35 million appropriated for the Navy had gone unaccounted for. The key suspect was the recent Treasurer of the Navy, a Whig MP named Robert Walpole. More on that story in Chapter 6.

Parties, Interests, and the Question of French Trade

It was in this antagonistic political atmosphere that Harley entered upon one of the most important endeavors of his young administration: bringing an end to the war with France. The peace question played squarely into Tory-Whig animosities. The Tories' desire for a swift peace and more insular foreign policy clashed with the Whigs' fiercely anti-French sentiments and their commitment not to end the war without ensuring Spain's safety ("No Peace without Spain"). From a commercial perspective, the stakes were not so black-and-white. Many Britons were optimistic that a well-negotiated peace could advance the nation's trade and usher in new prosperity. Yet any peace plan promised to affect numerous interest groups in different, entangled ways. The Tories and Whigs—not to mention the nation's myriad towns, trade groups, and companies—had different ideas.

By the time peace was being considered, trade between England and France had been curtailed almost continuously for fifty years. The prospect of reopening commerce had been mooted in 1697 and 1709, though negotiating impasses and renewed warfare thwarted those efforts. By 1712, the Tories, led by Harley and St. John, had identified the reestablishment of Anglo-French commerce as a critical objective. Leading Tories felt that the key to national prosperity lay in forging trade pathways linking cheap sources of goods to high-demand markets. Reexportation was seen as vital to Britain's commercial health. Opening the French trade would unveil lucrative targets for English manufactures and reexports while frustrating the Dutch, Britain's greatest commercial rival in Tories' eyes.[28] In June 1712, Queen Anne (strongly Tory in her personal politics) gave a lengthy speech proclaiming the commercial benefits promised by the imminent peace. In the following months, representatives of towns, companies, and trade groups addressed the queen and her ministers, thanking them for their attention to commerce and requesting special considerations in the peace negotiations. In early 1713, these pleas became increasingly heated in tone. West Indian planters expressed concern about potential restrictions the French might place on English sugar, for example, while silk and woolen manufacturers feared peace would increase competition from French industries.[29]

Faced with a rising chorus of complaints in spring 1713, Harley turned to the media to publicize the common benefits of renewed French trade. On March 31, the official peace and commercial treaties between Britain and France were signed at Utrecht in the Netherlands. Pro-Tory newspapers like the *Gazette, Post Boy,* and Defoe's *Review* reported stories of jubilant post-Treaty celebrations. Yet partisan tempers quickly began to flare. Though the treaties were signed, Parliament—which had control over any changes in taxation—still had to pass new legislation to enact the Commerce Treaty by lowering the tariffs charged on French imports. The Tory ministry began in May, proposing a bill to repeal certain taxes levied on French wine. This "Wine Bill" garnered heated opposition, especially from Iberian traders and woolen manufacturers. Critics feared a renewed French wine trade would jeopardize trade with Portugal, England's favored trading partner since the 1703 Methuen Treaty, by decreasing British appetites for port wine and thus limiting reciprocal demand for English woolens.

As the Tory ministers began drafting a more comprehensive "Commerce Bill," Whig spokespeople vigorously attacked the expansion of French trade. Many factors fueled the Whigs' fervor, including the party's traditional antipathy to France and the opportunity to defeat a major Tory project. More pointedly, the Whigs drew strong support from powerful commercial interests, like the Iberian and Levant traders. Many of these leading merchants were religious Dissenters, traditionally a strongly Whig constituency, and many had a lot to lose from any significant rearrangement of Britain's commercial policy. Broader political-economic principles were also at play. Whig thinkers like John Oldmixon and Henry Martin argued that national prosperity derived from the products of human labor, not the shuffling of goods. Accordingly, they feared direct competition with the much more populous France, whose vast reservoir of cheap labor meant they could undercut British prices.[30] While the Whigs for the moment lacked the official authority and access to data enjoyed by their Tory opponents, they had substantial political and informational resources of their own, which they began to marshal in early 1713. For example, the influential Levant merchant and Whig lobbyist Charles Cooke had built a remarkable private archive of commercial documents, containing not only canonical printed works but also an array of manuscript sources—Parliamentary petitions, minutes of corporate board meetings—all carefully ordered, annotated, and cross-referenced. This archive proved an invaluable resource for Cooke and his Whig colleagues as they disputed the Commerce Treaty in Parliament and the public press.[31]

The spring and early summer saw frantic debate and petitioning around the Commerce Treaty. This campaigning was emblematic of a new brand of special-interest politics that, alongside Tory-Whig partisanship, was also transforming political practice in Britain. Since 1688, Parliament was increasingly becoming a site where private interests—specific localities, trade groups, corporations, entrepreneurial "projectors"—systematically promoted their agendas. At a time when Parliamentary seats were often hotly contested, many MPs appealed to their constituents' interests by introducing "private" legislation seeking official support for local infrastructure projects, business ventures, special trade protections, and so on.[32] This new interest-based politics could be a messy, complex business, especially when the questions had national implications. The early controversy over the Tory Commerce Treaty made this only too clear. In early 1713, MPs and the reading public were inundated with competing appeals from different interest groups, each with different stakes in the outcome of the Commerce Treaty debate. While the Sheffield ironmasters pleaded for French markets for their cutlery, for example, cloth manufacturers in Colchester described the dire threats posed by competition from French clothiers. Any change to Anglo-French trade might further affect British trading prospects globally, from Spain to the Levant to the East Indies.[33]

In such unruly circumstances, interest-group and party politics intersected with—and ultimately reinforced—one another. Specific lobbies aligned themselves at the poles of Whig and Tory and drew upon partisan logics to tie their objectives to national goals. Party leaders, especially among the Whigs, tried to enroll interest groups under their party banner. These alignments were not always straightforward at first, even for influential interests like the Levant Company that would ultimately take a clear and vocal position on the Treaty question. When that Company met in late May to determine whether to petition against the Commerce Treaty, the vote was hardly unanimous: thirty-five members voted in favor of the Whig plan to petition against the Treaty, eleven voted not to petition, and eight abstained. (As it happened, these local sites of political conflict created new opportunities for calculation: the Tory-leaning minority in the Levant Company contended that, though outnumbered, their ranks included the biggest traders and thus represented a majority of the Levant trade by volume. Charles Cooke countered with a detailed manuscript calculating the total number of cloths exported by the two factions in 1712, showing the anti-Treaty group actually did the bulk of Company business: 17,203 cloths versus 11,639.)[34] In addition to these defined interest groups, the parties also tried to court independent and moderate Tory MPs who were likely to tip the balance in Parliament.[35] Partisan publicists had to appeal to specific interest groups and, at the same time, find captivating arguments showing why their position might benefit the entire nation. For all its rancor, Whig-Tory partisanship helped bring some order to what might have otherwise been an even more chaotic contest of fractured interests. The contest over the Commerce Treaty in 1713 was a signal example of the broader, organizing function played by party politics in the period, as recently described by historian Aaron Graham. During the early eighteenth century, Graham writes, "only political partisanship could impose sufficiently common priorities upon otherwise discordant public and private interest groups."[36]

It was in this political context that calculating the Anglo-French balance of trade took on new significance, specifically in spring 1713. Peace and trade negotiations had been ongoing with France since early 1712, but the question of the trade balance had attracted only passing attention among political leaders in those early days, including among the Tory ministers pushing for opening up trade between the two nations. For example, in a May 1712 letter to the Board of Trade reporting on the ongoing trade discussions with France, Tory Secretary of State Henry St. John had cited summary trade-balance data from one of Davenant's reports, but only in an offhand way. The data suggested that, in fact, England had had a positive trade balance with France of over £308,000 in the three peaceful years between 1698 and 1701, though at the time St. John did not offer much comment on that auspicious figure.[37] Rather, the quantitative question of the trade balance only came to the fore when the Commerce Treaty became a matter of public, partisan dispute.

The Whigs seem to have struck the first match that ignited the computational conflict. In their arguments against the Commerce Treaty in early 1713, some Whig critics began recalling the 1674 *Scheme of Trade* and its claim that England lost £1 million a year trading with France—a calculation that was nearly forty years old, but effectively the only numerical reference point that was easily available to the public at the time. These invocations of the 1674 *Scheme* kindled a surge of interest in trade data among members of Parliament. In May, MPs began to demand a massive volume of information from various government agencies, including updated trade data from the Customs Commissioners.[38] Many Whig critics probably believed the updated data would reaffirm the message of the 1674 *Scheme*. Demanding data also may have helped the Whigs stall the progress of the Treaty deliberations, and gave Whig MPs a chance to call Tory ministers and administrators to account—always a key use of numbers in the period. As numbers began to come in, MPs discovered that available data on Anglo-French trade yielded few easy answers. The Customs Commissioners tried to explain why: trade data for early years like 1663 and 1665 had been lost in London's Great Fire, and numerical records for ports outside of London were highly inconsistent. Such informational disorder was hardly atypical at the time, but it no doubt embarrassed a Tory ministry trying to claim that it understood what was in the nation's best commercial interest.[39]

Through May and early June 1713, procedural votes in Parliament suggested the Tory ministry still held the upper hand on the trade question. Yet on June 18, the first key vote on the Treaty, the so-called "Commerce Bill," was defeated 194 to 185 in the Commons, with over seventy Tory MPs defecting.[40] This surprising setback for the Tory ministry was barely the beginning of the public controversy over French trade, though. The possibility that the Treaty might be reintroduced later, plus the Parliamentary elections looming in the fall, kept the matter at the forefront of the nation's political attention. Tory journalists dismissed the defeat of the Commerce Bill as a momentary error of Parliamentary judgment engineered by Whig scheming; their Whig counterparts tried to convince nationwide voters of the commercial dangers that had been averted and to parlay their initial victory into future electoral success. Over the summer, Whig and Tory polemicists became locked in an intense, unrelenting battle. And both parties, for different reasons, wanted to make the fight about the numbers.

Party Epistemologies of Quantification; or, Two Reasons to Love Numbers

Tories and Whigs both saw virtues in numbers. But those virtues were different, stemming from the fundamentally different ways the two parties came to think about the nature of commercial knowledge in the early 1710s. Tory commentators, led by the *Mercator,* contended international trade was an immensely complex subject of inquiry. The inescapable role of self-interest in commercial affairs made it so. Claims about the dynamics of trade were inherently hard to trust, especially when advanced by merchants themselves, who could exploit the public's naiveté all too easily. In such self-serving hands, numbers themselves could be dangerous. But in the benevolent hands of honest authorities like the *Mercator,* skilled computation could protect the credulous public from calculated frauds. For the Whigs who confronted the *Mercator,* Britain's community of experienced merchants was its best source of knowledge about trade. To the Whigs, mercantile self-interest generated not deception, but understanding. They urged the public not to dismiss hard-earned mercantile wisdom—and indeed their own common sense—in favor of the supposed authority of pedantic experts. Plain-spoken numerical accounts were indispensable because they reinforced the shared intuitions of the merchant community. The remarkable "battle of figures" that began in 1713 was a collision of these two competing quantitative epistemologies.

Daston and Galison observe that "all epistemology begins in fear." Further, historian of medicine Harry Marks has observed how such epistemic fears—and the techniques of knowledge they engender, like the "randomized control trial" in twentieth-century medicine—can be targeted at specific kinds of people, what Marks calls "social objects of mistrust" (the unscrupulous pharmaceutical manufacturer, for example, or the glory-seeking medical researcher).[41] The formative role of fear in shaping epistemological attitudes was clearly evident in debates about commercial policy in the early 1710s. Tories and Whigs both harbored deep-seated fears about the security of commercial knowledge, and both identified clear epistemic villains. For Tories, the driving fear was a fear of deception, and their foremost objects of distrust were self-serving merchants. In September 1713, as the trade-balance dispute was raging, the Tory *Mercator* reminded readers of its mission: "The *MERCATOR*'s Business is to restore Men to their right Knowledge of these Things; to undeceive the abused People; to detect the abominable Falshoods that are spread about for private Designs among them; and to Inform them rightly in Things wherein they are abused and misled." The constant specter of deception was precisely why the nation needed the *Mercator*'s skilled guidance. The primary deception the *Mercator* intended to fight was, of course, the "Notion or Opinion which has been too generally received, *That the Trade with France has always been carried on to the Disadvantage of England.*" The *Mercator* made combating misrepresentation its foremost mission.[42]

Tories believed the threat of deception was especially heightened in matters of international trade. Since the very earliest emergence of the Tory party, Tory-leaning commentators like Josiah Child and Charles Davenant had explained that the dynamics of commerce on an international level were highly complicated, generating widespread confusion among the public. "It must be acknowledged to be something wonderful," *Mercator* no. 1 exclaimed, "that a Nation who drives the greatest Trade in the World . . . should be so very ignorant of their real share in one of the most considerable Branches." Britain's many different trades constituted a veritable "Knot," as Davenant had put it a couple of years earlier. Disentangling how those different trades affected national prosperity was a task that exceeded the narrow competence of practicing merchants, and was instead best left to specialists.[43] In his June 1713 *General History of Trade,* Tory polemicist and *Mercator* collaborator Daniel Defoe dubbed this field of knowledge "Universal Commerce." "Where are the Heads turn'd for the General Advantage of their Country, by seeing into the Scale of the World's Trade?" Defoe asked. "These are rarely to be found." Tory writers worried that their fellow Britons misjudged the complexity of commerce and thought it a much simpler subject than it truly was. This left many vulnerable to being misled, especially by merchants hiding their self-interested designs in the guise of simplistic commercial principles.[44]

The Whigs' attitudes about commercial knowledge stemmed from a different fear: a fear of overweening authority, like the unilateral claims to authority made by the *Mercator.* In a series of polemical pamphlets and the specialized *British Merchant* newspaper, an army of Whig pamphleteers fervently disputed the Tories' French Commerce Treaty and the elitist vision of commercial knowledge underpinning it. These writers, loosely coordinated by Whig political chiefs like Lord Halifax, represented a range of backgrounds and skills: the urbane journalist Richard Steele; the ruthless polemicist John Oldmixon; the affluent merchant Theodore Janssen; the cerebral lawyer Henry Martin.[45] The Whig commentators unanimously derided the arrogance of "the confident Authors of the *Mercator*" in trying to overturn well-established commercial truths, like the "General receiv'd Notion, that the Trade to France was always Prejudicial." "To contradict the Experience of fifty Years Commerce and to give no better Authority than his own bare Word for it," wrote Oldmixon in typically biting terms, "must end in the most terrible Mortification." What made the *Mercator*'s "Air of Confidence" (as the *British Merchant* once put it) especially threatening was that it came from a position of public trust, supported by governmental resources and privileged information. In 1721, *British Merchant* editor Charles King would later recount the fears that had driven the Whigs in 1713: "As [the *Mercator*] had a Knack of writing very plausibly, and they who employ'd him, and furnish'd him with Materials, had the Command of all Publick Papers in the Custom-House; he had it in his power to do a great deal of Mischief, especially amongst such as were unskill'd in Trade."[46]

For the Whigs, national commerce was no intractable riddle. It was a clear and comprehensible subject, understandable for anyone with mercantile experience. One anonymous pamphleteer argued that even seemingly abstract political-economic principles like the balance of trade were essentially intuitive to the nation's traders, known by any "trading Merchant of Common Sense or

Common Honesty" from their early days. "Don't every Boy upon the *Exchange* know," he asked, "that the One Article of Wines from *France,* will more than Ballance all that we can send them; and that for the Overplus . . . must be so much Money put into their Pockets?" For the Whigs, national commerce was, in fact, a matter of "Common Sense."[47] Recently, Whig political thinkers led by Anthony Ashley Cooper, the 3rd Earl of Shaftesbury, had come to celebrate Britons' shared "common sense"—or "Sensus Communis," as Shaftesbury titled an influential 1709 essay. More than just basic logic or good sense, Shaftesburian common sense described an instinctual awareness of the public good that held Britain's often discordant polity together. Whig political thinkers stressed that Britons needed to listen to their common sense and not cast it aside in favor of learned abstractions or authoritative proclamations. In matters of politics and morals, "honesty is like to gain little by philosophy or deep speculations of any kind," Shaftesbury proclaimed. "In the main, it is best to stick to common sense and go no further."[48] In 1713, Whig commentators argued that the primacy of common sense applied to commerce as well. Through their trading activities, Britain's merchants not only pursued their own private interests, but cultivated a common sense of what was in the nation's best interest as well—a far surer guide to future prosperity than any presumptuous philosophizing. For many Whigs, the Treaty question boiled down to this: "Is Theory better than Practice, and Sophistry stronger than Experience?" Of course not.[49]

The Anglo-French trade controversy was hardly the first time the parties had clashed over the politics of knowledge. A telling example came a year earlier, when Tory and Whig men of letters butted heads over the regulation of the English language. Early in 1712, prominent Tory critic Jonathan Swift penned a *Proposal for Correcting, Improving, and Ascertaining the English Tongue,* dedicated to Robert Harley. The pamphlet laid out a plan for appointing a panel of the "best qualified" experts to standardize the lexicon and curb linguistic abuses, like the "Infusion of Enthusiastick Jargon" and rampant "Manglings and Abbreviations." Whig critics swiftly rebuked Swift for his fawning praise for Harley, his desire to emulate the French in creating a national literary academy, and especially for his elitist attempt to, as Oldmixon put it, "Bully us into his Methods for pinning down our Language." Much as the debate over trade statistics would, the Tory-Whig linguistic quarrel became a dispute about civic epistemology, about where intellectual authority ought to lie in British society: in the knowing hands of select authorities or in the common sense of the people.[50]

While Tories and Whigs disagreed about how knowledge ought to be ordered in society, their competing epistemologies converged in certain ways as well. For one, both parties claimed to love *facts*. Yet they conceived of facts in subtly different ways. Tory writers described facts as incisive, antagonistic, and indomitable—the most potent arms for combating misrepresentations. The first issue of the *Mercator* described its agenda in militant terms. The delusion about the French trade, the authors explained, "is a strong Enemy, so it must [be] dealt with in a forcible manner . . . [with] Demonstrations, Matters of Fact, authentick Vouchers, clear Documents, and such like irresistible Weapons."[51] The *Mercator*'s epistemic arsenal was diverse and well-stocked: "demonstrations" echoed the logical deductions of philosophers; "Matters of Fact" evoked the new experimental philosophy and, more distantly, English common law; "Authentick Vouchers" connoted the paper accounts kept by honest merchants. Philosophy, natural science, law, accounting—all contributed to an emerging fashion for "facts" in the early eighteenth century. Critically, the Tories held that facts did not speak for themselves. As in the conception of "matters of fact" in the common law tradition, facts had to be justified by evidence. Skilled judgment was required to identify honest facts and distinguish them from abundant misinformation.[52]

Whigs also turned to the facts, but they valued facts for different reasons from their Tory adversaries. For Whig polemicists, a true fact was clear and accessible. It was the opposite of a privileged claim of authority, like "the *Mercator's* bare Word." Joseph Addison, famed author of *The Spectator* and frequent Whig campaigner, brought this Whig epistemology to life in a series of satirical pamphlets about "Count Tariff," a personification of the Tory position on French trade. In his *Late Tryal and Conviction of Count Tariff*, Addison chronicled the fictional courtroom showdown between "Count *Tariff*, Defendant" and the plaintiff "Goodman *Fact*." While Tariff accused Fact of disrespecting his betters, Fact claimed that he did nothing but "TO SPEAK THE TRUTH AND NOTHING BUT THE TRUTH." Addison especially lauded Goodman Fact for how he communicated. "Goodman *Fact* is allowed by every Body to be a plain-spoken Person, and a Man of very few Words," he wrote. "He affirms every Thing roundly, without any Art, Rhetorick, or Circumlocution." While Tory publicists warned that facts did not always speak for themselves, Whigs like Addison celebrated them for exactly that reason.[53]

The two parties not only converged on "facts" in general, but on numerical facts in particular. Yet again, the two parties saw different virtues in numbers. The opening of the chapter already mentioned how the *Mercator* had begun, in its very first issue, with a forceful computational promise: to "Take a Medium of any three Years for above forty Years past, and calculate the Exports and Imports to and from *France*," and thereby disprove the longstanding "false Glosses and Misconstructions" that had tricked the public into fearing the French trade. For Tory writers, the great political possibility of calculation lay in its ability to "undeceive." Writing as the official representative of the incumbent Harley ministry, the *Mercator* presented itself as the benevolent guardian of commercial truth, wielding advanced calculations to protect the ill-informed public. It was a slightly different epistemological position from the Country one that Davenant and Harley had pioneered in the 1690s, but not irreconcilably so. In their earlier exercises in public accounting and political arithmetic, those Country outsiders had used calculations to interrogate powerful government insiders. Now at the helm of government themselves, they used calculations to interrogate entrenched commercial interests. Their own level of authority had changed, but their vision of the political promise of calculation fundamentally had not. In a way, the *Mercator* was simply making good on a mandate Davenant had laid out in his 1698 *Discourses on the Publick Revenues*. "To perplex Things which have Relation to Trade," Davenant wrote, "is the Interest of so many." Political arithmetic had the power to combat this and to "shew the Links and Chains by which one Business hangs upon another."[54]

The Whigs, too, saw great potential in numbers, but not in the patronizing computations the *Mercator* advanced. John Oldmixon offered a telling, and vicious, critique of the Tories and their appointed expert Davenant. The Inspector-General was little more than a "Theorist, who talks of Trade in the *Park* as a Mathematician does of Navigation on *Hampstead* Hill," Oldmixon charged. "Yet because he can marshal up three or four Ranks of Numbers, which are like so many magical Figures to certain Politicians, because he has acquir'd a little Mercantile Court, he is made much of as a Person extreamly well acquainted with Trade." But an official title and some arithmetical acumen did not add up to true commercial understanding: "I will lay him as much Money as he is worth, that he cannot fit out a Ship for *Newfoundland*, nor make out an *Invoice* for *Virginia*."[55]

Though Oldmixon and his Whig associates railed against Davenant's "magical Figures," they did not dismiss calculation itself. Whig social commentators like Richard Steele, Addison's *Spectator* collaborator, contended that numerical reasoning was not an abstruse technical skill but a widely accessible one, essential for country gentlemen as much as hardened merchants. In an early edition of his *Guardian* in 1713, for example, Steele described a fictitious Northamptonshire family, the Lizards, whose eldest son Sir Harry exemplified such a calculating gentleman. Not only did calculation aid Sir Harry in the management of the Lizard estate, but "the same Capacity, joined to an honest Nature, makes him very just to other Men, as well as to himself." Facility in calculation, Steele suggested, was a foundation for honest discourse—a kind of common sense, perhaps.[56]

Scheming

In the summer of 1713, the Tory supporters of the French Commerce Treaty and their Whig opponents both seized upon the calculation of the Anglo-French balance of trade. Yet they had different reasons to believe in numbers, grounded in competing political and epistemological values. Critically, these divergent quantitative epistemologies manifest themselves within the calculations the two parties produced. The Tories promised that an expert calculation of the trade balance would help to undeceive the people and combat merchants' self-interested myths. This was reflected in the *Mercator*'s calculations. Davenant and company developed technically intricate analyses designed to correct potentially misleading data, and focused their calculations on showing their opponents were wrong—and untrustworthy—rather than making a positive case of their own. They also paid considerable attention to computational methodology, as a way to demonstrate their superior command over commercial complexities.[57] Their Whig opponents, on the other hand, favored a common sense approach to commercial knowledge, letting the "raw" numerical data speak for itself. To some degree, the two parties had to play the hands they were dealt. The Tories held ministerial power at the time and capitalized on the authority and the access to information to which that entitled them. The Whigs were forced to take an oppositional stance but had the *prima facie* numerical evidence and popular anti-French prejudice on their side, a position that allowed them to appeal to raw data and common sense. But the Whig and Tory epistemologies that became so starkly evident in 1713 were not just the product of rhetorical opportunism. The contrasting computational approaches reflected long-standing disagreements about civic epistemology, which were refined and intensified by the heightened partisan polarization of the early 1710s.

As noted, it seems to have been the Whigs who provided the first numerical spark in the balance-of-trade debate, by invoking the 1674 *Scheme of Trade* in early arguments against the Tory Commerce Treaty. But it was the next move by the Tories and the *Mercator* that stoked the flames, turning the debate into a full-fledged computational conflagration. In late May and early June 1713, as the crucial vote on the Commerce Bill loomed, the *Mercator* began an all-out assault on the 1674 *Scheme*, seeking "to expose the Arts and Contrivances" by which the British people had come "to receive it as an uncontroverted Truth, that we lost by that Trade." *Mercator* no. 2 contended that, in 1674, nefarious men had produced misleading figures about the nation's trade in order to derail King Charles II's efforts at closer diplomacy with France and sow discord between Charles and his subjects. Almost forty year later, this calculated deception was still invoked as "the Test or Touchstone of the French Trade, and as an unquestion'd Authority."[58]

The frauds in this "Touchstone," the *Mercator* claimed, were extensive. The *Scheme* relied on data from extremely misleading years, measuring exports (1668) and imports (1674) at different times. It excluded goods imported from Asia, Turkey, and American colonies that were reexported to France and thus should have counted positively toward England's trade balance. For particular items, the *Scheme* simply misstated the quantities grossly, as in the claims that the English imported 150,000 pounds of wrought silks or 11,000 tons of wine from France, numbers totally unjustified by the Customs House books. The *Mercator* offered a preliminary computation based on a first batch of Customs figures provided to MPs in May, suggesting that the *Scheme*'s £1,000,000 negative trade balance was far from accurate. Those numbers suggested England's balance was no worse than negative £269,000 in 1663 and £433,000 in 1669, and that was before several additional adjustments were made to the figures. In fact, "by a just and true Account," the *Mercator* argued, "the English Trade had the Gain."[59]

Over June and into July, the *Mercator* delivered an onslaught of specialized calculations aimed at undercutting specific Whig arguments against trade with France. None of the calculations presented a complete account of the Anglo-French trade balance. Rather, the *Mercator* used skillful arithmetic to critique various anti-French "Common-places."[60] For example, one common concern was that freer trade with France would ruin the English silk-weaving industry. *Mercator* no. 9 presented an account of the average silk imports from 1707 to 1709 from Germany, Italy, Holland, Spain, the "Streights," and Venice, showing that the total English market for imported silk textiles (under £33,000 per year) was not one-thirtieth of the volume of silks manufactured domestically. The English silk-weaving industry was far too robust to be threatened by French imports. A similar argument in the following issue attacked the idea that opening up trade with France would decrease English imports of Portuguese wines and thereby jeopardize Portuguese demand for English woolens. The numbers showed that Britain exported over £2 million in woolens and imported only £200,000 worth of wine. Portugal needed English woolens far more than England needed port.[61]

The *Mercator*'s strident critiques soon received some useful outside reinforcement. On June 15, the Commissioners of the Customs submitted a report to the Commons officially declaring that, after review of relevant records, "the said Scheme [of 1674] could not be taken from the Custom-house books."[62] The *Mercator* pressed its advantage, and landed its most decisive critical blow shortly thereafter. Seizing upon data on historical imports and exports with France that were newly made available by the Customs administrators, the *Mercator* published an incisive analysis showing that the 1674 *Scheme* had overstated French imports at the Port of London alone by over £721,000. Those new figures not only provided resounding evidence that the *Scheme* could not have been based on "exact Accounts from the Custom-House Books," but showed just how extensive was the deception perpetrated by the 1674 calculation.[63] The Whig opposition had not just been mistaken about the trade balance, they had been passing a veritable "forgery"—a charge the *Mercator* began to repeat regularly.[64] Just as a manipulated account-book or a counterfeit financial bill threatened networks of commercial trust, the counterfeit *Scheme* threatened the political trust of the nation.[65]

In its first three months, the *Mercator*'s computational tactics were persistently adversarial, using numbers to undercut the opposition's arguments and weaken public trust in the Whigs. The *Mercator* promised that, at some point, it would deliver an "exact Calculation" of the balance of trade that would conclusively show the *Scheme* was a "design'd Collusion and Prevarication."[66] But the *Mercator* delayed making any definitive claims about what the current Anglo-French balance of trade actually was. It was not until early fall that the *Mercator* began to approach that question directly, but only after taking an extensive excursion into methodology. *Mercator* no. 49, dated September 12–15, laid out a detailed procedure describing what it would take to produce a truly accurate balance-of-trade calculation—"all the just rules by which such a thing can be computed." Numerous corrections would need to be made to the Customs data: exports would have to be adjusted to include French-bound exports from London inaccurately recorded in the Customs books; valuations of imports would have to be corrected; reexports from the outlying ports would have to be added, as would the profits and employment generated for the English in shipping goods to and from France. This performance of methodological sophistication reminded readers just how complicated questions of "universal commerce" were, and why it was necessary to entrust them to proper authorities like the *Mercator*.[67]

For the Tory *Mercator,* the need to calculate was fueled by a desire to undermine political opponents and debunk deceptive claims, in this case by powerful commercial interests. This was typical of Britain's early quantitative age: political antagonism drove quantification. The same held true for their Whig opponents. The Whigs first began to cite the 1674 *Scheme* in order to critique the Tory Treaty plans. But what really inspired their calculative efforts were the officious claims being made by the *Mercator.* "My principal Care," announced the first issue of the *British Merchant* in early August, "shall be to detect the Frauds and Cheats of the *Mercator,* by which he imposes upon the People." For the Whigs and their *British Merchant,* calculation offered an ideal tool for cutting down the *Mercator* and its "Air of Authority." *British Merchant* no. 4 offered an early display of what the Whigs could do with numbers, taking apart a lengthy calculation offered in *Mercator* no. 26 that claimed that England had reexported £500,000 of foreign goods to France in 1686–1687. The computation borrowed the *Mercator*'s numbers regarding the quantities of goods exported and recalculated the total value of those exports using different prices for various products. The conclusion was that the value of the reexports in 1687 was less than £176,000, not even two-fifths of what the *Mercator* claimed. "Is this then one of the *Mercator*'s indisputable *Vouchers?*" the Whig paper asked. "Is this his way of proving Facts by *clear Documents?*"[68]

This became a common refrain for the *British Merchant:* the *Mercator* had failed to live up to its own lofty standard of knowledge, claiming to rely on facts while actually making unsubstantiated proclamations based on authority alone. The Whigs reveled in pointing out that the arrogant *Mercator* could not actually prove many of its central claims.[69] Most egregious was the *Mercator*'s inability to produce a complete balance of trade, for any year, showing the French trade actually benefited Britain. *British Merchant* no. 20 recalled the situation this way:

> The *Mercator,* at his first setting out, affirm'd, *That the Trade between* England *and* France *was ALWAYS beneficial to this Kingdom,* and that this should be prov'd by a CALCULATION of the Exports and Imports between both Countries, in which it would appear the Balance was always on the *English* side. . . . Tho his Paper comes out thrice every Week, and has done so for five Months together, has he ever yet produc'd the Account of any one Year? And till this is done, ought not all his Readers to throw him aside for an Impostor?[70]

Although it was the Tories who had first promised it, by autumn 1713 it was the Whigs crying loudest for new, comprehensive balance-of-trade calculations. "Upon the whole matter, I think, an exact Account of Exports and Imports is the only infallible way to shew whether we have gain'd or lost by our Trade with France," read *British Merchant* no. 12.[71] The Whigs did just this two issues later, in late September (see Figure 3.1). The four-page account, relying on Customs data for the years 1685–1686 presented to Parliament earlier that year, walked item by item through a dizzying array of goods, starting with ninety-four double barrels of anchovies imported into London (worth £70.10s. total) and continuing to much more massive items, like the 12,760 tons of wine (£223,300) imported into London or the 115,000 lbs. of wrought silks (just under £289,000) imported into England's "Out-Ports." The total amounted to over £1,284,000 in French imports for that year, set against only £514,000 exports.[72] The negative balance of about £770,000 was a retreat from the £1 million loss implied by the 1674 *Scheme*. But it was still a substantial number, and the Whig authors also provided several reasons, like the smuggling of clandestine French goods into England, why the deficit was likely even worse.[73] Though highly detailed, this "New Scheme" for 1685–1686 demonstrated the Whigs' plain-spoken, "common sense" approach to calculation. Though many readers may have balked at the sheer volume of numbers, the calculation itself was relatively transparent and did not depend on the intricate analyses or adjustments to the data favored by Davenant and the *Mercator*. It placed the burden of argumentation squarely back on the Tories. "It is time now for the *Mercator* . . . to falsify that Account," dared *British Merchant* no. 16.[74]

The Whigs had called the Tories' bluff. For months, the *Mercator* had contended that the Whigs preferred to argue from hearsay and popular myth instead of solid evidence. This position became hard to sustain after September 1713, once the *British Merchant* had taken up the *Mercator*'s challenge to "search the [Customs] BOOKS" themselves."[75] The Tory calculators were left with the task of trying to convince readers that the Whigs' figures—the sort of calculation the Tory paper had been promising for months but failed to deliver—could not be trusted. True to form, the *Mercator*'s first response was to try to out-calculate the Whigs, using more technically sophisticated analyses to show that the opposition's accounts were naive. In some cases, it was relatively easy to make the Whigs look amateurish. One of the Tories' most decisive calculative take-downs came in October 1713, in response to a highly speculative calculation that had been published in Richard Steele's pro-Whig *Guardian*. Following on the conclusions of the *British Merchant*'s "New Scheme," issue

Numb. 14.

The Britiſh Merchant ;

OR,

COMMERCE PRESERV'D:

In Anſwer to the *Mercator*, or *Commerce Retriev'd*.

From FRIDAY, September 18. to TUESDAY, September 22. 1713.

An ACCOUNT *of the* IMPORTS *and* EXPORTS *to and from* ENGLAND *and* FRANCE,

From *Michaelmas* 1685. to *Michaelmas* 1686.

Which was laid before the LAST PARLIAMENT by the PRESENT COMMISSIONERS of the CUSTOMS,

With a juſt *Valuation* of all the PARCELS,

Shewing the Loſs that *England* ſuſtain'd by our TRADE with *France* that Year.

An ACCOUNT *of Goods Imported into the Port of* London *from* France, *from* Michaelmas 1685, *to* Michaelmas 1686.

		l. s. d.	l.	s.	d.
ANchovies	94 Double Barrels	at 00 15 00 *per* Barrel	70	10	00
Ditto	354 Single Barrels	00 07 06 *per* Barrel	132	15	00
Annotto	2100 *lib.*	00 02 06 *per lib.*	262	10	00
Bugles Great	1241 *lib.*	00 04 00 *per lib.*	248	04	00
Small	100 *lib.*	00 06 08 *per lib.*	33	06	08
Lace	166 *lib.*	00 08 00 *per lib.*	66	08	00
Books Unbound	229 *Ct.*	01 00 00 *per Ct.*	229	00	00
Brandy	1568 Tuns, 215 Gallons	20 00 00 *per* Tun	31377	01	03
Basket Rods	921 Bundles	00 06 08 *per* Bundle	307	00	00
Bracelets or Necklaces of Glaſs	37 Small Groce	01 04 00 *per* Groce	44	08	00
Boulteel Raines	368 Pieces	00 10 00 *per* Piece	184	00	00
Buckrams	842 Pieces	02 10 00 *per* 12 Pieces	175	08	04
Cheeſe	47 $\frac{1}{2}$ *Ct.* weight	01 05 00 *per Ct.*	59	07	06
Cork	5104 $\frac{1}{2}$ *Ct.* weight	00 16 08 *per Ct.*	4253	15	00
Capers	86474 *lib.*	00 00 06 *per lib.*	2161	17	00
Dornix with Caddas	932 $\frac{1}{2}$ Pieces	01 10 00 *per* Piece	1398	15	00
Ditto with Silk	19 Pieces	02 00 00 *per* Piece	38	00	00
Fleams to let Blood	3876 Pieces	00 00 02 *per* Piece	32	06	00
Flax Undreſs'd	112 $\frac{1}{4}$ *Ct.* weight	01 00 00 *per Ct.*	112	05	00
Feathers for Beds	761 $\frac{3}{4}$ *Ct.* weight	06 00 00 *per Ct.*	4570	10	00
Martins Skins	3 $\frac{3}{4}\frac{1}{5}$ Timber	10 00 00 *per* Timber	38	15	00
Fans for Corn	360 Pieces	00 06 08 *per* Piece	120	00	00
Fans for Women	162 Dozens	02 00 00 *per* Dozen	324	00	00
Glaſs for Windows	1487 Caſes	01 10 00 *per* Caſe	2230	10	00
Glaſs-Pipes Great	92 $\frac{1}{4}$ *Ct.*	07 10 00 *per Ct.*	691	17	06
Raiſins Solis	81 $\frac{3}{4}$ *Ct.*	01 05 00 *per Ct.*	102	03	09
Looking-Glaſſes	2 Dozen	45 00 00 *per* Dozen	90	00	00
Glaſs Sights for Ditto	6 Dozen	30 00 00 *per* Dozen	180	00	00
Goods unrated		Value	13558	04	06
Gauls	100 *Ct.* $\frac{1}{4}$	03 10 00 *per Ct.*	350	17	06
Almonds Sweet	309 $\frac{1}{4}$ *Ct.*	04 10 00 *per Ct.*	1391	12	06
Anniſeeds	18 $\frac{1}{2}$ *Ct.*	02 00 00 *per Ct.*	37	00	00
Pepper	530 *lib.*	00 00 09 *per lib.*	19	17	06
Prunes	17256 $\frac{3}{4}$ *Ct.*	00 15 00 *per Ct.*	12942	11	03
Iron	160 *Ct.*	12 00 00 *per* Tun	96	00	00
Incle unwrought	932 *lib.*	00 02 06 *per lib.*	116	10	00
Britiſh Linen	200 Single Ells	06 13 04 *per* hundred Ells	13	06	08
Canvas Vitry	6145 $\frac{1}{4}$ hundred	05 00 00 *per* hund. of 120 Ells	30726	05	00
		Carry over	108786	17	11

Figure 3.1 Front page of *British Merchant,* no. 14 (Sept. 18–22, 1713), showing the beginning of the *British Merchant*'s "New Scheme." General Collection, Beinecke Rare Book and Manuscript Library, Yale University.

no. 170 of the *Guardian* produced a one-page analysis showing that, by 1713, the Anglo-French trade balance had likely gotten twice as bad (£1.45 million) as it had been in 1685–1686. In response, the *Mercator* crafted a clever arithmetical model designed to prove that the *Guardian*'s estimates of recent French imports were completely absurd—so absurd that, if correct, then the Customs taxes on those French imports alone would have exceeded all of the Customs duties the British government currently received. If that were true, the *Mercator* quipped, then the *Guardian* ought to be "praised then for the best Projector of the Age"![76]

The *British Merchant*'s New Scheme itself provided a much more serious challenge, though. Even the Tories had to admit that the Whig calculations drew upon legitimate Customs data and did not egregiously falsify import-export figures like the 1674 *Scheme* had. Instead, the *Mercator* argued, the New Scheme relied on more subtle forms of deception, which could only be made evident through detailed technical critiques. In issues no. 60 and 61, the *Mercator* argued that the choice of the year 1685–1686 prejudicially inflated imports from France. It had been a rare year during which English tariffs were reduced, prompting English consumers to rush for newly available French items. The Tory paper also contended that, while the New Scheme had honestly represented the amounts of goods traded between England and France, it had severely misrepresented those goods' average prices. *Mercator* no. 66 included correspondence from several specialist traders—"Wholesale Grocer" (on French prunes), "Silversmith" (on French silver plate), "Calf-skin" (on leather)—elaborating on these pricing errors. Together, the *Mercator* argued, the biased choice of year and myriad mispricings artificially depressed England's annual trade balance by over £600,000. The Whig calculation had also overlooked another £500,000 of benefits to Britain due to shipping profits. Finally, the *Mercator* argued that the Whigs' New Scheme overlooked abundant English exports to France that were shipped through other countries or fraudulently reported to the Customs Office as intended for other destinations; an estimated £300,000 of English goods officially destined for Holland and £100,000 aimed for Flanders ultimately ended up in French markets. The *Mercator* composed a series of tables summarizing the aggregate impact of these various mistakes (see Figure 3.2). According to the *Mercator*'s analysis, rather than losing by trading to France, the English in fact stood to profit by almost £900,000 a year.[77]

It was a powerful conclusion. But it was based on many lines of complicated computational adjustments that were difficult to explain in simple terms. The Whig reaction was swift and unforgiving. They assailed their Tory opponents for

		Source
The *British Merchant*'s "New Scheme": Year ending September 25, 1686		
Imports from France (*negative*)	(£1,284,420)	*B.M.* no. 14
Exports to France	514,137	"
Total Balance of Trade, per *British Merchant* (*negative*)	(£770,283)	
The *Mercator*'s Corrections to the "New Scheme"		
Imports from France, per "New Scheme"	(£1,284,420)	*B.M.* no. 14
Plus: Biases related to choice of year and mispricing	374,867	*Mercator* no. 61
Plus: Other adjustment	19,649*	
Corrected Imports from France (*negative*)	(£889,904)	
Exports to France, per "New Scheme"	514,137	*B.M.* no. 14
Plus: Biases for mispricing of goods	227,940	*Mercator* no. 63
Plus: Trade to France via Holland	300,000	*Mercator* no. 65
Plus: Trade to France via Flanders	100,000	"
Plus: Additional exports of corn	50,000	"
Plus: Profits on retailing in France not accounted for	64,200	"
Plus: Profits on shipping	500,000	"
Corrected Exports to France	£1,756,277	
Corrected Total Balance of Trade, per *Mercator*	**£866,373**	

*There is a discrepancy of £19,649 between the *Mercator*'s figures in issue nos. 61 and 65.

Figure 3.2   Table comparing the "New Scheme" for Anglo-French imports and exports for the year ending September 25, 1686, as proposed in *British Merchant (B.M.)* no. 14 and *Mercator* nos. 61–65.

taking advantage of their privileged political position to impose interpretive judgments on the public, even hinting that the Tory government might have falsified records.[78] The *British Merchant* turned the *Mercator*'s authoritative voice against the Tory paper, showing how its artful computations contradicted commercial experience and common sense. The *Mercator,* for instance, had sternly criticized many of the import estimates given in the *Guardian,* like the claim that English imports of French linen textiles would likely amount to £600,000 per year. The *British Merchant* tried to defend the *Guardian*—and discredit the *Mercator*—by using a basic demographic calculation showing that such a projection was entirely within common-sense bounds. "It is generally believ'd there are seven Millions of People in *Great Britain,*" the Whig paper explained, "and will the Value of 600000 *l.* in Linen (not above 1 *s.* and 8 *d.* ½ for every Head in *England,* for Shirts, Sheets and all sorts of Linen) be thought extravagant?"[79] This kind of rough-and-ready exercise in political arithmetic exemplified the Whigs' common-sense calculative style.

As the pressure of the Whig critiques mounted, the Tory *Mercator* turned to a new—and somewhat ironic—tactic: to question whether the import-export data collected by the Customs House could really reflect the state of Anglo-French commerce at all. In effect, Davenant and the *Mercator* revived the skeptical critique about the balance of trade Davenant had expressed in his 1711 report as Inspector-General. Exasperated by the Whigs' persistently outlandish claims about how much the English supposedly imported from France, *Mercator* no. 64 exclaimed: "Do they think the Gain or Loss of a French Trade is to be Ballanced in the Custom-House Books! Well may the Authors acknowledge they do not understand what they are doing!"[80] Late in 1713, the *Mercator* spent much of its effort detailing all of the factors that compromised the integrity of the Customs data, like the various tricks merchants played to evade tariffs. Merchants frequently shipped English goods to neighboring countries like Holland and Flanders before reexporting them to France, the *Mercator* argued, or falsified documentation regarding the destination of export goods to avoid trade prohibitions.[81] Charles Davenant, of course, had extensive experience with such deceptions from his position as Inspector-General. Initially, the *Mercator* argued that this "Masquerade by the Merchants" was a major reason why expert analysis was necessary to correct the superficial Customs data.[82] Over time, though, the *Mercator*'s elitist skepticism about the Customs data hardened into outright doubt about whether a true trade balance could ever be calculated. A later issue put it bluntly: "No Accounts from the Custom-house can give a true Scheme of the Trade to France, or so much of a Scheme as to make a rational Conjecture from."[83]

To the Whigs, the *Mercator*'s turnaround was sheer hypocrisy. "What a strange Creature is the *Mercator?*" reflected the *British Merchant*. "One while all for the *Custom-House* Accounts, and for proving the *French* Trade beneficial by those infallible Vouchers: This has been promised in many of his Papers. At another time, *it is impossible to make any Conjecture of the* French *Trade by any Accounts from the* Custom-House." What could explain such a reversal? The Whigs thought the matter was simple: the *Mercator* rejected the numbers because they "are found now to be against him," proving the Whigs' case instead of the *Mercator*'s.[84] The Tories' about-face on the possibility of a conclusive balance-of-trade account represented something beyond just hypocrisy or poor sportsmanship, though. Rather, it marked their begrudging acknowledgment of the strategic realities of quantitative politics. At the end of 1713, the thinkers behind the *Mercator* probably remained committed to the belief that England's balance of trade with France was positive. They had provided several compelling reasons—the challenges of pricing, misreporting in the records, reexportation—why the superficial evidence of the Customs books was misleading. But the Tory calculators found that, while poking holes in their opponents' case about the balance of trade was one thing, calculating the numbers themselves was another. The authors of the *Mercator* believed that a truly comprehensive balance-of-trade calculation—one that would conclusively prove their case—would need to be extremely complex, based on an array of arcane adjustments to the raw Customs data. But given the political environment, such a calculation proved extremely difficult to carry out—too time-consuming to put together under the pressure of the two-to-three-day "news cycle," nearly inscrutable to most readers, and an incredibly easy target for Whig mockery.

After October 1713, Tory journalists stuck consistently to the position that the balance of trade was not reliably calculable from Customs House data. But they did not abandon numbers altogether. Rather, they explored an alternative computational strategy seemingly better suited to their political and media environment. Beginning in *Mercator* no. 68, in late October, the authors began regularly presenting new "raw" data describing the specific English products currently being exported to France. *Mercator* no. 76, for example, concluded with a simple chart showing the English goods sent to France on the single day of November 13.[85] The paper continued to publish these unadorned figures through the very final issue. The purpose of the figures was to demonstrate how, even without the benefits of the Commerce Treaty, France represented a vibrant market for English products. This strategy was a new one for the *Mercator*. Instead of presenting complicated, heavily interpreted computations that demonstrated their authors' competence, the paper opted to present fresh, uncomplicated data that required little explanation. This change in tack may have been related to a change of leadership, as Daniel Defoe may have taken greater control of the publication at some point in late 1713 due to Davenant's failing health.[86] But it also showed the effects that relentless partisan pressure had on the computational ambitions of the *Mercator* and the Tories. In effect, the Tories started to calculate more like the Whigs.

In-Conclusion

Eventually the trade-balance battle stopped. In spring 1714, the Tory ministry was still pushing the Commerce Treaty and making efforts to renegotiate better terms with France. They appeared to be making cautious progress. As summer came, though, a confluence of misfortunes finally ended the Tories' Treaty aspirations. In June, one of the ministry's key negotiators, Arthur Moore, was excoriated in Parliament for reaping personal financial rewards from his official duties. Diplomatic relations between France and Britain broke down over disagreements about Newfoundland and the Spanish West Indies. When Parliament was prorogued July 9, the future of the Commerce Treaty appeared bleak. Issue no. 181 of the *Mercator,* dated July 17–20, announced the paper would be taking a hiatus until Parliament reconvened.[87] That would prove to be the final issue. The final blow came with the death of Queen Anne on August 1, 1714. King George I soon ascended to the throne, and new Parliamentary elections in 1715 brought a sweeping victory for the Whigs, who predictably killed the Tories' trade project. Anglo-French trade would remain closed off by prohibitory tariffs until 1786.

Later Whig publicists recast this political victory as a sign of the superiority of their party's economic ideas. In his massive 1722 chronicle of the reign of Queen Anne, Abel Boyer recounted how the *British Merchant* had thoroughly "unravell'd the Fallacies" perpetrated by the Tory *Mercator.*[88] A year earlier, the editor Charles King compiled the issues of the *British Merchant* along with other key Whig writings from the 1713–1714 debate and reprinted them in three volumes. King's compendium would long stand as a monument to the victory of an imagined Whig consensus and commercial common sense, reissued three times in English (1743, 1748, and 1787), translated into Dutch (1728) and French (1753), and cited by Lord Shelburne in 1785 as "a book which has formed the principles of nine tenths of the public since it was first written."[89]

To whatever degree the Whigs "won" the war over the Commerce Treaty, though, they did not win the computational battle over the balance of trade—at least not decisively. That battle ended, but it never reached a conclusion. As historian Perry Gauci puts it, "neither side could provide a satisfactory case to disprove the claims of their rivals."[90] The Anglo-French balance of trade remained—perhaps remains—an open question. The *Mercator* had initiated the fight with a resounding numerical challenge, stating that a conclusive balance-of-trade calculation was going to settle the trade question once and for all, but never came particularly close to that goal. Their Whig opponents got further in actually computing the magic number, but never achieved a clear victory. Whig calculators were forced to back off from their long-established slogan that trading with France cost the nation £1 million a year, revising that estimate down by about a quarter (£770,000), and that was before taking into account any of the numerous methodological complications pointed out by their Tory adversaries.

Time has not been kind to Charles King's claim that it was the Whigs who "set the State of our Trade in a clear Light." In 1951, economic historian Margaret Priestley retrospectively analyzed Anglo-French trade during the era of the 1674 *Scheme of Trade*, arguing (as the *Mercator* had) that the *Scheme* had presented the situation "from England's point of view, in the gloomiest possible light." Priestley called into serious question whether the balance was even unfavorable to England during that period. Subsequent quantitative studies based on the ledgers of the Inspector-General's office also paint a much brighter picture. One historian has recently stated that England had an "endemically positive balance of trade with France" throughout the early eighteenth century. But, as Priestley notes, there are many factors that would need to be taken into account in assessing the historical trade balance which, given extant data, simply "cannot be estimated on a quantitative basis." The conclusive account of the Anglo-French trade balance that the *Mercator* promised may well never be calculated.[91]

In understanding the birth of Britain's quantitative age, the most striking lesson of the trade-balance battle is that highly polarized, two-party politics proved incredibly amenable to quantitative thinking in public life. Yet ironically, the very thing that energized the quantitative conflict in the first place—Whig-Tory partisanship—also made it nearly impossible for either side to achieve the conclusive calculation they sought. Under such antagonistic conditions, there was no time for compromise and no room to be wrong. Incorrectness was fraud, wrong numbers were forgeries, and any small victory for the opponents seemed an almost existential threat. Under such merciless conditions, both parties always seemed to—and always *had to*—find a new move to play: a new methodological critique, a new piece of data, a new conceptual wrinkle to add to existing models of international trade. This meant the argument could never really be finished. Irregularities in key sources of data like the Customs accounts contributed to this open-endedness, because there was always sufficient "give" in the data to allow each side to make another plausible calculative move. The competing computational styles and numerical epistemologies of the two parties also served to make the argument more indeterminate. Each party had its own standards for computational success and its own reasons to keep up the numerical fight. The Tories always had the superficial quantitative evidence, the "raw" data, against them. But for the Tories, the fact that the public had been taken in by superficial arguments was precisely why expert calculations were necessary "to undeceive the people." The Whigs had "common sense" and raw data on their side, but saw themselves threatened by the authoritative claims of the *Mercator*. So long as the Tories held official power, the Whigs could not rest. Indeed, it was precisely when the Tories lost that power, and the structure of partisan conflict was rearranged, that the two sides stopped calculating.

As noted in the Introduction, Robert Boyle had shied away from quantitative methods in his vision for experimental philosophy in the mid-seventeenth century, fearing that too much reliance on numbers and calculations would promote uncivil arguments and yield intractable disputes. Based on the evidence of 1713–1714, Boyle was right. But this hardly meant that all calculation was a waste of time, producing abundant partisan heat and little analytical light. The balance-of-trade debate was highly consequential even if inconclusive. Scholarship on intellectual controversies, particularly in the history of science, has been keenly interested in processes of *closure*—on how disputes get settled, experiments ended, and winners and losers decided.[92] Less attention has been paid to openness, indeterminacy, and nonclosure—to why some arguments seem to resist meaningful conclusion and what the consequences of such in-conclusion may be. In explicitly political contexts, competing investigators often never have the luxury of seeing a dispute closed; debates end, participants move on, but the questions at hand remain frustratingly open to debate. Yet such controversies can produce a lot even as they fail to produce answers.

In many ways, the balance-of-trade debate advanced contemporary discussion of the dynamics of international trade to unprecedented heights, making it conceptually, methodologically, and empirically richer. The stress of computational combat forced combatants to complicate their working models of international trade and to wrestle with difficult conceptual problems, like the value of labor employed in shipping, the dynamics of reexportation, and the national costs of smuggling. It prompted them to refine their methodologies for handling trade data. And it drove them to seek out new sources of economic knowledge, like the comments by the *Mercator*'s correspondent "Wholesale Grocer" on the pricing of French prunes, and new quantitative data, like the *Mercator*'s daily reports of export shipments from London. In fact, the open-endedness and inconclusiveness of the balance-of-trade debate—its resistance to closure—only made it more productive in these adjacent ways.

In some ways, this unrelenting partisan pressure also brought a certain order to the trade-balance debate. The threat of an immediate counterattack from the other party limited how much either side could stretch the data. As they continued to trade numerical blows back and forth, the two parties found themselves converging on certain common premises. While they remained staunchly opposed about the size of the Anglo-French trade balance, they agreed more and more about how that balance might ultimately be calculated—it would need to be highly granular, treating different import-export goods separately; it would need to be based on historical data from a fair and representative period; it would need to adjust available Customs data to compensate for fraudulent practices and to normalize the pricing of imports with exports. Even their computational styles seemed to converge somewhat in the later stages of the debate: the Whigs' "new" schemes of the balance of trade became increasingly elaborate and detailed—more Tory-like—in order to defend against Tory criticisms; the Tory *Mercator* came to rely evermore on Whig-style "raw" data about exports to France. The computational game between Tories and Whigs was vicious, but it was not without a certain kind of order, and it actually became more orderly the more it was played. In this sense, the debate exemplified the epistemological dynamics of the post-1688 public sphere. As Mark Knights has described, this new sphere was not kind, well-mannered, or reasonable. But the highly contentious, disingenuous, and sometimes irrational practices of argumentation that prevailed nonetheless helped forge the notion "of a rational nation capable of discerning truth amid the lies it was being told, and a set of rules by which judgement of public discourse could be made."[93]

Over time, the partisan combat over the trade balance helped to make the public debate about Anglo-French trade more nuanced, technically sophisticated, and well-informed—in some ways. But what of the central concept of the balance of trade itself? All the wrangling over the trade balance—and especially the fact that neither side seemed to be able to answer the question conclusively—might have served to undermine confidence in the balance of trade as a model, or at least in the possibility of ever calculating accurate trade balances. For some of those who were closest to the numbers, such skepticism did indeed arise. We have seen how, as the dispute dragged on, the *Mercator* came to reject the possibility of calculating a comprehensive trade balance on the basis of the Customs accounts (though it did not reject the value of numerical trade data, nor the balance-of-trade model, overall). Other veterans of the balance battle, including on the Whig side, came to have similar misgivings. Most notable was Henry Martin. After Charles Davenant died in late 1714, he was replaced as Inspector-General of Imports and Exports by his quantitative adversary Martin, probably the most computationally sophisticated of the *British Merchant*'s contributors. As Inspector-General, Martin undertook the daunting task of providing, at long last, a consolidated account of the nation's trade balances from the beginning of the office in 1697 to 1714. In this new position, he expressed a new attitude about the possibility of calculating a precise balance for English trade. "We can never come at the knowledge of the values bought and sold by England by the values imported and exported according to the [Customs] entrys," he explained in a 1718 manuscript commentary to the Board of Trade. "And if so the balance of trade . . . is not possible to be known."[94]

And yet—though many of those closest to the numbers left the trade-balance battle with new awareness of the challenges in that calculation, and new doubts about its feasibility, the overall effect of that controversy on the broader public seems to have been very different. The Tories' and Whigs' ceaseless number-slinging constituted a remarkable advertisement for the balance-of-trade concept, which reinforced, rather than undermined, the position of that idea in popular economic thinking. Each round of quantitative sparring made it harder and harder to imagine that the problem of Anglo-French trade could be adjudicated in any other way. This was arguably the most significant product of the computational contest: an ever stronger and more pervasive confidence in balance-of-trade calculations. An analysis of the frequency of the phrase "balance of trade" in eighteenth-century printed texts—an imperfect metric, to be sure—tells a suggestive story. (See Figure 3.3.) Beginning in about 1715, discussion of the "balance of trade" became increasingly common, rising consistently through the 1740s before leveling off in mid-century. Texts from the 1720s were two-and-a-half times more likely to feature the expression "balance of trade" than texts from the first decade of the century; texts from the 1740s were almost five times more likely.

This evocative word-frequency data is corroborated by how British economic commentators spoke about and deployed balance-of-trade reasoning after 1714. Many commercial writers took the mechanism of the trade balance as the obvious starting point for discussions of international trade. Exemplary was Joshua Gee, whom Charles King listed as an original contributor to the *British Merchant*. Gee's highly influential *Trade and Navigation of Great-Britain Considered* went through at least seven English editions between its publication in 1729 and 1767. For Gee, the fundamental unit of analysis in the study of national commerce was the bilateral trading relationship between two nations. *Trade and Navigation Considered* began with twenty separate chapters sketching Britain's past and present balance of trade with various places, from Turkey to Pennsylvania to Africa.[95] In 1752, David Hume would recall how "The writings of Mr. Gee struck the nation with an universal panic" about the nation's negative trade balances.[96] It was not just members of the Whig *British Mer-*

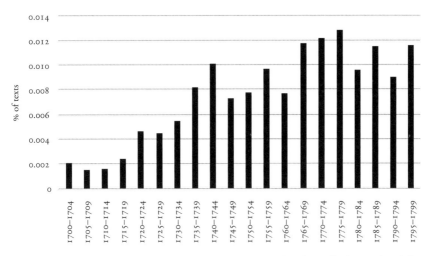

Figure 3.3 Percentage of British printed texts containing the phrase "balance of trade" at least once, by five-year increments. Based on analysis of the Eighteenth-Century Collections Online (ECCO) digital text database as of November 4, 2015. A Google nGrams analysis for the phrase "balance of trade" tells a similar story. Graphic generated via Google Books Ngram Viewer.

chant circle who focused on the balance of trade. Far more iconoclastic economic writers, like Anglican clergyman Josiah Tucker, also took it as their point of departure. "In the Exchange of Commodities, if one nation pays the Other a Quantity of *Gold* or *Silver* over and above its Property of other Kinds, this is called a BALANCE *against* the Nation in *favour* of the other," Tucker wrote in 1750. *"And the whole Science of gainful Commerce consists in bringing this single Point to bear."*[97]

The balance-of-trade battle thus had two different consequences for thinking about the balance of trade, which at first glance seem quite contradictory. On the one hand, for those closest to the numbers, the conflict made them look upon the balance of trade with greater nuance, caution, and even suspicion; on the other hand, within the broader sweep of political-economic thinking, it further entrenched the idea that balance-of-trade calculations ought to be the ultimate authority for making political judgments about foreign trade. But these two products were not really contradictory at all. They were understandable products of antagonistic, open-ended numerical controversy. The unforgiving exchange of computational blows in 1713–1714 *both* forced calculating specialists to realize many of the intricacies and inadequacies in the balance-of-trade model *and* provided tremendous publicity for that calculation among nonspecialists in the public. That wider audience probably followed little about the technical disputes that occupied the *Mercator* and the *British Merchant,* but nonetheless left believing that trade balances were a really, really important thing to calculate.

Perhaps the most striking thing about the quantitative battle over the trade balance is that it did not actually harm the standing of the balance of trade in public esteem, but helped it—despite the fact that, from a certain perspective, the calculation had proven itself incapable of settling the question at hand. This suggests another vital lesson about the quantitative age: the public authority of calculations did not depend on those calculations producing consensus answers. It was the practice of argument and not the production of answers that mattered. As with the Equivalent, Britons disagreed widely about what calculation to believe in, but through their disagreements many converged on the notion that *some* calculation held the answer. What is more, these disagreements extended not only to the results of specific calculations, like the balance of trade, but to calculation in general. Not only did the Tories and Whigs disagree about the correct number for the trade balance, they disagreed about what made numbers valuable to the British polity. Tories celebrated their incisiveness, their ability to debunk deceptions and misconceptions; Whigs celebrated their clarity, their ability to speak for themselves. As with the numerical question of the Anglo-French trade balance, these broader arguments *about* numbers did not get answered by 1714. But there did not have to be a consensus about why to trust in numbers. The quantitative age was not built on consensus. It was built on conflict.

CHAPTER FOUR

The Preeminent Bookkeepers in Christendom: Personalities of Calculation

FOR THE MOST PART, the calculative combat that became a fixture of post-1688 politics was not an especially elegant or erudite affair. But every so often, amid the barrage of fiscal figures and arithmetical insults, a calculator would venture a lyrical sentence or two reflecting on the quantitative enterprise in which he (and it was, admittedly, always he) was engaged. One of the most expressive commentaries—and one of the epigraphs for this book—appeared in 1718, buried several pages into a highly technical pamphlet on the historical growth of the national debt. Its author was a former commercial bookkeeper-turned-Whig polemicist named John Crookshanks, who was entangled in a heated computational dispute about public debt with an MP named Archibald Hutcheson. "Truth and Numbers are always the same," Crookshanks wrote. "Tho' the first may be obscured, yet it can never entirely lose its Lustre; and tho' the latter may be transpos'd, they can never lose their Denomination and true Value, when ranged in their proper place."[1] Though few of his calculating contemporaries praised numbers in such forthright or flowery terms, Crookshanks's sentiment was not entirely uncommon. He was giving voice to an attitude that had, over the previous thirty years, become more and more pronounced in British civic epistemology: in reckoning with public problems, numerical knowledge warranted special authority, and bore a special relationship to the truth. The first three chapters have described how the foundations of numbers' public authority were laid. This chapter uses Crookshanks's ode to "Truth and Numbers" as the starting point for investigating another aspect of Britain's quantitative age, which has gone somewhat overlooked so far: the people who did the calculations.

Who was John Crookshanks, and who was his opponent Hutcheson? How did they learn to calculate? How did they end up devoting their time and talents to crafting pugilistic calculations about public finance? Britain's quantitative revolution was wrought by the hard computational work of individuals. Some of these calculators—Harley and Davenant, Paterson and Gregory—would attain some degree of historical notoriety, though rarely just for their political calculations. But much of the numerical work behind Britain's emerging calculating culture was carried out by more marginal figures, like Hutcheson and Crookshanks. These calculators discovered that, in the new political environment that had developed after 1688, skill with numbers could be a good way to win political friends and influence people.

The calculating lives led by Crookshanks and Hutcheson were, in part, enabled by a broader social process: the general expansion of numeracy in Britain. Over the course of the early modern period, more and more Britons learned to work with numbers in increasingly skillful ways. New sites of mathematical education arose, which made it possible for individuals to develop the rudimentary skills necessary to undertake more creative calculations. As a rising number of Britons learned how to think with numbers, they constituted an audience—or at least a perceived audience—for the numerical publications those calculators produced. Following the paths navigated by Crookshanks and Hutcheson provides new clues to the social history of numeracy in early modern Britain. Those two men were not representative of the numerical skills of the average Briton, of course. But their lives are illustrative in a different way. As a social-historical phenomenon, numeracy can be hard to handle and difficult to measure.[2] Examining two individuals who made remarkable numerical lives allows us to glimpse the era's numerical possibilities as they opened up, to see where computational skills were learned and refined, and to watch as numerical talents became personal opportunities.

The biographies of Hutcheson and Crookshanks share some telling commonalities. Both hailed from Britain's geographic periphery, and both came from "middling" families. They both had to make their own way in life, and both began their careers overseas. Both held brief positions in military administration. Both used their numerical skills to attract powerful patrons and used those connections to establish a foothold in political life. Most critically, they both discovered that political calculation was a tough game to play, finding their careers threatened by unforgiving rivals and their characters by unsavory rumors. Yet the differences between their respective calculating paths are as telling as the similarities. They were trained in very different occupations: Crookshanks through overseas trade and Hutcheson through law. While both men had economic resources to support themselves, Hutcheson was the more secure and prosperous, thanks to the massive dowry he received upon his 1715 (third) marriage to the daughter of a wealthy West India merchant. While Crookshanks had few connections to elite scientific and literary culture, Hutcheson was a Fellow of the Royal Society, a friend to leading wits like John Arbuthnot, and later the first employer of famed writer Samuel Richardson.[3] Both used calculation to advance their political careers, but in different ways. Crookshanks attempted to convert his computational skills into a coveted "place" in the executive government; Hutcheson used calculation to garner Parliamentary influence as a critic of the government machine.

For both Crookshanks and Hutcheson, knowing how to calculate helped them to become the craftsman of their own fate—a *faber fortunae,* as contemporaries put it. But making a calculating life was an adventurous undertaking, and Crookshanks and Hutcheson went about it in quite different ways. By these different paths, they eventually converged on a similar political and intellectual role.[4] That role had no clear label. "Bookkeeper," "accountant" (or "accomptant"), and even "mathematician" were sometimes used to described them. The most precise term—and the one generally used throughout this book—was probably "calculator." While hardly universal, the term "calculator" was used with some frequency in contemporary texts and was a favorite of the prolific Daniel Defoe.[5] Notably, the designation was often used pejoratively by one calculator to describe another.[6] What bound the era's "calculators" together into some kind of a shared identity was the bonds of antagonism, more than shared purpose or mutual respect. In the early eighteenth century, the essence of being a calculator was fighting against other calculators with numbers.

Observing how Crookshanks and Hutcheson forged their respective calculating personalities also sheds historical light on one of the most crucial virtues often attributed to quantitative reasoning: its apparent *impersonality.* The question of how certain forms of knowledge have come to be seen as impersonal has been one of the formative questions in science studies. For example, historians of early modern science like Steven Shapin and Simon Schaffer have sought to explain how the kinds of facts generated by experiments—fabricated by individuals, acting in specific times and places, using unusual and contrived apparatuses—came to be seen as timeless, placeless, and impersonal indications of how nature worked. As Shapin and Schaffer put it, "the solidity and permanence of matters of fact reside in the absence of human agency in their coming to be." As Shapin has shown through the case of Robert Boyle, making that experimental knowledge seem impersonal was a highly personal process. In the seventeenth century, this depersonalization depended heavily on the gentlemanly credibility of the experimenter. Boyle's father, for instance, was an earl.[7]

Scholars have also observed that, in many modern contexts, quantitative techniques have proven essential to this process of detachment whereby knowledge becomes untethered from the conditions of its creation. Compared to nonnumerical forms of knowledge, numerical data and calculations seem better at resisting the whims, biases, and agendas of the people who create them. As Theodore Porter puts it, "a decision made by the numbers (or by explicit rules of some other sort) has at least the appearance of being fair and impersonal."[8] This impersonality is critical to the success of quantification as an instrument for generating trust across geographic, institutional, or political distance. But what makes reasoning "by the numbers" seem more impersonal than other ways of knowing? This apparent impersonality cannot be explained simply as an intrinsic feature of numerical reasoning itself, or the consequence of purely instrumental needs. It is something people have come to believe and expect about numbers, part of certain moral economies of knowledge. As Lorraine Daston notes, "impersonality and impartiality are cultivated by quantifiers as much for moral as for functional reasons."[9] And they have not always done so; recall how Robert Boyle worried that too much quantification would exacerbate interpersonal conflicts and hinder the collective pursuit of truth. So how did impersonality become a recognized epistemic virtue associated with numerical reasoning?

Porter, Mary Poovey, and other scholars have sought the roots of an answer in the development of heavily regimented quantitative practices in commercial life, notably accounting techniques like double-entry bookkeeping. The disciplined rules of double-entry yield numerical inscriptions that, though calculated by individuals, appear to have been mechanically produced and free from human discretion. (Though accountants will be the first to explain that the practice of accounting includes many subtle judgments and choices.) Double-entry accounts have thus been an especially useful mode of recording and communication for businesspeople trying to cultivate credibility with their counterparties, or for government agents trying to do the same with their citizens. Scholars argue that the apparent impersonality of accounting numbers in particular may have contributed to a broader, cultural sense that numbers *in general* were an impersonal way of making knowledge. Poovey has traced, for example, how English merchant-writers in the seventeenth century converted double-entry's "effect of accuracy" into a rhetorical device in their writings on political economy.[10]

The history of Britain's post-1688 quantitative age, though, suggests that there is more to this story about how numbers became impersonal. Accounting was definitely an important source of contemporary attitudes about numerical knowledge, and the public financial calculations that circulated during the period were often treated as a kind of "accounts." But the perceived impersonality of numbers was no straightforward result of the mechanical "effect of accuracy" conveyed by double-entry accounts. For one thing, the act of keeping accounts was intensely personal for early modern merchants. Rigorous procedures like double-entry imposed discipline on the individual, but they were not intended to negate an accountant's individual personality. Rather, well-kept books performed a merchant's virtue, and so bookkeeping was closely bound up with matters of reputation, self-fashioning, and even spiritual well-being (as was mathematics in general during the early modern period).[11] Fourteenth-century Tuscan merchant-banker Francesco Datini, for example, famously began all of his double-entry account ledgers with the phrase, "In the Name of God and Profit." Literary scholar Adam Smyth has recently argued that, in early modern England, bookkeeping was a critical early genre of personal "life-writing," which helped lay the foundations for later forms like diary-keeping and autobiography.[12]

Likewise, public contests over financial calculations during the early eighteenth century were highly personal, morally laden affairs. When they attacked one another's numbers, dueling calculators like Crookshanks and Hutcheson made things personal. They regularly tried to undercut their opponents' numerical "accounts" by casting moral aspersions on the individual who produced them, framing their calculating conflicts as trials of personal virtue. We have already seen this at play: in the 1690s, critics questioned Davenant's honor for publishing numerical estimates without sufficient "vouchers," while calculators throughout the Anglo-French trade debate denounced their opponents' computations as forgeries. This was especially evident in the conflict between Crookshanks and Hutcheson over the national debt. Neither man could validate his calculations—as the genteel Robert Boyle did with his pneumatic experiments—by calling upon his personal reputation. If anything, the opposite was almost true. Both men had spotty reputations, dogged by lingering accusations of cynicism, corruption, even treason. Their disreputable character, in turn, threatened to stain the very numbers they calculated.

Yet Crookshanks and Hutcheson—and the personal, political, often petty ways they calculated—also helped to contribute to the perceived *impersonality* of numbers. In critiquing their opponent's calculations, and trying to justify their own, both men grasped at a putative domain of numerical truth that transcended any individual calculation or calculator. Their logic went roughly like this: your calculation is wrong, because you are partisan and dishonest; my calculation might be imperfect, because my data is incomplete or irregular; but the correct numbers exist somewhere, and they are the ultimate authority. Public calculators, regularly assailed for their individual character (and regularly assailing the character of other calculators), came to suggest that the numbers could be better than the people who made them. The use of numbers as an instrument of dispute played a crucial part in the historical process by which impersonality became one of numbers' signature epistemic virtues.

One of the most crucial features of Britain's emerging quantitative culture was that it was grounded on a collective confidence in calculations, not in calculators. To put it another way: people like Crookshanks and Hutcheson did not function as *experts* in the context of British politics. They resembled experts in some ways.[13] Both were acknowledged by others for their exceptional technical abilities. Crookshanks, for example, was once described as someone "who disputes the Preheminence with all *Book-keepers* in *Christendom.*"[14] Both tried to convert those abilities into professional advancement and personal authority. But they were definitely *not* experts in the way experts function within modern civic epistemologies. Experts are authorized by the rest of the community to make judgments about complicated problems, to fix things, to determine what is safe. As Sheila Jasanoff writes: "Experts are the people to whom publics turn for answers."[15] People defer to experts. They believe what experts are going to say simply because they are experts. By that standard, Crookshanks and Hutcheson were not experts at all in 1718. As individual calculating personalities, neither man would have seemed especially trustworthy to a majority of Britons. Yet somehow the numbers that they peddled became more trustworthy than they ever could have been.

Making Mathematical Fates in Early Modern Britain

Rising numeracy across the British population was a crucial factor enabling the emergence of the quantitative age over the course of the long eighteenth century. It is difficult to measure how many truly numerate people there were—or even to define what "numerate" might exactly mean. (Is to be numerate to be able to add or multiply? To use the "Rule of Three"? To compute compound interest?)

Despite these ambiguities, the assessment of historian Keith Thomas is apt: "there can be little doubt that numerical skills were more widely dispersed in 1700 than they had been two centuries earlier." Over the course of the early modern period, more and more people adopted a recognizably modern form of numeracy based on Arabic numerals and pen-and-paper calculations, aided by a proliferation of new didactic books and new sites of mathematical instruction.[16] The range of practical applications for mathematics expanded, producing an array of different "mathematical practitioners," especially in urban areas. The social range of mathematics expanded as well. Calculation went from being a specialized skill, cultivated primarily by a narrow set of merchants, navigators, and tradesmen, to being recognized as something with practical utility in various walks of life.[17]

Beginning in the first half of the sixteenth century, the resources available for learning basic mathematical calculations began to expand markedly. The most tangible evidence comes from new mathematical textbooks, starting with Cuthbert Tunstall's first Latin text in 1522. Vernacular arithmetic books soon followed, first the anonymous *An Introduction to Learne to Rekyn with the Pen and Counters* (1537) and then the immensely influential *Grounde of Artes* (1543) by Robert Recorde, which went through at least forty-six editions by 1699. Recorde's benchmark text presented arithmetic as a series of rules: first "numeration," or counting; then addition, subtraction, multiplication, and division; then more specific calculations like converting weights and the indispensable "Rule of Three," a technique for calculating the missing fourth term in a proportion. More advanced rules explained how to deal with "compound" proportions and how to divide the profits of a commercial partnership (the "Rule of Fellowship"). Subsequent editors, like Elizabethan mathematician and occultist John Dee, regularly updated Recorde's text, adding sections on fractions, square and cube roots, and decimals. Eventually, new textbooks emerged offering streamlined alternatives to Recorde's cumbersome primer, most notably *Cocker's Arithmetic: Being a Plain and Familiar Method Suitable to the Meanest Capacity* (1677), which would run through over a hundred editions. Writers increasingly targeted lessons to the novice calculator—to the "meanest capacity"—through illustrative exercises, easy-to-use estimation techniques, and "ready reckoning" tables.[18] Textbooks did not just pile up on shelves. Jessica Otis's recent study of 365 surviving early modern textbooks finds that roughly three-quarters of those texts contain marginal annotations, and over half contain substantial marginalia showing significant engagement with the mathematical content. Readers sometimes copied textbooks by hand, reproducing and studying a text simultaneously. (Not every student got to the end of the book, of course. One manuscript textbook in the Huntington Library stopped before chapter eight, on "Addition of Fractions & Mixt Numbers.")[19]

At the same time, mathematical educators produced more specialized and advanced instructional texts. Books on accounting and financial calculation were an important early segment. John Mellis's *A Briefe Instruction and Manner How To Keepe Books of Accompts* (1588) and Richard Dafforne's *The Merchants Mirrour* (1635) instructed English readers in double-entry bookkeeping, while Richard Witt's *Arithmeticall Questions: Touching the Buying and Exchanging of Annuities* (1613) and William Purser's *Compound Interest and Annuities* (1634) provided rules and tables for solving problems about simple and compound interest. William Leybourn's *The Compleat Surveyor* (1653) set the standard on land surveying, and Richard Collins's *Country Gaugers Vade Mecum* (1677) for "gauging" or calculating the volume of vessels (like ale casks, essential for excise taxation). Similar specialist texts addressed carpentry, gunnery, navigation, and "dialling" (the making of sundials). The trade in practical mathematical books continued to grow in the eighteenth century. At least fifty different guides were published on gauging alone between 1700 and 1750.[20]

As vital as textbooks were, learning to calculate was not only a solitary and bookish endeavor. Alongside new texts grew new sites of mathematical instruction, notably in and around London. Textbook writers were usually professional tutors as well. Humfry Baker drafted a prominent textbook, *The Welspring of Sciences,* that went through at least twenty-four editions between 1562 and 1687, and also operated a school "on the Northside of the Royall Exchange" in London in the late sixteenth century, advertisements for which still survive.[21] A century later, Edward Cocker offered private lessons, too, teaching for nearly two decades in Northampton and then Southwark. Quantitative instruction often came not from mathematical specialists but rather generalist "writing" tutors, who taught the pen-and-paper skills of arithmetical calculation along with reading and writing. These complemented another key site of quantitative learning: artisanal and commercial apprenticeships, in which novices would learn arithmetic and bookkeeping in the normal course of their training.[22]

But the growth of numeracy was not a smooth, even, or inevitable process, either. Many people ignored numbers and resisted learning how to use them. As discussed in the Introduction, Britons had long disparaged calculation as too solitary, mercenary, or trivial for a true gentleman, or likened it to a form of spell-casting. While written arithmetic grew in prominence across the early modern period, other traditional "numeracies" persisted as well, including those based on hand gestures, Roman numerals, and physical objects like tally sticks and the abacus. Editions of Robert Recorde's famed textbook, for example, taught calculation by physical counters—alongside pen-and-paper methods—until 1699. Though many new sites of numerical learning opened up, they were mostly small and scattered operations.

For many—perhaps most—Britons, learning the basics of calculation was thus a somewhat irregular affair. Many individuals who went on to bright mathematical careers first learned arithmetic in odd or even accidental ways. John Wallis, appointed Savilian Professor of Geometry at Oxford in 1649, learned arithmetic only when he was fifteen, not at the prominent Felsted School he attended, but by happening upon the textbook of his brother, an aspiring merchant, while home on holiday. John Flamsteed, who would become the first Astronomer Royal in 1675, claimed not to have learned arithmetic, fractions, or the Rule of Three until age fourteen, taught by his father after dropping out of school from poor health. The famed astronomer Edmund Halley, Wallis's successor as Savilian Professor, reportedly learned arithmetic from an apprentice of his father, a soap-boiler. Perhaps most famously, Samuel Pepys had to hire a former ship's mate to teach him basic mathematics at age thirty, because a new administrative post with the Royal Navy demanded it, and he had never learned at St. Paul's School or Magdalene College, Cambridge.[23]

Stories like Wallis's and Pepys's have often been cited as evidence of the relative mathematical backwardness of England's educational establishment: the petty schools, grammar schools, and universities. Recent research has shown, though, that these traditional sites of learning were not quite as disparaging toward mathematics and natural philosophy as critics (including contemporaries like Wallis) have claimed. Some "petty schools," aimed at children aged five to seven, started offering basic numerical lessons in the late sixteenth century. By the seventeenth century, about a quarter of all petty schoolmasters' licenses listed arithmetic teaching as a specific skill.[24] More advanced "grammar schools" often required students to have basic numerical competence, and coordinated outside mathematical lessons for students needing special instruction. The universities of Oxford and Cambridge were also not the mathematical backwater that stories like Pepys's suggest. Some arithmetical competency was expected for all university graduates in the seventeenth century, and enterprising students would have had ample opportunity for more advanced mathematical education in libraries, lectures, and tutorials. Yet these orthodox educational institutions probably did not prioritize quantitative skills, either. Many students would have been able to pass through every level of traditional schooling, from petty school to grammar school to university, without learning much more than the first few pages of Recorde or Cocker.[25]

The best way to describe the topography of mathematical education through the middle of the seventeenth century, therefore, is that it was decidedly uneven. There were many places to learn to calculate, but they were diverse, disparate, and impermanent. The place of numerical knowledge in school and university curricula varied widely by schoolmaster and student. The unevenness in mathematical education mirrored the unevenness in contemporaries' attitudes toward numerical thinking. This was the numerical world from which John Crookshanks and Archibald Hutcheson emerged.

During the latter half of the seventeenth century, arguments for the usefulness of numerical skills grew louder and more frequent. Educational reformers like Francis Osborne and John Aubrey lobbied strenuously for an expanded role for mathematics at all phases of education. In a manuscript on education begun in 1669, John Aubrey contended that a young student should take up "arithmetic as soon as he could speak," such that "arithmetic and rules will be so bred in the bone . . . that it will be habitual."[26] Mathematicians like Wallis and David Gregory (as we saw in Chapter 2) worked hard to expand the presence of mathematics in the universities. New institutions like the Mathematical School at Christ's Hospital, founded in 1673 by Pepys and others, and chartered by Charles II to educate sailors in mathematical navigation, gave a measure of official validation to the usefulness of calculation. One key strategy adopted by mathematical promoters like Aubrey was to recount the stories of those curious calculators who had used mathematical learning to fashion successful careers. Exemplary of these new *faber fortunae* ("maker of his own fortune") narratives was a brief autobiography that William Petty appended to his own will in 1685. Petty proudly made his early mathematical learning a foundation of his subsequent prosperity. He recounted how by age fifteen he had mastered "the whole Body of Common Arithmetick, the practicall Geometrie and Astronomie," skills which earned him his first job in the Navy and helped him, by age twenty, to have saved up "about threescore pounds."[27] Advocates also stressed that the capacity to calculate could aid traditionally genteel pursuits including military leadership and estate management. Some students got this message as well, like the young Thomas Foley, nephew of the would-be Commissioner of Public Accounts Paul Foley and later brother-in-law of Robert Harley. Writing to his father in February 1689, at age fifteen or sixteen, the younger Foley argued that he needed to leave Shropshire and relocate to London to advance his education, and particularly to "perfect my self in Mathematicks." As he put it, "the Mathematicks . . . are necessary for the accomplishment of a gentleman."[28]

The fact that Foley even had to make this argument was, of course, telling. He could not take it for granted that the value of an expanded mathematical education was well understood by his father—or society at large. In his 1701 *Essay on the Usefulness of Mathematical Learning*—probably the period's most successful piece of mathematical boosterism—John Arbuthnot lamented that the field still "hath been generally neglected, and regarded only by some few persons, whose happy Genius and Curiosity have prompted them to it, or who have been forced upon it by some immediate subserviency to some particular Art or Office."[29] Around the time Arbuthnot published, though, an evolving political environment afforded new opportunities for those "few persons" who developed intensive calculating skills. The remainder of this chapter is dedicated to two such figures. As with the career of calculation in British culture in general, the paths they trod were rough and jagged.

"A Very Knowing Able & Experienced Accomptant"

If you had to choose one character to narrate the story of Britain's early quantitative age, there is perhaps no better candidate than John Crookshanks. He was not the most influential nor especially typical—there was no "typical" calculating life, after all. Rather, his story is illustrative precisely for its odd twists and turns. He can be seen wandering through practically every chapter in this book. He managed military payments during the War of Spanish Succession, helped reform the Scottish Customs after the union, and engaged in shady stock deals during the South Sea Bubble. He was one of the only people who could explain how the Equivalent worked. He calculated for Tory leader Robert Harley and for Whig leader Robert Walpole. He was both perpetrator and victim of the era's ugly numerical politics. His calculating adventures took him from Tuscany to Edinburgh to London to Seville. He left countless traces of his calculating career: in manuscript letters and calculations in the Treasury archives, in the personal papers of Harley, Walpole, and others. Yet he was, for the most part, an anonymous figure. He has no modern biographies or encyclopedia articles. Were it not for his public showdown with Hutcheson in 1718, he might never have made his way into this story of Britain's quantitative age.[30]

John Crookshanks began his life in the 1670s and "had the happiness to be born in *Scotland,* to be Educated in *Ireland,* brought up in *France,* and Finished in *Italy.*" While few details are available on his early life, two features are especially noteworthy. First, he was Scottish. Many of the most enterprising public calculators of the era, like William Paterson and David Gregory, were also Scottish and used their technical skills to build careers in political and intellectual centers like London and Oxford. Second, Crookshanks seems to have learned to calculate at sea, at least metaphorically speaking. By the mid-1690s, he had become a trusted accountant for the well-reputed trading house of Alexander Rigby & Co., based in Livorno, Italy, where he earned a reputation as one of the world's "preheminent" bookkeepers. International commerce was unquestionably one of the most important sites for quantitative learning in the period. Maritime trading enterprises required numerate officers to navigate their ships and skilled bookkeepers ("pursers") to manage the ships' intricate business activities. Many calculating pioneers of the seventeenth century also began their careers at sea: William Petty was barely a teenager when he set off as a cabin boy on a merchant ship; financial mathematician and mathematical "intelligencer" John Collins was a purser in the Venetian trade; one important author of early textbooks in financial mathematics was a former Royal Navy mate and Bristol-based merchant, aptly named William Purser. Their nautical excursions were often harrowing—Petty broke a leg, Purser lost a hand in a naval battle. In the seventeenth century, accountancy was rarely a boring career.[31]

Crookshanks's early career did not lack adventure, either. In 1697, he became embroiled in a diplomatic scuffle when a former company partner, William Plowman, was arrested by the Grand Duke of Tuscany. The charge was piracy, perhaps trumped up as retribution for Plowman's attempts to muscle in on the Egyptian coffee trade. The accusations sank the traders' business interests, and Plowman was imprisoned for three years. Crookshanks became Plowman's primary agent, responsible for negotiating his boss's release with Tuscan and French officials, including the Grand Duke himself. The bookkeeper was not immune from the turmoil. When Rigby & Co. later tried to recoup some of their losses by appealing for damages from the Grand Duke, the trustees for the English factory at Livorno, loath to alienate the Tuscans, hung Rigby & Co. out to dry. Instead, they blamed the company's failure on "the Intriegues and Unfair Practices of *Plowman* and *Crookshanks* their Book-keeper, who neglected his proper province, keeping up no Books for several Years, that so he might better serve his own and *Plowman's* Secret Designs." Those critics intimated Crookshanks had cooked the books to conceal Plowman's contraband trade. Whether there is any truth to the charges may be impossible to know. Here we see another vital feature of quantitative life in the early eighteenth century. Because financial calculation was an inherently sensitive activity—in which secrets were managed, testimony was made, and the credit of an enterprise was constantly at stake—skilled calculators always occupied a precarious position, haunted by the specter of deceit and corruption.[32]

Crookshanks returns to historical view in 1706. By then he had migrated from the world of commerce to that of military administration (potentially just as lucrative), assisting with payments and recruitment under Lieutenant General Richard Ingoldsby, a trusted deputy of the Duke of Marlborough during the War of Spanish Succession. In the following year, he was promoted to Comptroller-General of the Customs in Scotland, recently unified with England, where he was tasked with drafting an instruction book on accounting procedures for Scottish Customs officers. Crookshanks's book ran to over a hundred pages. Contents ranged from prosaic best-practices—"do not clear any Accompts or Dispatches in Taverns or Ale-Houses"—to a lengthy appendix containing dozens of example forms, accounting templates, and calculations. Among his challenges in the Customs Office were managing the fiscal complications of the union and combating the smuggling and tax evasion rampant in Scotland, particularly in tobacco. Given the tense post-union environment in Scotland, the Customs Commissioners saw him as an agent of political surveillance as well, urging him to "have watchfull Eyes on any fomenters of misunderstanding" and to "observ[e] any open or secret Contrivances."[33]

Crookshanks's work caught the attention of key politicians in London, including Lord Godolphin, then Britain's leading minister. By late 1708, the Customs Commission in London had become frustrated with the "great Backwardness and Confusion" in Scotland's Customs records. They requested the Lords of the Treasury promote Crookshanks, "who hath the Character of a very knowing able & Experienced Accomptant," into a new London-based position as Accountant General of the Customs for all of unified Britain. That promotion never materialized due to petty jealousies; Lord Chief Justice John Holt stymied the move in order to protect the political position held by his brother, Rowland. But the nomination bolstered Crookshanks's reputation in administrative circles in London.[34] Among those impressed was Robert Harley. As Lord High Treasurer in the early 1710s, Harley called upon Crookshanks for insight on fiscal and commercial matters, notably regarding Scotland and France. He synthesized data on Scottish trade and customs revenue, calculated the costs of illicit Scottish trading in wine and salt, and passed along intelligence regarding anti-union sentiment in Scotland. In one especially remarkable bit of calculation in June 1713, he produced an extended account of the "happy consequences of an Union with Scotland," probably intended to aid Harley in propaganda efforts. This veritable balance sheet of the union included both the "liquid advantages" the Scots had gained—the £398,085.10s. of the Equivalent plus £70,000 of Scottish debt absorbed by Britain—as well as a series of other, more abstract benefits, like an estimated £52,500 benefit to Scottish trade from being able to trade certain "Goods Duty Free." Crookshanks was not subtle about his desire to parlay the connection with Harley into a better job.[35]

Politically speaking, perhaps Crookshanks's most valuable asset was his understanding of the Equivalent—something that, as Crookshanks recalled, "at that time very few understood."[36] As noted in Chapter 2, paying the Equivalent proved even harder than calculating it, leading to various political problems in the two decades following the union. Many Equivalent creditors were slow to get paid, and new computational controversies emerged as time passed. One especially thorny problem related to the forward-looking nature of the Equivalent calculation. The Equivalent calculations were based on projections about likely future Scottish taxes; what if they did not come to pass? The Union Treaty made a provision for this possibility, promising that if Scottish excise and customs revenues turned out to exceed the levels projected (£33,500 and £30,000 per annum, respectively) in the first seven years, the surplus would be refunded to the Scots. This was termed the "arising Equivalent." The outcome was the most confusing one possible: Scottish excises exceeded predictions, but customs fell well short. Some argued that the customs shortfall would simply cancel out the excise surplus. Others contended the Scottish should repay some Equivalent money because the customs proved deficient. Yet dozens of Equivalent creditors, and the Parliamentary commission that represented them, felt they were entitled to the value "arising" from the excise surplus. The Commissioners met with Harley in early June 1714 to demand the arising funds, which they calculated to be £11,882 (and later revised to over £65,000). Harley and the Tory government were already under considerable fiscal strain and were hesitant to commit precious funds. Amid this acrimonious—yet highly arcane—dispute, Crookshanks's calculating skills proved invaluable. He produced a battery of detailed analyses that helped Harley make sense of the controversy and defend against the creditors' agitations.[37]

Harley was not the only one impressed by Crookshanks's numerical acumen. Around 1714, the Earl of Halifax, a major Whig powerbroker, took notice of his handling of the Equivalent, too. It was a timely connection. In 1715, following Queen Anne's death, a massive electoral victory for the Whigs ushered in nearly half a century of Parliamentary dominance for Halifax's party. Crook-shanks had little hesitation in switching allegiances. He soon began advising Whig politicians, including John Aislabie and the Earl of Sunderland, on Scot-tish affairs and technical matters of "Trade & Revenue." While he maintained his official position with the Scottish Customs, he also spent much of his time at a home in Twickenham, near London, staying closely connected to West-minster politics and providing regular advice and reports to the Lords of the Treasury. It was in this capacity as an informal adviser that Crookshanks took his calculating skills public in the dispute with Archibald Hutcheson in 1718.[38]

Crookshanks's story shows the hustle it took to make a calculating life in this early quantitative age. Crookshanks bounced around, from place to place and job to job, trying to use his calculating skills to get noticed. His reputation as a "very knowing able & experienced" calculator opened up opportunities in military logistics, the Customs administration, and informal political consul-tancy. Critically, it was not just the administrative demands of the state that made his calculating skills useful, but the demands of politics. Calculations had become a vital instrument of political dispute. Leaders like Harley and Halifax needed someone like Crookshanks, who could use numbers to explain com-plex policy issues, analyze proposed actions, and formulate public arguments. Yet the kind of calculating life this afforded was complicated and sometimes treacherous to navigate. Because of economic and political rivalries, he was accused of piracy in Livorno and had his path to promotion blocked in London. There was nothing impersonal about the numbers in which Crookshanks traf-ficked. He used his calculating skills to make his fate, but it was a constant gamble.

From London to the Leeward Islands and Back Again with Archibald Hutcheson

Archibald Hutcheson's early life in calculation was no less eventful than that of his eventual nemesis.[39] Like Crookshanks's, Hutcheson's family hailed from outside England. Archibald was born around 1660 to Ulster-Scottish par-ents from County Antrim in northern Ireland. Compared to Crookshanks, Hutcheson's path to a calculating life was even less direct. The story seems to begin in the law. In the early 1680s, Hutcheson enrolled at the Honourable So-ciety of Middle Temple, one of the four surviving Inns of Court, the centuries-old institutions responsible for training and regulating English barristers. In the early modern period, the Inns served a mix of professional and social functions. Many students enrolled to undertake the peculiar aural curriculum, learn the common law, and begin legal careers. Others saw the Inns as a passage-point in the life of a well-to-do gentleman, a place to make connections, expand cultural horizons, and learn a few skills for managing an estate. Hutcheson was squarely in the former category. But he also would have benefited from the vibrant intellectual life that grew up around the Inns, where curious students could explore new languages, poetry, drama, astronomy, anatomy, theology, and even mathematics. One historian writes that, among all educational institutions in the period, "the Inns came closest to fulfilling the role of humanistic academies for young English gentlemen."[40] In the late seventeenth century, the Inns of Court had also come to be a key nexus between London professionals and the city's burgeoning scientific community. John Aubrey, for example, used Middle Temple "as a convenient base of operations," though he undertook little legal training. The Inns even served as a "recruiting ground for the Royal Society." While Secretary of the Society from 1677 to 1682, Robert Hooke signed up sixteen members schooled at the Inns, half from Middle Temple. Hutcheson probably learned little about numbers in any Middle Temple classroom, but his time there was likely formative in stimulating his curiosity for calculation.[41]

Hutcheson was "called to the bar" in 1683. He practiced law in London and Ireland, before his career took an adventurous turn in March 1688. Recommended by two Catholic noblemen, the Earls of Middleton and Carlingford, Hutcheson was appointed by the Colonial Governor of the Leeward Islands to be the new Attorney General. The Leeward Islands were a group of island colonies at the northern tip of the Lesser Antilles chain. The first island colonized by the English was St. Christopher (St. Kitts) in 1623. The French set up their own colony a couple years later, leading to a joint massacre of the native population and an unstable partition of the island between the two European powers. In 1628, the English settled on nearby Nevis, which would become the seat of colonial governance (and, a century later, the birthplace of Alexander Hamilton). Antigua and Montserrat were settled the next decade. The economic promise of the Leewards lay in their potential for intensive agriculture, first in tobacco, which grew well in St. Christopher's volcanic soil, then later indigo and sugar. By the 1640s, Leewards colonists had begun to transport enslaved Africans to provide involuntary agricultural labor, beginning the islands' development into a slave society.[42]

The Leewards were a violent and insecure place. Though less lucrative than nearby Barbados, the islands nonetheless became objects of desire for England's imperial rivals. Due to their precarious geographic location, they played host to regular military conflict, notably during the second Anglo-Dutch War (1665–1667). England's hold on the distant Leewards remained tenuous throughout the century, due both to foreign threats and the English colonists' own distaste for metropolitan oversight. Leewards residents had always enjoyed mostly unrestricted trade with neighboring colonies and foreign traders, particularly the Dutch, and stringently resisted any efforts by the English government to regulate trade more carefully. Hutcheson's appointment as Attorney General was likely driven by the desire of James II and his ministers to bring the Leewards, and other parts of the nascent empire, under increased Crown control.[43]

Shortly after Hutcheson arrived in the Caribbean, Governor Nathaniel Johnson reported back to England that he had already "proved worthy of the recommendations." Yet Johnson worried about how Hutcheson would take to the job. "I can do little to induce a gentleman educated as a lawyer to spend his time in the Colonies," he fretted, "where little of pleasure or the improvement of knowledge can be expected." Hutcheson found the Leewards in legal and political disarray. Laws and regulations were widely disregarded, and Johnson's government held limited sway over the residents. When Johnson asked the islanders to lend their support to the new Attorney General, he found "them unwilling to comply with anything." Such disorder seems typical of the Caribbean colonies at the time. Hutcheson was given a broad mandate for reform, and he took to the task with vigor. His job combined roles as financial manager, public auditor, administrative reformer, and justice of the peace, plus political fixer for the colonial governors.[44]

Among Hutcheson's first orders of business was to tour several of the islands to survey administrative conditions and ascertain the legal status of local estates. What he found was shocking. Weak legal infrastructure, opportunistic colonists, and intermittent military conflicts had made it extremely difficult to know who legally owned what land. "Altogether," Hutcheson concluded in a report on the matter, "I believe that nearly all the titles in the Leeward Islands are insecure." Many Leewards colonists, with dubious claims to their own estates, liked things that way. Hutcheson was especially frustrated by one Joseph Crispe, who for eight years had held the position of "Escheator General" in St. Christopher. In that position, Crispe was responsible for administering estates that, due to the death or disqualification of their proprietors, were legally due to revert or "escheat" back to the Crown. (Crispe was also apparently "a persistent smuggler of negroes and sugar to and from the Dutch Islands.") When Hutcheson made his initial tour, Crispe was unable to furnish any documentation for the lands he received as Escheator General, dismissing "the receipts as trivial." Hutcheson's early experience in the Leewards mirrored what the Commissioners of Public Accounts would endure a couple years later. Tasked with bringing order and accountability to a world that knew little of either, they found themselves fighting against entrenched interests who had thrived in the absence of public oversight. Hutcheson quickly proposed a creative plan to tackle the chaotic land-title situation and simultaneously improve the colonial government's dire fiscal condition. He proposed that the king solidify the legal status of Leewards estates by granting new "patents" to legitimate colonial proprietors. In exchange for this legal privilege, the Crown would take a rental fee of ½ to 1 percent on each estate's annual produce. Hutcheson projected that the resulting "revenue would be considerable," enough to fund the salaries of the colonial officials (including Hutcheson himself).[45]

Hutcheson's other duties were varied and vexing. These included managing finances and logistics for the Leewards' military forces, who were regularly embroiled in conflicts with the French over St. Christopher and neighboring islands, especially after a new Governor, Christopher Codrington, arrived in 1689. Military administration was a contentious business, as soldiers were rarely paid on time and got restless as a result.[46] Hutcheson's varied responsibilities brought him considerable power, and new enemies. In May 1697, for example, a disgruntled Leewards resident using the pseudonym J. Johns Sonn wrote a colorful complaint to Admiral Nevill describing Codrington as "an unhappy, covetous and unprofitable governor, an oppressor of the poor." Sonn specifically cited the fact that the Governor "employs one Hutcheson, a lawyer, in all his intrigues, a confirmed Jacobite."[47] A planter named Edward Walrond subsequently leveled a sustained attack against Codrington, accusing him of corruption and profiteering in illicit trades. Hutcheson, who seems to have returned to London at various points during the late 1690s and early 1700s, lobbied strenuously to clear Codrington's name with the Board of Trade and other English officials. In the course of doing so, he got a first taste of the hostile battles over public accounts that were becoming a fixture of English politics. In fall 1699, Hutcheson was called before the Lords of the Treasury to testify regarding irregularities in the accounts of customs revenues raised in the Leewards under Codrington. There, Hutcheson had to answer meticulous questions about potential collusion in sugar sales and about suspicious losses of public revenues attributed to "over rates, wastage, agency, & com[issi]^{on}" (accounted down to the quarter-penny).[48] Hutcheson managed to avoid getting caught up in the accusations swirling around his former boss, but the experience contributed to his ongoing political—and computational—education.

After about 1702, Hutcheson became less and less involved in Leewards affairs, though he may have held his official position as late as 1704. Back in London, he resumed his legal practice. He also reestablished connections among the city's scientific circles and was elected a Fellow of the Royal Society in 1708. At some point, Hutcheson developed a professional relationship with the immensely powerful James Butler, 2nd Duke of Ormond, a leading military commander and holder of massive estates in Ireland. Hutcheson's earliest work for Ormond was legal and administrative; in 1710, for example, he served as a trustee for the sale of some of the duke's property. Hutcheson subsequently took on increasing responsibility for managing the duke's affairs and began to benefit from his patron's political clout. Aided by Ormond's patronage, Hutcheson earned election to the House of Commons in 1713, representing the seaside town of Hastings. There, he began to build a new calculating personality.

Hutcheson's Search for "Exact Estimates"

In his early years in Parliament, Hutcheson's political identity was hard to pin down. Historians have termed him "a well-meaning oddity": "variously described as a Whig or a Tory, but who seemed comfortable with neither party"; friends to many Jacobites but not clearly one himself; "Country" in principles but unmatched in his knowledge of city finance.[49] At first, Hutcheson's affiliations straddled partisan divisions. The patronage of Ormond, a Tory, got him elected. But he also curried favor with the presumptive next King George, then Elector of Hanover, and his Whig supporters, even traveling to Germany in summer 1713 and visiting the Hanoverian Court.[50] He made a sufficiently good impression on Whig leadership to earn a spot on the Board of Trade in December 1714. In his early years in Parliament, he championed some relatively nonpartisan economic projects, encouraging domestic woolen manufacturing and the African slave trade. He also quickly showed himself unafraid to speak out against major party causes on both sides. He stridently opposed both the 1714 Schism Act, a Tory bill requiring all schoolmasters be licensed by the Church of England, and the 1715–1716 efforts to repeal the Triennial Act, a Whig project that would have reduced the frequency of elections.[51] When Ormond was impeached by Parliament in 1715 for his involvement in that year's Jacobite rebellion, Hutcheson's continued loyalty to his patron further alienated him from Whig leaders. He consequently resigned from the Whig-dominated Board of Trade in early 1716. After that, his loyalties became squarely Tory, no doubt largely a reflection of his growing distaste for the particular Whigs in power.[52]

If anything, Hutcheson's primary allegiance was to the principle of opposition—to an almost obsessive distrust of those wielding government power. His politics were suffused with the Country ideals that had energized the Accounts Commissioners in the 1690s. He can be seen as the main heir to the critical, Country approach to calculation first formulated by Charles Davenant.[53] Hutcheson shared Davenant's concerns about governmental ignorance, fiscal waste, and ministerial corruption, concerns heightened by his own administrative experience in the Leewards. Like Davenant, he embraced calculation as a means of overcoming informational asymmetries and combating secrecy. Yet Hutcheson's calculative style differed from Davenant's in a crucial way. While Davenant embraced the political value of calculations that were merely probable, Hutcheson craved certainty. He saw such calculations as imperfect, stopgap measures, necessary for dealing with the crippling uncertainties that plagued British politics. Hutcheson's distinctive approach to political calculation would become evident in his most notable computational innovation: what we might call the arithmetical "spreadsheet" model.

The political problem that worried the new MP Hutcheson most was the national debt. Since 1688, outstanding government debts had surged from perhaps £1 million to more than £40 million by 1715.[54] Almost immediately, Davenant and others had begun warning against the specter of permanent national indebtedness, contending that chronic interest payments and high resulting taxation threatened to destabilize England's finely balanced social order, impoverishing landowners and empowering a new "class of professional creditors."[55] Hutcheson's first published pamphlet, *A Proposal for the Payment of the Publick Debts*, from January 1715, articulated this argument forcefully. He lamented how all Britain's standard taxes had been "Mortgaged and Sold for ever, or for long Terms of Years" to finance the nation's mounting debt. This necessitated burdensome new imposts on malt and landed property as well as excessive customs burdens, which proved "great Obstructions to Trade." He encouraged swift and decisive remedies, including a 10 percent, one-time levy on all property (a "decimation"), an apparel tax, a decrease in the maximum allowable interest rate, and a rigid schedule for debt retirement.[56]

Hutcheson's 1715 text contained only a single numerical table (see Figure 4.1). But that small array of numbers said much about Hutcheson's political thinking,

The Annual Sum to pay off 45 Mill.	The Number of Years according to the Rates of Interest			
	6 p. C.	5 p. C.	4 p. C.	3 p. C.
A Million and [a] half	18	19	21	22
A Million	23	25	27	29
Half a Million	32	35	39	45
250,000 *l.*	43	48	54	63
100,000 *l.*	58	65	76	91

Figure 4.1   The lone numerical table in Archibald Hutcheson's 1715 *A Proposal for the Payment of the Publick Debts, and an Account of Some Things Mentioned in Parliament on that Occasion* ([London]: *s.n.,* 1714 / 15), 21. The table shows the projected amount of time it would take to repay £45 million of national debt given various annual payments (£100,000 to £1.5 million) and various interest rates (3–6 percent).

and the role of calculation therein. The table showed how long it would take to repay the national debt of £45 million, assuming different initial annual payments (£100,000 to £1.5 million, over and above what was already paid for interest) and different interest rates (3–6 percent). Two features were especially remarkable. First, Hutcheson's projections implicitly assumed that the debt would be paid off at an exponential, rather than a linear, rate. He projected that, as Parliament paid off capital, the interest saved would be used to pay down even more capital in subsequent years, effectively harnessing compound interest to pay down the debt.[57] This was the exponential logic of a "sinking fund," a financial technology that would become increasingly popular over the next decade and a half.

The second notable feature of this table was its computational structure. His table was not informational; it was not filled with discrete pieces of empirical data. Rather, it consolidated the results of different calculations—or, rather, a single computation, repeated twenty times over various inputs (in this case, five hypothetical values for the annual payment size and four hypothetical interest rates). Each of the twenty different entries in the table was the outcome of a distinct calculation; the table itself represented a range of potential futures under unknown conditions. In short, it was a *model*. He admitted that such a model was only a starting point, an imperfect reflection of a complex problem. "In Order to the forming any Scheme," Hutcheson explained, "'tis absolutely necessary to have an exact State of the entire Debt of the Nation, and of the clear yearly Income of the Revenues appropriated to the Payment of the same."[58] But, until such ideal information was available, modeling different scenarios using arithmetic was a valuable way to think through an uncertain future. This style of calculation—akin to a modern spreadsheet model—would become Hutcheson's trademark.

He called upon that spreadsheet style again in 1717, in a new pamphlet analyzing the national debt. He had better data to work with this time, courtesy of new Treasury accounts offered to Parliament in March. Yet much remained unknown. His *Computations Relating to the Publick Debts* included a massive table (Figure 4.2) of twenty-four rows by ten columns, showing ten different features of the national debt at that point, including original sums, amounts repaid, current interest, and the impact of changes in interest rates. He also described ten different types of outstanding government debt: eight different forms of "redeemable" debt—debts that could easily be repaid by the government at any time—plus two different groups of annuities. The annuities required the government to make fixed yearly payments to creditors—for example, £100 per year every year for ninety-nine years—but did not carry a fixed principal value at which the government could pay them off. It was thus a complicated question how much those annuities added to the national debt as a whole. To deal with this complexity, Hutcheson calculated three different "valuations" for the annuities. (For example, outstanding ninety-nine-year-long annuities were assigned values of 17, 17½, and 19 times the annual payment.) Once again, he used multiple parallel calculations to show a range of outcomes for an unknown parameter—in this case, annuity valuation. Hutcheson's table modeled both an economic problem and an informational one. It synthesized everything that was known about the national debt in order to frame how much was unknown about it.[59]

Following the table, Hutcheson offered a lengthy exposition of the additional fiscal data that were needed to formulate an effective plan for conquering the debt. As he explained, he hoped his figures would "incite those who know a great deal more, to rectify not only my Errors in Calculations, but in my way of reasoning on the same." This was more than feigned humility. Behind his dizzying tables was a conviction that there were other people who knew more than he did. For example, after providing detailed "Estimates" of what various tax revenues were expected to produce, he conceded that those estimates would likely "fall short or exceed" actual outcomes. Such variances would alter the future trajectory of the national debt. But, he continued, whether he was exactly right or not was a secondary point, because the matter would "easily be seen when the exact Estimates are given into Parliament." Hutcheson's calcula-

The Redeemable Funds which have above 5 *l. per Cent. per Annum* Interest, *Viz.*	1. *Sums originally advanced.*	2. *The Sums to be repaid more than advanced on Redem^n. or Repurchasing.*	3. *The Sums repayable on Redemption, and at which the Repurchasing is computed.*	4. *Sums repaid.*	5. *Sums still due.*	6. *The Annual Sums for which the Funds were given.*	7. *The Yearly Interest as proposed to be reduced.*	8. *The present sinking Fund for the Discharge of the Debt remaining due.*	9. *The Encrease of the sinking Fund by the Reduction of Interest.*	10. *The whole sinking Fund after the Reduction of Interest.*
1. The Debt to the *Bank of England.*	8,051,840 7 10	— — —	8,051,840 7 10	115,787 10 —	7,936,052 17 10	535,063 16 —	396,802 12 10	14,991 13 2	123,269 10 —	138,261 3 2
2. The 4 Lotterys of 1711, and 1712.	7,100,000 —	2,114,500 —	9,214,500 —	451,875 —	8,762,625 —	657,676 —	438,131 5 —	131,918 10 —	87,626 5 —	219,544 15 —
3. The *South Sea* Company.	10,000,000 —	— — —	10,000,000 —	— — —	10,000,000 —	608,000 —	500,000 —	— — —	108,000 —	108,000 —
4. The Bankers Debt.	664,263 —	— — —	664,263 —	— — —	664,263 —	39,855 16 —	33,213 3 —	— — —	6,642 13 —	6,642 13 —
Funds at above 5 *l. per Cent.*	25,816,103 7 10	2,114,500 —	27,930,603 7 10	567,662 10 —	27,362,940 17 10	1,840,595 12 —	1,368,147 — 10	146,910 3 2	325,538 8 —	472,448 11 2
The Redeemable Funds which have 5 and 4 *l. per Cent. per Ann. viz.*										
5. The *E. India* Comp. at 5 *per Cent.*	3,200,000 —	— — —	3,200,000 —	— — —	3,200,000 —	160,000 —	160,000 —	— — —	— — —	— — —
6. The *Bank* Annuities at 5.	1,069,000 —	— — —	1,069,000 —	— — —	1,069,000 —	54,600 —	54 600 —	— — —	— — —	— — —
7. The Civil List Lottery at 4.	500,000 —	133,010 —	633,010 —	33,820 —	599,190 —	35,000 —	23,967 12 —	11,032 8 —	— — —	11,032 8 —
8. The Lottery 1714. supposing the 64,300 *l.* paid off, be proportionable to the Blanks and Prizes, then 1,117,694 *l.* is at 5 *per Cent.* and 694,406 *l.* at 4 *per Cent.*	1,400,000 —	476,400 —	1,876,400 —	64,300 —	1,812,100 —	116,573 12 —	83,660 18 9	32,912 13 3	— — —	32,912 13 3
Funds at 5 and 4 *l. per Cent.*	6,169,000 —	609,410 —	6,778,410 —	98,120 —	6,680,290 —	366,173 12 —	322,228 10 9	43,945 1 3	— — —	43,945 1 3
Funds at above 5 *per Cent.*	25,816,103 7 10	2,114,500 —	27,930,603 7 10	567,662 10 —	27,362,940 17 10	1,840,595 12 —	1,368,147 10	146,910 3 2	325,538 8 —	472,448 11 2
Total of the Redeemable Funds.	31,985,103 7 10	2,723,910 —	34,709,013 7 10	665,782 10 —	34,043,230 17 10	2,206,769 4 —	1,690,375 11 7	190,855 4 5	325,538 8 —	516,393 12 5
The first Computation.	A COMPUTATION of the Repurchasable Terms of 99 and 32 Years at the following three Rates of Purchase; *Viz.* 17. 17½, and 19 Years Purchase for the 99 Years. And 13. 13½, and 14 for the 32 Years. The Purchase Money to be repaid with Interest, at 5 *l. per Cent. per Annum* on the two first Valuations, and at 4 *l. per Cent. per Annum* on the third Valuation.									
9. The Annuities for 99 Years.	9,859,617 7 1	1,491,374 10 3	11,350,991 17 4	— — —	11,350,991 17 4	667,705 8 1	567,549 11 10	— — —	100,155 16 3	100,155 16 3
10. The Annuities for 32 Years.	2,400,000 —	408,000 —	2,808,000 —	— — —	2,808,000 —	216,000 —	140,400 —	47,520 —	28,080 —	75,600 —
	12,259,617 7 1	1,899,374 10 3	14,158,991 17 4	— — —	14,158,991 17 4	883,705 8 1	707,949 11 10	47,520 —	128,235 16 3	175,755 16 3
The second Computation.										
9. The Annuities for 99 Years.	9,859,617 7 1	1,825,227 4 3	11,684,844 11 4	— — —	11,684,844 11 4	667,705 8 1	584,242 4 6	— — —	83,463 3 7	83,463 3 7
10. The Annuities for 32 Years.	2,400,000 —	516,000 —	2,916,000 —	— — —	2,916,000 —	216,300 —	145,800 —	47,520 —	22,680 —	70,200 —
	12,259,617 7 1	2,341,227 4 3	14,600,844 11 4	— — —	14,600,844 11 4	883,705 8 1	730,042 4 6	47,520 —	106,143 3 7	153,663 3 7
The third Computation.										
9. The Annuities for 99 Years.	9,859,617 7 1	2,826,785 6 4	12,686,402 13 5	— — —	12,686,402 13 5	667,705 8 1	507,456 2 1	— — —	160,249 6 —	160,249 6 —
10. The Annuities for 32 Years.	2,400,000 —	624,000 —	3,024,000 —	— — —	3,024,000 —	216,000 —	120,960 —	47,520 —	47,520 —	95,040 —
	12,259,617 7 1	3,450,785 6 4	15,710,402 13 5	— — —	15,710,402 13 5	883,705 8 1	628,416 2 1	47,520 —	207,769 6 —	255,289 6 —
	A COMPUTATION of the Annuities for two and three Lives, of which several Lives being drop'd at the Rate of 6 *l. per Cent. per Annum* Interest, may be Valued one with another at 12 Years Purchase. And also a Computation of the Annuities, with the Benefit of Survivorship at the Value of the Money originally paid for the same, being something above 14 Years Purchase.									
11. The Annuities for two and three Lives.	192,152 6 3	52,724 18 9	244,877 5 —	— — —	244,877 5 —	20,406 8 9	12,243 17 3	5,713 16 1	2,448 15 5	8,162 11 6
12. The Annuities with the Benefit of Survivorship.	108,100 —	— — —	108,100 —	— — —	108,100 —	7,567 —	5,405 —	1,081 —	1,081 —	2,162 —
*	300,252 6 3	52,724 18 9	352,977 5 —	— — —	352,977 5 —	27,973 8 9	17,648 17 3	6,794 16 1	3,529 15 5	10,324 11 6

Figure 4.2 Large table summarizing the state of the national debts, from Archibald Hutcheson, *Computations Relating to the Publick Debts, Taken from the Abstract Deliver'd into Parliament the 14th of March, 1716* (London: Printed for H. Clements, 1717), after p. 4. Courtesy of the Kress Collection of Business and Economics, Baker Library, Harvard Business School. (HOLLIS: 007393968).

tions were, to his mind, merely educated guesses. The "exact Estimates," which he presumed were known by leading government officials in the proper revenue departments, offered the certain knowledge that was truly needed.[60]

Hutcheson believed that those who had more exact numerical knowledge owed it to the public to make it available. Getting those individuals to disclose their privileged numbers became his chief political cause. In 1717–1718, he set his sights on military spending, long a target of suspicion for Country partisans who feared the growth of a "standing army." Amidst ongoing debates in the Commons, he published two pamphlets on military spending: In his first pamphlet, he offered calculations detailing various dubious expenditures being made for military purposes, from peacetime recruitment costs to the improper allocation of officers' personal expenses. He also constructed detailed tables measuring the costs of ill-considered pension policies that granted "half pay" to retired officers. As in the case of the national debt, Hutcheson stressed that all of his calculations were mere estimates, and urged military insiders to fill in the gaps in his data. "When this matter is fully explain'd to the House, by the proper Officer," he wrote, "it will be then seen whether there be, or be not, any Prejudice to *Britain* therein." In a subsequent pamphlet he proposed a new solution to this persistent public ignorance: revitalizing Parliament's Commission of Public Accounts, which had fallen dormant.[61]

In Hutcheson's mind, his role as public calculator was almost a begrudging one. This became especially clear in a 1718 pamphlet entitled *Some Calculations and Remarks Relating to the Present State of the Publick Debts and Funds*—the text that would spark his showdown with Crookshanks. The core of the pamphlet was a series of eight numerical "States" elaborating different aspects of the national debt. Five of those States described key features of the debt as of the current moment, including the amount of redeemable and irredeemable debts and an estimate of the gross and net revenues appropriated for debt repayment. The "Fourth State" dealt with the recent past, showing the "State of the Increase of the National Debt, since the Peace concluded at Utrecht." By Hutcheson's calculations, the national debt had increased by some £11.8 million over the preceding five years. The final two States dealt with the future, projecting the ameliorative effects of two remedies he proposed for addressing the debt problem: a reduction of interest rates and a more potent sinking fund. Yet it was as if Hutcheson wished he did not have to do all these calculations. He had already spoken out on such matters some fifteen times, he explained, and he worried that such computational demands were weighing upon his "sickly Constitution." "The several Accompants for the respective Duties, or some proper Officer for the whole, should have taken this trouble upon them," he complained at one point, "and not left it upon the Gentlemen of the House of Commons." The nation needed a far better informational infrastructure, with dedicated revenue officers, standardized accounting procedures, and consistent reporting to Parliament. Only then, he explained, would Parliament, and the nation, "be informed with certainty, and not need rely on the conjecture of a *Medium* calculated in any manner."[62]

For Hutcheson, the very kind of conjectural calculation he was performing—estimating "Mediums," modeling uncertain variables—was a palliative remedy for a deeper political disease. The nation needed certainty, not conjecture. Though he stressed that his own calculations could never be perfectly accurate because he lacked access to the proper data, Hutcheson made a powerful argument about the possibility of—and indeed the need for—certain numerical truth. Certain numbers existed, but for the moment they lay concealed in the hands of those "proper Officers" who refused to reveal them out of laziness, secrecy, or corruption. Hutcheson saw numerical truths as highly personal things, held by specific individuals. But the more he invoked these ideal numbers, and the more calculating he himself did, albeit imperfectly, the more he effectively argued that the answers to the nation's problems lay in the numbers—numbers that, perhaps, transcended any individual.

Truth and Numbers Are Always the Same

Hutcheson's 1718 pamphlet, *Some Calculations and Remarks,* rankled the Whig establishment. They took greatest offense at Hutcheson's "Fourth State," the historical calculation showing that the national debt had risen by nearly £12 million since the 1713 Peace of Utrecht. Contemporaries immediately understood those figures as a critique of the Whig government that had taken power in 1715. The zealous Abel Boyer published a pamphlet in rebuttal, in which he claimed "several *Mistakes* are rectify'd; *Misrepresentations* exploded; and the *Present Ministry* Vindicated." John Crookshanks offered another—and here the calculating courses charted by Hutcheson and Crookshanks would collide. It was David Dalrymple, an influential Scottish Whig, who presented Crookshanks with Hutcheson's pamphlet and enlisted him to "examin[e] some accounts therein." The calculator responded with a thorough letter to the top Whig minister of the moment, Lord Sunderland, who urged Crookshanks to publish a formal response.[63]

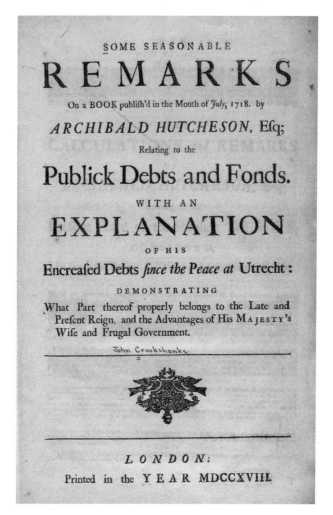

Figure 4.3 Title page of John Crookshanks, *Some Seasonable Remarks on a Book Publish'd in the Month of July, 1718* (London: *s.n.,* 1718). Courtesy of the Kress Collection of Business and Economics, Baker Library, Harvard Business School. (HOLLIS: 007391979).

Crookshanks's rebuttal ran a dense fourteen pages, plus an elaborate fold-out table as an appendix. His antagonistic agenda was evident on the very cover of his pamphlet (see Figure 4.3). The date Crookshanks gave—November 5, Guy Fawkes Day and, by some accounts, the date of William of Orange's landing in 1688—was just one of many signals to readers that the questions at hand were of great importance. The beginning of the pamphlet's title was anodyne: *Some*

Seasonable Remarks on a BOOK published in the Month of July, *1718.* . . . But the next line revealed—in large type, italicized and capitalized—the true object of attention: *ARCHIBALD HUTCHESON,* Esq. Throughout the title page, and indeed the pamphlet as a whole, Crookshanks framed the argument as a test of his opponent's skill and honor, not simply an honest dispute over an analytical question. Strikingly, Hutcheson's name was on the cover of the pamphlet and Crookshanks's was not, with the "Esq." in plain view, a reminder of that adversary's supposed gentility and legal credentials. The title page further explained that the pamphlet would offer an "Explanation of his Encreased Debts since the Peace at Utrecht," the "his" intimating that those "Debts" were matters of Hutcheson's making.[64]

Crookshanks cleverly framed the opening of the pamphlet as a story of personal disenchantment. "Having for many Years entertain'd a favourable Opinion of Mr. *Hutcheson's* Capacity, Learning, and good Qualifications," Crookshanks recalled, "I was in hopes of receiving some advantageous Instructions from his last Publication." Having read Hutcheson's July pamphlet with "Respect and Attention," he set about trying to understand Hutcheson's calculations better by comparing them to other sources, including original exchequer accounts. But his initial admiration soon gave way to astonishment when he read Hutcheson's "Fourth State," which claimed that the national debt had risen "amazingly and unaccountably" since the Peace of Utrecht, at least £11.8 million and perhaps as much as £15.4 million according to Hutcheson's reckonings. It was a shocking and dangerous calculation, which it seemed Hutcheson had formed "without a perfect Knowledge of the Particular Articles" involved. Crookshanks feared these negligent numbers provided ammunition for every scheming person who might seek to weaken the current ministry or even the Crown.[65]

The remainder of the pamphlet was a masterwork of the quantitative critique genre, drawing upon the full arsenal of combative tactics to discredit Hutcheson's "Encreased Debts." His first move was to point out very specific, though not especially large or significant, irregularities in the numbers Hutcheson had presented: a £44,150 difference between one of Hutcheson's accounts and the Exchequer account he cited as the source of his data; a roughly £33,080 inconsistency between the figures given for certain outstanding lottery debts on two different pages within Hutcheson's pamphlet. "These are certain Errors," Crookshanks pointed out, though he stopped short of saying they were intentional. "My Opinion of Mr. *Hutcheson's* Probity, restrains me from censuring his Designs." But the implication was clear enough; in matters of accounting, any hint of inaccuracy cast doubt on the entire account, as well as the accountant. It was provocative even to suggest that Hutcheson might have "Designs," a consistently pejorative term in the eighteenth-century political lexicon. It was not Crookshanks's last such innuendo. Shortly thereafter, Crookshanks dropped a glancing reference to the Jacobites and their "late Rebellion," lest readers forget Hutcheson's questionable political loyalties.[66]

Crookshanks moved easily between pointing out Hutcheson's accounting errors, alluding to his treasonous past, and challenging him on substantive questions of financial analysis. He next embarked on a series of highly sophisticated technical arguments, designed to undermine three calculations central to Hutcheson's claims about the "Encreased Debts." On all points he strayed into murky and unsettled conceptual territory. The first disputed calculation concerned the valuation of government annuities, in particular whether an increase in the market price of annuities constituted an increase in the national debt. Hutcheson had argued it did. The total market value of one large group of government annuities had appreciated from about £12.7 million in 1715 to nearly £17.0 million in December 1717. For Hutcheson, this amounted to an additional £4.3 million burden, because if Parliament tried to pay off the national debt immediately by buying up outstanding annuities at market rates, they would have to pay that higher price tag.[67]

In fact, Hutcheson worried that the problem was even worse than that. He suggested that the value of the annuities was likely to appreciate even further, due to impending plans on the part of Parliament to lower the nation's maximum "usury" rate. Contemporary financial thinkers had observed that the market value of fixed-income assets like annuities, bonds, and landed property was closely tied to interest rates: as interest rates went down, the fixed annual return on an annuity would become relatively more valuable, driving the price of annuities up. This posed quandaries when it came to public money. Reducing the maximum interest rate promised major benefits: it could reduce interest costs on the national debt, stimulate trade, and potentially increase the value of the nation's landed estates. In July 1714, toward the very end of Queen Anne's reign, Parliament had in fact passed such a rate reduction, dropping the maximum allowable rate from 6 to 5 percent. That change had broad political support, including from the recently elected MP Hutcheson himself. Hutcheson continued to support such rate reductions, but also realized that if Parliament reduced rates even further, the market price on many outstanding debts would continue to rise. Hutcheson's 1718 pamphlet pointed out this troubling consequence: If Parliament dropped the usury rate further to 4 percent, annuity prices could be expected to rise almost £3.6 million—so much more added onto the national debt. This was the second major calculation with which Crookshanks took issue.[68]

Together, these two calculations—the £4.3 million increase in market values from 1715 to 1717, plus the £3.6 million expected from the interest-rate reduction—implied a £7.8 million increase in the national debt, entirely because of the changing market value of existing debts. Crookshanks rejected this £7.8 million entirely, giving three reasons why it did not qualify as a legitimate increase in the national debt. First, the increased value of government annuities and corresponding declines in interest rates were good things! As Crookshanks wrote, "Mr. *Hutcheson* could not give a more authentick Proof of the Advantages of his Majesty's Reign, nor a brighter Encomium on the Administration." To accuse the Whig government of mismanagement on account of such a beneficial change was preposterous. Second, the claims about the increased value of government debts were unjustifiable because, he wrote, "Mr. *Hutcheson* is not yet appointed Arbitrator between Parliament and the Annuitants, to fix a Value on their Fond [*sic*]." In other words, Hutcheson did not have the authority to determine how much government creditors would demand in exchange for their annuities. That was a matter for Parliament and the annuitants to negotiate, not a matter for Hutcheson's calculations.[69]

The third reason was, perhaps, of a rather different sort from the first two. It had to do with Hutcheson's own personal role in the matter. "Mr. *Hutcheson* is as blameable as any Person for the advanc'd Value of those Annuities," Crookshanks wrote, "if he remembers the Part which he acted, and, I think, commendably, in the Reduction of the National Interest." Hutcheson could not push for lower interest rates and then saddle the Whigs with the ill consequences. The £7.8 million was partially his responsibility! Collectively, the three reasons Crookshanks gave to dispute Hutcheson's £7.8 million "increase" show how very different modes of argumentation became entangled in quantitative conflicts over public money. His first reason was essentially political-economic—about what was best for the nation's wealth. The second reason was methodological or even epistemological—about the nature of financial value and who had the authority to determine it. The third reason was ethical and deeply personal—about Hutcheson's responsibility for his own actions.[70]

In addition, Crookshanks rejected a third purported "Encrease" in the national debt that Hutcheson had calculated: £2.4 million that arose because Parliament had rerouted certain revenue away from debt repayment and toward Crown expenses (the "Civil List"). In total, this amounted to £10.2 million that

Hutcheson considered an increase in the national debt since 1713, but which Crookshanks claimed was simply a calculated illusion. By Crookshanks's reckoning, nearly two-thirds of Hutcheson's "Encreased Debts"—£10.2 million out of £15.4 million—were completely illegitimate, leaving at most £5.2 million in legitimate new debts that had been contracted since Utrecht.[71]

He spent the next few pages whittling away even further. He gave considerable attention to questions of chronology, arguing that many of the burdens Hutcheson implicitly attributed to the new Whig ministry were really the result of earlier actions by the preceding Tory government. Throughout, he embroidered his technical critiques with a running commentary on his adversary's character. He often reminded readers of Hutcheson's reputed skills in accounting and calculation. Crookshanks introduced one critical calculation by remarking: "It is very commendable in Mr. *Hutcheson,* whose Time has been employ'd in studying Matters of greater Consequence than Accounts, to have such a Proficiency in them, as to out-shine many Professors of that Art." These ironic references to Hutcheson's skill made errors by that "Masterly Accomptant" all the more contemptible.[72]

As the pamphlet progressed, Crookshanks dropped the courtesies and eventually gave Hutcheson the lie. One assertion by Hutcheson—"pag. 31, very astonishing"—gave Crookshanks particular offense. Therein, Hutcheson computed that the national debt under the previous monarch, Queen Anne, had actually been lower than usually thought, £42.5 million, implying that even more of the responsibility for the current debt fell on Whig shoulders. Crookshanks found this calculation deeply hypocritical, because it directly contradicted prior numerical claims Hutcheson had made. In his January 1715 pamphlet, which was fashioned as a letter to the new King George, Hutcheson himself had stated that the debt under Anne was £45 million, not £42.5 million. Crookshanks's exasperated response is worth quoting at length:

> Has this worthy Gentleman forgot his ingenious Letter to the King in *January,* 1714[15], reprinted in his Publication, making them upwards of 45 Millions? Did he venture to speak at Random to his Sovereign? Or pretend full Knowledge of a Matter, while in the Dark concerning the same? Or is it possible, that Zeal for his present Argument should make him forget the Reputation he had acquired?[73]

Crookshanks felt he had found incontrovertible proof of his opponent's dishonorable intentions. There was no legitimate justification for why Hutcheson's account of the national debt under Anne ought to have changed in the intervening years. One of his two accounts had to be flawed: either he had published irresponsible and misinformed figures in 1715, or he was manipulating the numbers in 1718.

In Crookshanks's mind, Hutcheson's computational inconsistency was not just an error. It was a false account, a profound ethical failure. What made his transgression so clear were the numbers themselves; Hutcheson could not escape the legacy of his own figures. Here we return to the lyrical words with which we began this chapter:

> Truth and Numbers are always the same; tho' the first may be obscured, yet it can never entirely lose its Lustre; and tho' the latter may be transpos'd, they can never lose their Denomination and true Value, when ranged in their proper place; and 45 in Figures, must always be read Forty five in Words.[74]

Crookshanks's reverent equation of "Truth and Numbers" arose in a distinctive argumentative context. Throughout most of his critique, he had not been talking about abstract, impersonal, absolute numbers at all. He had been talking about very specific, highly personal *accounts:* Hutcheson's computations of the national debt. The point of Crookshanks's computational criticisms had been to discredit Hutcheson personally, as well as his calculation. Traditionally, accounts were only as reputable as the accountant who produced them. But in doing so, Crookshanks stressed that those personal accounts were subject to another—*impersonal*—set of rules: the arithmetical logic of the numbers themselves. Hutcheson could try to pass off speculative estimates as true accounts, or manipulate his assumptions toward political ends, but his numbers would always be accountable to themselves. While one individual's accounts might be idiosyncratic and unstable, numbers were universal and stubborn. Numbers could be transposed and transfigured, but they would not lose their true value. Hutcheson could lie, but his numbers would always confess.

Crookshanks's celebration of the impersonal honesty of numbers was, therefore, a strategic maneuver in the course of a deeply personal argument. It came as he was trying to catch his opponent, Hutcheson, in the midst of a numerical lie. For Crookshanks, elevating "Numbers" to such an exalted status went hand-in-hand with trying to knock down his computational opponent. Arguably, his ode to "Truth and Numbers" was part of a rather dirty trick, trying to undermine his opponent by overstating the significance of a small inconsistency (£42.5 million versus £45 million) between two calculations made more than three years apart. This was in keeping with Crookshanks's overall objectives. He persistently sought to transform computational questions into personal tests of Hutcheson's character, and to use numerical truth as a standard for ethical judgment. In his own way, Hutcheson did this, too. In Hutcheson's case, the targets were the government officers who had failed to provide the public with vital financial information, those "exact Estimates" he so coveted. In the course of these political maneuvers, the two calculators helped to effect a process of epistemological detachment essential to the quantitative age. They both argued for a domain of impersonal, objective numbers that transcended and governed the highly personal, subjective realm of *accounts*.

The Ordeal of John Crookshanks

The story of how numbers came to be impersonal is also, fundamentally, a story about the personalities who made the numbers. Both Hutcheson and Crookshanks used calculation to fashion a new kind of professional and political identity, a process that was unpredictable and often unpleasant. They found opportunities in their ability to argue with numbers, but engaging in such numerical contests meant subjecting themselves and their reputations to harsh scrutiny. Both accountants lived under suspicion, and neither was trusted as an "expert" in a modern sense. Rather, they both took advantage of, and helped to foster, an emerging civic epistemology in which numbers and calculations carried far more authority than the individual calculators who produced them. As they both harangued their adversaries for their dishonest accounting, they cast doubt on those individuals who produced the numbers while exalting the numbers themselves. They drove a wedge between the untrustworthy calculator and the trustworthy calculation. STS scholars have observed that, across many modern polities, trust in quantification and trust in expert judgment are often inversely correlated, representing competing impulses in civic epistemology; political communities demand rule-bound, quantitative procedures for decision-making when they do not feel comfortable delegating those decisions to individual experts. In the calculating contest between Crookshanks and Hutcheson, we can see that antithesis coming into being.[75]

The irony was that, even as they helped to make the position of numbers increasingly secure in British political life, those calculators did not manage to create the same kind of security in their own political lives. After their initial showdown in 1718, they each continued to wander precarious paths. It should be noted that their calculating contest did not stop with Crookshanks's pamphlet. In a new pamphlet of his own, Hutcheson meticulously addressed Crookshanks's computational concerns, composing a series of detailed accounts comparing the two men's calculations and attempting to show exactly where differences arose. Crookshanks answered back, and the game went on. But their political trajectories diverged from there. Within a couple years, Hutcheson would go on to greater renown, thanks to the fateful events of 1720. That is a story for Chapter 5. For now, it suffices to note that however prominent he would become, and however great his computational triumphs, his career would always remain shrouded in jealousy and suspicion. Opponents would continue to challenge him on every calculation, and he continued to fight off accusations of Jacobite treason and even electoral tampering. Hutcheson's numbers were always trusted more than he was.[76]

The rest of Crookshanks's calculating life would be far less exceptional. At around the same time he was grappling with Hutcheson over national debt accounts, he became entangled in another conflict. It was also ostensibly about accounts and was arguably even pettier. But it proved to be more threatening to Crookshanks himself. In 1719, Crookshanks lost his job as Comptroller-General of the Scottish Customs. His breaches in accounting protocol were the official reason. But there were complex political forces at play. Around 1717, there had been a shake-up among Whig factions at the Treasury: Robert Walpole was forced out as Chancellor of the Exchequer and a new coterie, led by the Earl of Sunderland, James Stanhope, and John Aislabie took over key positions. This new administration seems to have brought with it a "new whiff of 'reform.'" Those new leaders accused Crookshanks of failing to follow the standardized accounting procedures that had been specified at the 1707 union. Crookshanks believed he had been sacked for more cynical reasons, because "he had declared himself to be a Creature & Dependent of Mr Walpole's" and was therefore a rival of the new ministerial faction. What his superiors pointed to as accounting failures were, to Crookshanks, part of his own attempt to improve Scottish accounting methods. That was what made the calculating life so treacherous. One man's methodological reform was another's calculated deception.[77]

Jobless, Crookshanks went back to hustling, peddling his calculating skills in search of new political opportunities. The Earl of Sunderland, though a leader of the Whig faction that had ejected him from office, continued to appeal to him for numerical advice and to make vague promises of a new position. In early 1720, Sunderland apparently asked him for thoughts on two competing projects to refinance the national debt, one proposed by the South Sea Company and the other by the Bank of England. (Several years later, Crookshanks would proudly recall he had preferred the Bank's scheme.) In return for this service, he received some kind of insider deal from Sunderland for £3,000 of

South Sea Company stock. But some stock transaction he made in that fateful year seems to have gone awry, leaving him entangled in debt and legal controversy. He continued to hawk his computational services in the 1720s, especially to his former patron, Walpole. Once the latter became Britain's leading minister in 1721, Crookshanks advised him regularly on matters like tobacco customs, national debt accounting, and the "sinking fund," and offered commentary on the polemics written by opponents. All the while, Crookshanks pleaded with his patron for a permanent job. In 1727, he wrote Walpole with a blow-by-blow account of his professional misadventures, detailing all his undeserved setbacks and all the political promises made to him still unfulfilled.[78]

Eventually, Crookshanks's appeals were heard. It took about a decade. Around the early 1730s, Walpole granted Crookshanks his coveted governmental position—in Spain. It was a vital diplomatic outpost, to be sure, but far afield for someone who had spent such effort trying to establish himself in London. That distance was emblematic. Just as he began his career far from the center of British economic and political life, keeping (or cooking) books in Livorno, he would end it fighting over calculations in another distant city, this time Seville. On January 1, 1733, he wrote to his patron Walpole to offer his thoughts on recent polemics critiquing Walpole's new excise reform plan, including one critique by his old foe, Hutcheson. Crookshanks admitted that the "solitary climat" weighed on him, but he remained committed to supporting Walpole and the nation through his computational exertions, "at whatsoever distance my lott may fall." He enclosed a detailed "Reply to M^r. Hutcheson Page 13." Like many of Crookshanks's writings and calculations, that paper would make its way to the center of British power, to be carefully preserved in the archives of political history—detached from the person who created it, who never made it there himself.[79]

CHAPTER FIVE

Intrinsic Values:
Figuring Out the South Sea Bubble

This shews the little power that reason and truth have over the passions of men, when they run high. In the late revolution in the Alley, *figures and demonstrations would have told them,* and the directors could have told them, that it was phrenzy; that they were pursuing gilded clouds, the composition of vapour and little sunshine.

JOHN TRENCHARD and THOMAS GORDON, "How Easily the People
are Bubbled by Deceivers," *Cato's Letters* (1720)

IN DECEMBER 1720, the political essayist "Cato"—a pseudonym for the influential republican thinkers John Trenchard and Thomas Gordon—looked back on the events of the preceding year with consternation. It was the infamous year of the "South Sea Bubble," an unprecedented crisis that ignited Britons' passions in a way no financial event ever had before, and perhaps none has since. The primary object of "phrenzy" had been the stock of the South Sea Company, a peculiar corporate hybrid established nine years earlier to refinance government debt and trade slaves to the Americas. At the height of the Bubble, in the middle of the summer, the Company's Stock touched £1,000 per share, nearly eight times its market price at the beginning of the year (£128). On paper, the total market value of the South Sea Company's stock was worth hundreds of millions of pounds, perhaps upward of £300 million at its peak—a truly mind-boggling aggregation of financial capital for the time. (To put that in perspective, in 1696, political arithmetician Gregory King had estimated the total value of the wealth *in all of England* at £650 million.)[1] By the time Cato wrote, though, the price had fallen well below £200 per share. Subsequent generations would long remember the rise and fall of South Sea Stock as history's paradigmatic financial "bubble" and a cautionary tale about "the little power that reason and truth have over the passions of men."[2]

For the people who lived through it, though, the 1720 Crisis was more than a fable about the dangers of avarice and the limits of reason. It also offered a powerful, if costly, testament to the virtues of one specific form of reasoning: mathematical calculation. As Cato pointed out, one thing that made the South Sea calamity so devastating was that it could have been avoided—if people had listened to the numbers! "Figures and demonstrations would have told them," Trenchard and Gordon recalled, "that it was phrenzy; that they were pursuing gilded clouds." Following the collapse of the Bubble in autumn 1720, others echoed this refrain, lamenting that investors had neglected clear mathematical warnings about what South Sea Stock was really worth. One anonymous commentator stated around October 1720: "The Frenzy rose so high, that even Mathematical Demonstrations . . . were slighted, and the Authors of them treated as if they had been Quacks or Ballad-Singers."[3] The South Sea Bubble offered a devastating lesson about the dangers of ignoring the math. In fact, 1720 was perhaps the single most important moment in the ascent of calculative thinking in British civic epistemology.

No calculator did more to warn Britons about the mathematical dangers of South Sea Stock than Archibald Hutcheson. Almost monthly throughout 1720, Hutcheson published diligent analyses exposing the flaws and uncertainties in the South Sea Company's financial strategies. These cautionary calculations were remarkable feats of skill and ingenuity. One especially remarkable pamphlet from March 1720 offered one of the earliest attempts at using the exponential "discounting" of future profits as a technique for stock valuation—a technique that remains foundational in modern financial analysis. Though long cast as a rather peripheral character in histories of the South Sea Bubble, Hutcheson has recently attracted keen attention from economic and financial historians, who have praised his seemingly prescient insights about the Bubble and even granted him the honor of the "father of investment analysis."[4]

Though they contained many recognizably modern financial insights and techniques, Hutcheson's calculations were built around distinctly early modern values and assumptions about finance. And though they presaged modern techniques of investment analysis, his project was not that of a profit-seeking investor at all. His 1720 calculations were—like so many others in this early quantitative age—fundamentally instruments of political dispute. They grew out of his long-held fears about national indebtedness, "designing ministers," and the dangers of political secrecy. His South Sea analyses applied his favored computational tool, the "spreadsheet" model, to a new problem: calculating the value of stock. Prior to 1720, there was little agreement among Britons about how to understand what a stock was worth. Shares in "joint-stock" companies were among the era's most untested financial technologies, and many technical, conceptual, and legal questions remained open to debate. What made a joint-stock valuable: the assets the company held, its potential future profits, its political support and privileges? Was a stock's value even something that could be calculated? Hutcheson developed his own distinctive perspective on these questions, combining cutting-edge computational techniques with the traditional economic and ethical concept of "intrinsic value." Hutcheson reconceived that ancient notion in terms of the politics of knowledge. In his formulation, the intrinsic value of financial objects was a secret fact known only to privileged insiders— insiders who had the power to deceive. Though outsiders in the general public could not necessarily ascertain a stock's intrinsic value with certainty, they could use the tools of financial calculation to shed light on a company's mysteries and pressure its leaders to reveal their secrets. That was what Hutcheson's calculations during the early months of the South Sea Bubble were intended to do.

At first, Hutcheson's tireless computational work did little to dissuade eager investors. But the Bubble's collapse seemed to confirm his persistent warnings, thus giving his numbers new significance and authority in the eyes of the public. In the Bubble's aftermath, Hutcheson repurposed his calculations from a tool of protest into a tool of prosecution. In the winter of 1720–1721, he crafted retrospective models of what the Stock's value had been earlier in 1720 and used these models to argue that the South Sea Directors had violated the public trust by knowingly selling Stock for more than its intrinsic value. These post hoc valuations offered mathematical evidence of the Directors' guilt, and became central to how many Britons explained what went wrong in 1720. As a member of the Parliamentary committee tasked with investigating the Crisis, Hutcheson wrote these calculations into official Parliamentary reports. In the press, the Company's opponents echoed Hutcheson's claims, while Company defenders tried desperately to debunk them.

The chaos of the South Sea Bubble proved to be a triumphant moment for calculative thinking. This may seem like a rather strange, even paradoxical, tale to tell about a financial bubble. After all, aren't financial crises a testament to the limitations, even the deceitfulness, of calculations? Recent financial history seems to suggest as much. Undue confidence in faulty calculations, like the probabilistic valuation models used to price complex derivatives or the mathematical risk management protocols employed by investment banks, is often cited as a key factor behind the global financial crisis that began in 2007.[5] Many modern historians have assumed that a similar story held in the eighteenth century: the 1720 crisis revealed the failures of calculation, not its triumphs.[6] But the Bubble actually had the opposite effect on public attitudes toward calculation. For Britons who lived through that year, the South Sea Bubble was an argument for putting more faith in calculation, not less. It made Britons appreciate numbers more than ever.

The Arcanum of the South Sea Scheme

The story of the South Sea Bubble began as another episode in the ongoing saga of Britain's national debt. In 1719, the most pressing and perplexing question with regard to the debt concerned what to do about a large volume of long-term, "irredeemable" annuities that had been a popular mechanism for government borrowing in preceding years. These "irredeemables" carried relatively high interest rates (up to 9 percent) and, worse still, could not be paid off ("redeemed") by the government without their owners' consent.[7] Projectors came up with an array of creative, sometimes bizarre, solutions for retiring these stubborn debts. Humphrey Mackworth proposed that Parliament float a new "Temporary Species of Money" in paper. Richard Steele touted a plan to restructure the entire national debt in an elaborate configuration of "tontines" (a kind of group annuity). Perhaps the most radical plan called for nationalizing all of Britain's trading and manufacturing companies, mines, and colonies, and paying public creditors with shares in those monopolies.[8] These imaginative schemes were indicative of the dizzying atmosphere of financial invention that pervaded the later 1710s.

The convoluted solution advanced by the South Sea Company was somewhat less zany than some of these alternatives, but its technical workings would have confounded many contemporaries all the same. The Company was guided by John Blunt, a shoemaker's son who began his career as a scrivener before rising to become an influential financial adviser to Robert Harley's Tory government. The South Sea Company had been born in 1711, a combined product of Harley's policy ambitions and Blunt's financial ingenuity. Following a model pioneered by the Bank of England in 1697, the South Sea Company was created by "engrafting" £11 million of government debt into the Company, giving Stock in the new Company to government creditors in exchange for their government bonds. This financial operation provided the initial capital for what many imagined to be an unparalleled commercial opportunity: trading enslaved African workers to Spanish America. The British government was in the process of negotiating with Spain a special *Assiento* ("Contract") to carry out that brutal trade, a deal that would be finalized in 1713. Company promoters touted involuntary African labor as a source of potentially limitless profits and a new foundation for Britain's financial system. For Harley and the Tories, the Company also offered a way to harness the prosperous overseas trades that would open up following peace with France and to loosen the Whiggish Bank of England's stranglehold on public finance.[9]

By 1719, worsening diplomacy with Spain and operational missteps had stymied the Company's trading ambitions. Seeking to revitalize Company fortunes, and envious of the state-of-the-art financial projects recently undertaken in France by itinerant Scot John Law, Blunt and other Company Directors looked to the national debt as a strategic opportunity. After engineering a small stock-for-government-debt swap early in 1719, Blunt presented a far more ambitious plan to key Whig political leaders in the closing months of the year. The Company proposed to "engraft" the majority of the outstanding national debt, about £31 million, into the South Sea Company's capital. (Added to the nearly £12 million the Company already possessed, this would make the Company the owner of over 80 percent of Britain's long-term government debt.) Under the Scheme, the Company would offer to exchange new shares of South Sea Stock for various outstanding government annuities, including the vexing irredeemables, at prices to be determined by the Company at a later time. They would also pay a substantial fee to Parliament for the privilege: originally around £3 million, later rising to over £7 million.

The Scheme had evident appeal for the British government. The government would earn a hefty cash fee and reduce the interest rate on most of its debt to 5 percent. Most importantly, the Scheme offered a way to retire the irredeemables, as long as the investors who owned those government annuities agreed to take part. The Scheme's appeal to the Company itself is somewhat less obvious at first glance.[10] One crucial attraction was that, through its intricate financial engineering, the Scheme promised to afford the Company access to an extraordinary volume of new capital. At the time, the ability of joint-stock companies to issue stock was closely governed by their governmental charters. For the South Sea Company, the quantity of Stock it could issue was strictly limited to the amount of government debt the Company possessed. Through the 1720 Scheme, the Company stood to absorb up to £31 million in new government debt, meaning the Company could issue £31 million (par value) of new Stock. Even better, the Company was not required to give exactly one £100 share of Company Stock for every £100 of government debt it absorbed. If the Company could convince debtholders to accept the Company's Stock at less than one-to-one (above the £100 "par" price), the Company would create a surplus pool of Stock, which could be sold to private investors for cash.[11]

To illustrate: imagine a government creditor who held government annuities worth £3,000 in total. By absorbing that debt, the Company would earn the right to issue thirty new "shares" of Company Stock, each with a par value of £100. If the Company offered the creditor Stock priced at par, or £100 per share, the Company would have to give the creditor all thirty shares. But if instead the Company could convince the creditor to accept Stock priced at £300 per share, the Company would only have to give ten shares of Stock for the £3,000 in debt, leaving twenty surplus shares. At a price of £300, those surplus shares could be sold to the public for £6,000. The new capital raised by selling surplus Stock could be used to cover the £7 million government fee and to fund lucrative commercial opportunities, including a revived American trade. These innovative financial mechanics were both tantalizing and mystifying. To some, they seemed to promise a kind of perpetual financial motion, where raising capital made it possible to raise even more capital.

When the proposal was first presented to Parliament in late January 1720, it failed to gain immediate assent. Some MPs wished to see alternative proposals, and shortly thereafter the Bank of England pitched an engraftment scheme of its own. A two-week bidding war ensued. The South Sea Company countered by increasing its governmental fee from £3 million to approximately £7.5 million. That may have been enough to tip the balance in favor of the Company. Its plan earned Parliamentary approval on February 2, shepherded by key Whig leaders like Chancellor of the Exchequer John Aislabie. The next two months saw a flurry of conversation about the relative merits of the Scheme, in letters, print, and no doubt countless coffeehouse discussions.[12] Pro-Company pamphlets trumpeted the large dividends the Company could offer to investors, praised the Scheme's value as a remedy for public indebtedness, and chided opponents as unpatriotic. Some proponents specifically advertised the dynamic, innovative nature of the Scheme, arguing that it was engineered to create a virtuous cycle of rising Stock prices, expanding trade, and decreased national debt.[13] This optimism was reiterated in private correspondence, like that of James Brydges, 1st Duke of Chandos, one of the era's most influential financial players. Writing to his broker on February 8, Chandos eagerly speculated that South Sea Company Stock would likely "become one of the best Funds in Europe to lay one's Mony out in."[14]

Yet the Scheme also attracted swift and abundant criticism. Some objections came from ideological opponents like John Trenchard, who feared the corrupting consequences of investing one corporation with so much power over the nation's finances. Others came from partisan and commercial rivals promoting alternative plans, especially the Bank of England's. Critics took various approaches: they called attention to the Scheme's many uncertainties, insinuated that the Company had corrupt intentions, and explained why the Bank's proposal would have been more effective. Several of these pamphlets were highly quantitative. One prominent argument was that new investors in the South Sea Company were destined to take a loss, because calculations showed that the Company's financial assets could not support the above-par prices they were likely to charge new investors for shares of South Sea Stock. Other critiques contended that the Scheme could not possibly work because holders of irredeemable annuities would never relinquish their prized securities except at astronomical prices the Company could not afford to pay. The cynical John Trenchard worried that, once the Scheme got going, investors would quickly grow weary of the plentiful Stock, driving its price down and thwarting the Company's grandiose designs. For "what imaginary Hopes can there be," Trenchard mused, "that their Stock will keep at the advanced Price?"[15]

Events quickly proved Trenchard wrong, of course. The South Sea Scheme received royal assent on April 7, and the Company made its first public offering of Stock a week later, with the first debt-for-Stock exchanges following shortly thereafter. The Company initially priced its Stock at £300 per share, three times its "par" value and more than twice the market price of the Stock at the beginning of the year (£128). The Company marketed its Stock energetically from the start, promising to pay a 10 percent dividend in additional Stock at midsummer.[16] It also began to lend money generously to prospective investors to make it easier for them to purchase Stock. At the end of April, the Company issued another £1.5 million of new Stock, priced at £400 per share. By May, the mania was on. South Sea Stock sold in the London markets at £342 at the end of April, £595 by the end of May, and £950 by the end of June. In mid-June, the Company made a huge third offering of new Stock to the public: £5 million par value, or 50,000 shares, priced at £1,000 per share. The offering was massively oversubscribed, as eager investors actually signed up to buy over 112,000 shares. A fourth subscription at the same £1,000 price followed in August. In early September, as investors' appetite for the Stock seemed to be waning, the Company made a spectacular promise to pay a dividend of 30 percent (based on par value) at Christmas and 50 percent dividends every year for the following twelve years.[17]

For some observers, the Stock's spectacular ascent seemed to defy logic. Writing on April 20, leading financier and South Sea Company critic James Milner called the early surge in South Sea Stock to be "an Arcanum that nothing but the Philosophers Stone can make out."[18] Milner was a supporter of the rival Bank of England, and some amount of bluster could have been expected from him. But many others shared his bewilderment, and the sense that the South Sea Scheme amounted to a kind of alchemy. The Company's marvelous feats of financial engineering posed similar problems of knowledge to many other tantalizing technical "projects" circulating in the London marketplace at that very same moment, like the famed perpetual-motion machine being promoted by German engineer Orffyreus.[19] Britons simply did not know how to evaluate such spectacular claims, or who could be trusted to do so.

This collective puzzlement is essential to understanding the Bubble as an economic and political event. Financial economists have observed that stock markets are—almost by necessity—sites of disagreement, because buyers and sellers in a given transaction are effectively expressing different opinions about the value of the security being traded. This is especially true in times of rapid change (and potential crisis).[20] Early in 1720, Britons had many reasons to disagree about South Sea Stock. Contemporaries had little experience assessing complex financial enterprises with so many moving parts. The Scheme comprised various financial operations—stock-for-debt swaps, sequential sales of stock for cash, stock dividends—with unpredictable consequences. The Company also had a convoluted corporate ownership structure with various stakeholders, including original South Sea stockholders, those who traded in government debts, and new cash purchasers. Making things even more complicated, the Company was a monstrosity, combining two very different kinds of businesses: its financial operations, which involved large capital resources and stable income; and its trading operations, which involved few measurable "assets" and uncertain—but potentially massive—future profits.

Consequently, the innovative Scheme produced an array of difficult technical questions, which strained the limits of established financial knowledge. Early modern financiers had built up a relatively robust corpus of technical knowledge in some areas of finance, notably the valuation of "fixed income" securities like bonds and annuities. But when it came to understanding stocks, contemporaries had little to go by. The "jobbers" of Exchange Alley had considerable experiential know-how in trading stocks, but this was mostly unsystematic and unwritten. Commercial writers like John Houghton had published cursory lessons about the stock market but little about the technicalities of valuation.[21] Britons did not even know who among them had the knowledge and integrity to navigate such unfamiliar financial terrain. Exchange Alley was a shadowy world, and those with the most experience were not necessarily the most trustworthy. When it came to mechanical projects like Orffyreus's perpetual-motion wheel, careful British investors increasingly sought out the guidance of scientific experts, like the members of the Royal Society and its experimental demonstrator J. T. Desaguliers, to help sort technical fact from projectors' fiction.[22] When it came to strictly financial projects like the South Sea Scheme, by comparison, there were no obvious authorities to whom Britons could turn—at least not in early 1720. This was the context in which Archibald Hutcheson began to calculate his assessments of the South Sea Scheme.

Reckoning with Intrinsic Value

Behind the discordant conversations about the South Sea Scheme was an even deeper and more confounding question: What did it even mean to speak of the value of a stock? Hutcheson would soon provide a distinctive and powerful answer to this question, grounded in calculation and organized around the idea of "intrinsic value." That idea was ancient, but it had undergone consistent reinterpretation throughout the early modern period, especially in the economically tumultuous century preceding 1720. Sketching the genealogy of intrinsic value both helps to illuminate the discordant state of financial thinking at the onset of the South Sea Bubble and to recover the conceptual precedents Hutcheson built upon—and transformed—in his calculations during that infamous year.

Attention to the "intrinsic" in Western thinking dates at least to Plato and Aristotle. For those classical philosophers, and centuries of their followers, the distinction between the (superior) "intrinsic" value of things or actions that were good in themselves, and the (inferior) "instrumental" or "extrinsic" value of things or actions that were useful merely as means to an end, was foundational to understanding goodness and value. As one twentieth-century philosopher writes: "These philosophers took it for granted that, if there is anything that is good, then there is something that is intrinsically good or good in itself."[23] This emphasis on intrinsic goodness became a crucial guide for thinking about ethics, aesthetics, and economics. Among medieval Christian thinkers, the notion that goods held an intrinsic value based on the materials and labor that went into making them was central to the complex calculus of "just price."[24] By the late medieval period, the logic of intrinsic value had come to organize economic life in Europe in specific and powerful ways. For example, historian Bert De Munck has shown that, from the fifteenth to seventeenth centuries, urban guilds in the Low Countries relied heavily on the concept to explain the distinctive quality of guild-made products, and thus to justify guilds' exclusive commercial privileges. Guild "hall marks" were used to guarantee that a product contained raw materials of the highest intrinsic value, like quality textile fibers and pure metals.[25]

By the seventeenth century, the language of intrinsic value was in use in many different economic domains. Its meaning varied, depending on the object: for a manufactured good, intrinsic value might describe the value of the raw materials that went into it; for a unit of land, its natural agricultural yield; for a coin, the content and quality of precious metal it contained.[26] During the seventeenth century, the precise meaning and function of the concept was becoming less and less stable. Various economic forces—the rise of global trade, mounting commercial competition between nations, changing patterns of consumption and luxury—strained traditional logics of value. In the case of manufactured goods, De Munck argues that "the seventeenth and eighteenth centuries witnessed a shift from intrinsic value to design and decoration as important ingredients of product quality."[27] In *The Order of Things,* Michel Foucault contends that, during this period, the traditional concept of intrinsic value was eclipsed by an alternative understanding of economic value based on exchange, a seminal shift in economic thinking indicative of the broader transition from the Renaissance to the Classical *episteme.*[28]

In England, one issue above all stimulated new thinking about economic value: coinage. By the early seventeenth century, English coins had reached a troubling state, degraded by clipping and counterfeiting and depreciated in comparison to their foreign counterparts. The loss of specie due to hoarding and exportation was an especially acute problem for England, which lacked direct access to gold or silver mines. Many feared the nation would soon find itself without enough currency to facilitate commerce. As this currency crisis became a persistent topic of conversation throughout the seventeenth century, so did the topic of intrinsic value. In his 1623 *Circle of Commerce,* Edward Misselden explained how coins always carried two different kinds of value: the "*Intrinsique*" value, which was determined by the coin's metal content or "inward fineness"; and their "*extrinsique* or outward valuation," the official value stamped on coins by government proclamation. As Misselden noted, these different registers of value prevailed at different times. Sometimes intrinsic value was what mattered, as when merchants transacted business overseas using physical coins; at other times, extrinsic value took priority, as in many domestic transactions or those carried out on paper. In the 1620s, the big question was what to do when the intrinsic and extrinsic values of the nation's coins became misaligned with one another. Commercial commentators disagreed. Gerard de Malynes argued that the Crown ought to intervene to reset extrinsic valuations at intrinsic pars, a proposal that Misselden and Thomas Mun sternly opposed. Even though they argued about how to address the coinage problem, Misselden, Mun, Malynes, and many others at the time shared the belief that the concept of intrinsic value was key to solving such quandaries. As Andrea Finkelstein writes of Thomas Mun, "intrinsic (or objective) value was the cement holding the finite universe together."[29]

In the 1690s, another massive coinage crisis brought yet another flurry of contentious conversations about money and its value. As Deborah Valenze explains, many Britons still "cherished the principle of intrinsic value as the polestar within a vast, crowded vista of possibilities."[30] The most prominent defender of intrinsic value as the foundation of commercial order was John Locke, who argued that it was essential to stable commerce that the value stamped on English coins reflect their intrinsic value precisely. In order to rectify the nation's coinage, Locke advocated a heavily deflationary recoinage project whereby citizens were forced to bring eroded coins into local mints to be melted down and recast based on weight.[31] The "Great Recoinage" that Locke advocated was ultimately enacted in 1696, but not without considerable opposition. Locke's opponent William Lowndes proposed an alternate, inflationary policy whereby coins were to be reminted but with lower amounts of precious metal than the previous standard. More sweeping critiques came from merchant-writers like John Cary, James Hodges, and Nicholas Barbon, who disputed Locke's notion that intrinsic properties were the ultimate determinant of value. Hodges pointed out that "silver, considered as Money, hath, speaking properly, no real intrinsick value at all." The most radical critic of the intrinsic view of value was Barbon, who contended that the value of *all* things—money included—ultimately derived from their usefulness to people, as reflected in market prices, not from any intrinsic properties. As Barbon put it in 1690: "things are just worth so much, as they can be sold for."[32]

Coinage was not the only issue causing Britons to rethink intrinsic value. So too was the ongoing revolution in "paper" finance. For many conservative, land-owning Britons, the new forms of paper wealth circulating in Exchange Alley disturbed their "aristocratic notions of inherent worth and the preservation of hierarchy."[33] For others, though, traditional concepts of intrinsic value became a vital tool for explaining and justifying the era's new financial innovations. Promoters of new banking projects—like "Lombard banks," which issued paper credit instruments backed by mercantile goods, and "land banks," backed by landed property—pointed to the intrinsic value of those underlying assets as the source of their banks' security.[34] Corporate leaders adopted the language of intrinsic value in lobbying for their joint-stock companies. A notable early example came in 1681, amid debates over the East India Company monopoly on Asian trade. A pamphleteer for the EIC named "Philopatris," probably Josiah Child, argued that the EIC was not an exclusive monopoly because anyone could share in the company's profits by buying into its joint stock. Some EIC critics had disputed this point, noting that anyone who wanted to partake in EIC stock "must pay 280*l*. for 100*l*."—that is, had to pay a lofty price of £280 for a £100 share. Child countered this criticism by explaining that EIC stock cost such a high price because "the *intrinsic value* is worth so much," a fact he claimed "is as true as 2 and 2 makes 4."[35]

As London's stock market continued to grow in the early eighteenth century, Britons continued to ponder what made stock valuable and to debate Child's assertion that stocks had any kind of "intrinsic value." Carl Wennerlind has reconstructed two such positions, exemplified by a Whig-Tory dispute in 1710 over recent Tory political tactics and their effect on stock prices. Whig thinkers like Benjamin Hoadly disputed the seemingly antiquated notion that the value of stocks had to derive from something intrinsic, arguing instead that the value of stocks was a function of society's collective confidence in their worth and in the nation's commercial well-being. (In 1710, this had a political message: the Whigs insinuated that a recent fall in key stock prices was a sign of Britons' lack of confidence in the new Tory government.) Tory writers like Simon Clement, on the other hand, disputed Hoadly's assertion that "financial wealth was grounded in immaterial and abstract future-oriented notions—like trust, confidence, and opinion."[36] Clement, a veteran of the 1690s coinage debates, argued that stocks had a measurable intrinsic value that was the ultimate anchor of their market price, just like coins. For coins, intrinsic value was based on metal content; for stocks, it was based on the fundamental properties of the joint-stock company that issued the stock. When it came to stocks, he explained, "People ought never to value them by the Rates they may go at in Exchange-Alley, but to inform themselves truly of the certain Sum that has been paid into the Stock, and of the Dividend that is constantly made, together with the probable Success of the Management." With both coins and stocks, market prices might rise and fall due to the momentary manipulations of coin dealers or the fleeting fancies of stock jobbers. Occasionally, stocks could even be "run up to an imaginary Value much above their real Worth." But these imaginary variations were fleeting; intrinsic value always won out.[37]

Early eighteenth-century Britons, therefore, espoused a variety of different attitudes about stocks and their value. On one extreme were those who held a very traditional understanding of value and wondered how paper promises could bear any real worth at all; on the other extreme were those who believed that stocks derived their value purely from intangible properties like public trust and future potential, best expressed in their market price. In between were those like Clement who believed that stocks did carry a solid, measurable intrinsic value. Even among that intermediate group, though, there was considerable disagreement about what constituted that intrinsic value. This range of different attitudes about stock value was in full view in early 1720. Some observers called upon intrinsic value in trying to make sense of the South Sea Company's strange refinancing plan. The skeptical John Trenchard questioned the intrinsic value of the South Sea Company and wondered why anyone would want to buy its Stock. He observed that any "additional Rise of this Stock above the true Capital, will be only imaginary." After all, "One added to One, by any Rules of Vulgar Arithmetick, will never make Three and a Half." Trenchard and likeminded critics construed intrinsic value in a narrow way: for a stock, it meant the precise amount of capital a joint-stock company controlled. As another anonymous pamphleteer put it, a stock's "real Value" amounted to the "clear Amount of the Capital Fund belonging to any Company . . . legally incorporated to trade upon a joint Stock."[38]

Others in 1720 took a less restrictive view of stock value. Some moderate commentators contended that a stock's value had both intrinsic and non-intrinsic, or "imaginary," components. "I don't mean . . . that the Stock is always intrinsically worth full as much as it sells for," one pro-Company pamphleteer admitted frankly in early 1720. "For I am not at all afraid to own that Part of the Value is imaginary." Indeed, buying stock meant embracing some uncertainty, as intrinsic value only explained so much. Ultimately, stock value was "not capable of direct Proof either way." Yet others embraced the "imaginary" aspects of stock value with even greater gusto. Fortunes might be made by riding the ups and downs of public sentiment. The moralistic author of the *Letter to a Conscientious Man* fretted that "many of those who deal in this Stock, know well enough the Fallacy of it" and were only seeking to make quick profits "by taking the Advantage of the strong Fascination [of] the People." Later in 1720, a bookdealer-*cum*-stock-trader named T. Johnson put the matter even more bluntly: "Tho the funds have no great solidity any of 'em, yet there's money to be got in Stock jobbing while it lasts."[39]

Another line of financial thinking essentially jettisoned the intrinsic-imaginary framework altogether. Though many Britons thought that the rise and fall of stock prices was highly capricious, others believed that the price of a joint-stock company's shares was actually subject to a high degree of control by corporate and political leaders. It was not uncommon for observers to suggest that the South Sea Company's Directors and their Parliamentary supporters could effectively set the Stock's market price. For many, like the Scottish Duke of Montrose, this was precisely the reason to be optimistic. "There is this reason indeed to support that belieff that ye Company is not only supported by the Government w[ith] a vue to ye speedy payment of ye Publick debts," Montrose wrote in late March, "which will be one consequence of ye Stocks riseing high, but [also] all the individuals who have [an] interest in it wish the stock wrise high." Parliament did exert considerable regulatory control over joint-stock companies, with the power to dictate key variables like the maximum amount of stock a company could issue. Consequently, some Britons saw financial value not as a matter to be judged by calculation or the market, but by legislation. In the early months after the Bubble began to deflate, one pamphleteer reflected on this as he lashed out against a pro-Company opponent. "This Author . . . pretends to set a *Real Value* upon Stock," the critic smirked, "(tho' I do not know of any such Thing unless it be done by Act of Parliament)."[40]

Out of this morass of different thoughts on valuation that prevailed in early 1720, one would come to dominate collective understanding after the calamity of the Bubble year: Archibald Hutcheson's version of "intrinsic value." His formulation bore important similarities to earlier ones like Simon Clement's. But Hutcheson's achievement would be to transform and update the well-trodden notion in two essential ways. One vital revision was to give the amorphous concept of intrinsic value a rigorous mathematical meaning. His second essential contribution came in the way he reformulated intrinsic value in distinctly political and ethical terms. Many Britons thought of the value of stocks as something within human control, particularly within the control of political authorities like Parliament. Hutcheson combined that focus on financial responsibility with the seemingly more material notion of intrinsic value.

Hutcheson's first serious engagement with the intrinsic value concept seems to have come not from thinking about stocks but, as for Simon Clement and others, from thinking about money. In late fall 1718, after trading salvos over the national debt with John Crookshanks, Hutcheson went to France. There he connected with English and French diplomatic officials, including the British ambassador to the French Court, Lord Stair, and reported back observations to state officials in England. He was particularly attentive to the worrisome condition of the exchange rate between British and French currencies in Paris. British merchants in France were gravely concerned that those exchange rates had become artificially out of line, such that merchants bearing bills of exchange denominated in British pounds were getting far less than fair value when they cashed their bills in Paris for French livres. In a November 1718 letter, Hutcheson forwarded a manuscript by "An Eminent Merchant," which spelled out that the foreign exchange problem was so bad that British merchants were effectively "bubled out of ¼ of the Intrinsick Value of their Money" when trying to do business in France using bills of exchange.[41]

What could explain such an "extravagant" imbalance in the sterling livre exchange rate? Hutcheson contended that it could not have been a natural consequence of trade, but must have stemmed from "some other artfull Managements"—in other words, manipulation by skilled political agents. Hutcheson was clear on the culprit. It was "the very ingenious Gentleman who is now the sole Director of the Bank of Paris": John Law.[42] The high exchange value for the French livres was not just an ephemeral alteration in market prices. It was a deliberate effort by a powerful and devious agent to drive prices up to "imaginary" levels, above their intrinsic value. This was not the last time Hutcheson would confront such malfeasance.

Archibald Hutcheson and the Mystery of the South Sea Company

By early 1720, Hutcheson had returned from France and resumed his position as one of Parliament's most outspoken and critical voices on financial matters. Given his long-standing concerns about the national debt—and his long-standing fear of secrecy and corruption in public affairs—it is little surprise the South Sea Scheme attracted his scrutiny. Hutcheson published his first pamphlet on the Scheme in early March 1720, shortly after the South Sea Company had outbid the Bank of England. He quickly updated that text, adding a more extensive "Preface," dated March 31. He stressed to readers that he did not oppose aggressive attempts to combat the nation's debt problem, and even admitted that he initially preferred the South Sea Company's refinancing plan to the Bank of

England's. (After all, the South Sea Company had started as a Tory project and was built around principles Hutcheson could easily get behind: conquering debt through the horrific promise of the slave trade.) But, Hutcheson explained, he quickly had become wary of the Company's massive undertaking, which entrusted a tremendous amount of financial and political power to one privileged cabal. "To risque, at once, the Liberties of *Britain,* by making the Path to Arbitrary Power plain and easy, is a Measure which I never can come into," Hutcheson wrote. "For so great a Company, under the Influence of an ill-designing Ministry . . . may load the Nation with heavier chains, than the Debts we are endeavouring to Discharge."[43]

It was the duty of Parliament, Hutcheson urged, to stop that threat and "prevent the Ruin of many Thousands of Families." To aid this cause, Hutcheson took it upon himself to illuminate the Scheme's risks through calculation. Despite the grand promises made concerning the South Sea Scheme, Hutcheson observed, the Company had actually provided little information to Parliament or the public about its specific commercial plans and done little to explain the "solid Foundation of the real Value of their Stocks." "And," he explained, "it was this"—this uncertainty, this secrecy—"which put me upon the making the following CALCULATIONS." By using calculation to highlight how much was unknown about the Scheme, Hutcheson sought to compel the Company's leadership and political supporters to reveal further details about their plot.[44]

The core of Hutcheson's dense, thirteen-page pamphlet was an intricate arithmetical model, assessing the possible intrinsic value of South Sea Stock. It offered another example of the "spreadsheet" style that had featured in his earlier pamphlets on the national debt, though far more complex. His overall message was highly skeptical: the Company probably could not justify selling its Stock, or offering it in exchange for government debts, for prices above about £150 per share. Hutcheson was not alone in this general conclusion; several other calculators had come to a similar, skeptical result in early 1720. What made his calculation distinctive was how it was built—and the political message that computational structure conveyed. He did not set out to calculate a single, specific number indicating the intrinsic value of South Sea Stock (as some other calculators did). Instead, he calculated the level of future trading profits the Company would have to deliver in order to justify selling its Stock at various different prices. Hutcheson wanted to know: *if* the Company sold new Stock to the public at a certain price, then what profits would the Company have to deliver to make good on that price? The price of the Stock was thus an input to his model, rather than the output. In a vivid demonstration of his spreadsheet technique, he executed this calculation for sixteen different scenarios, created by flexing two key variables: whether or not the Company would follow through and buy the irredeemable annuities (in addition to other "redeemable" debt instruments); and what price the Company set for its stock (£100 or "Par," £125, £150, £175, £200, £300, £400, or £500).

Examples of this calculation are shown in Figures 5.1 and 5.2. In the specific case examined in Figure 5.2, Hutcheson assumed that the Company *would* take in the irredeemable annuities, and that it would price new Stock at £175 per share, a premium of £75 over par. Hutcheson determined that in order to justify a price of £175 (annotated by X in Figure 5.2), the Company would need to generate just over £6.5 per share in trading profits (Y, precisely £6.10s.9d.). Otherwise, purchasers of that Stock would suffer a loss. Hutcheson began by calculating the total funds that the Company would raise by selling new Stock to the public through the Scheme at the given £175 price (A, £23.9 million). Then, he deducted certain expenses the Company would incur, like the fee the Company owed to the government for the right to carry out the Scheme (B, a total of £10.9 million). This determined how much of a "Dividend" the Company's stockowners would collectively gain from the sale of new Stock (£13.0 million in total, or £29.15s.3d. per share, C). For new investors who paid £175 for their shares, this dividend partially compensated them for the markup they paid to buy the Stock. So did the value of the comparatively high interest the Company would earn on the government debt it absorbed, which Hutcheson valued at £6 per share (D). Subtracting these benefits, new investors were still effectively overpaying by more than £39 per share (E). This amount, Hutcheson argued, would have to be made up by the Company's trading profits. Hutcheson imagined these future trading profits as a kind of annuity paid to investors over the coming years. Using compound-interest discounting—the same technique William Paterson had touted for calculating the Equivalent in 1706—he determined what level of annual income would equate to a present value of £39 per share. The answer was £6.10s.9d. per share (Y) or, as he would show later, £2.8 million total. A later table (Figure 5.3) summarized these key figures for all sixteen different scenarios.

Hutcheson's model gave the slippery concept of a stock's intrinsic value a precise mathematical meaning. It also found a rigorous way to relate the Company's present financial assets to its potential future trading profits, thus making sense of the South Sea Company's confounding identity as a hybrid financial firm and trading firm. His key innovation lay in using the mathematics of compound-interest discounting to put a "present value" on the Company's

N° III. *The Purchase at* 150 l. per Cent.

THE Profits of the Additional Stock, at 150 *l. per Cent.* is   - - -   *l.* 15,904,000
  But the Money to be paid the Publick, and the Proprietors of the longTerms, is only 10,891,000
  Which leaves a Dividend on the whole Capital of   - - - - ———— 5,013,000

	l.	*s.*	*d.*
The Advanced Price, as aforesaid, is	50	00	00
But the Dividend of 5,013,000 *l.* is, *per Cent.*	11	10	02

	l.	*s.*	*d.*
Which reduces the Price to	38	09	10
But out of this Deduct the present Value of 1 *l. per Ann.* for Seven Years, which is	6	00	00
Then the Price, over and above the Value, is	32	09	10

Which will require, to make it a Saving Bargain, a Yearly Profit on Trade of 5 *l.* 8 *s.* 3 *d.*

N° IV. *The Purchase at* 175 l. per Cent.

THE Profits of the Additional Stock, at 175 *l. per Cent.* is   - - -   *l.* 23,856,000
  But the Money to be paid the Publick, and the Proprietors of the longTerms, } 10,891,000
  is only   - - - - -
  Which leaves a Dividend on the whole Capital of   - - - ———— 12,965,000

	l.	*s.*	*d.*
The Advanced Price, as aforesaid, is	75	00	0
But the Dividend of 12,965,000 *l.* is, *per Cent.*	29	15	3

	l.	*s.*	*d.*
Which reduces the Price to	45	4	9
But out of this Deduct the present Value of 1 *l. per Ann.* for Seven Years, which is	6	0	0
Then the Price, over and above the Value, is	39	4	9

Which will require, to make it a Saving Bargain, a Yearly Profit on Trade of 6 *l.* 10 *s.* 9 *d.*

N° V. *The Purchase at* 200 l. per Cent.

THE Profits of the Additional Stock, at 200 *l. per Cent.* is   - - -   *l.* 31,808,000
  But the Money to be paid the Publick, and the Proprietors of the long Terms, } 10,891,000
  is only   - - - - -
  Which leaves a Dividend on the whole Capital of   - : ———— 20,917,000

	l.	*s.*	*d.*
The Advanced Price, as aforesaid, is	100	0	0
But the Dividend of 20,917,000 *l.* is, *per Cent.*	48	0	5

	l.	*s.*	*d.*
Which reduces the Price to	51	19	7
But out of this Deduct the present Value of 1 *l. per Ann.* for Seven Years, which is	6	00	0
Then the Price, over and above the Value, is	45	19	7

Which will require, to make it a Saving Bargain, a Yearly Profit on Trade of 7 *l.* 13 *s.* 2 *d.*

C   N°

Figure 5.1 Examples of Archibald Hutcheson's analyses of South Sea Stock at various offer prices (March 1720). From *Some Calculations Relating to the Proposals Made by the South-Sea Company, and the Bank of England, to the House of Commons; Shewing the Loss to the New Subscribers* (London: Printed and Sold by J. Morphew, 1720), 5. Courtesy of the Kress Collection of Business and Economics, Baker Library, Harvard Business School. (HOLLIS: 007388124).

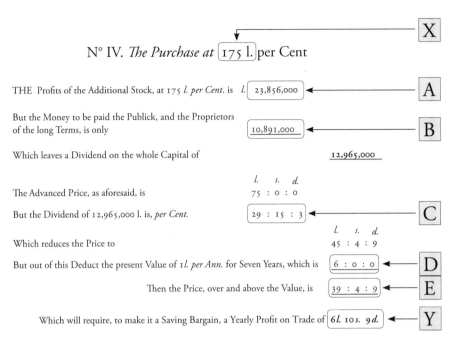

$N°$ IV. *The Purchase at* 175 l. per Cent

THE Profits of the Additional Stock, at 175 *l. per Cent.* is *l.* 23,856,000

But the Money to be paid the Publick, and the Proprietors of the long Terms, is only 10,891,000

Which leaves a Dividend on the whole Capital of 12,965,000

| | *l.* | *s.* | *d.* |
The Advanced Price, as aforesaid, is 75 : 0 : 0

But the Dividend of 12,965,000 l. is, *per Cent.* 29 : 15 : 3

| | *l.* | *s.* | *d.* |
Which reduces the Price to 45 : 4 : 9

But out of this Deduct the present Value of 1 *l. per Ann.* for Seven Years, which is 6 : 0 : 0

Then the Price, over and above the Value, is 39 : 4 : 9

Which will require, to make it a Saving Bargain, a Yearly Profit on Trade of 6 *l.* 10 *s.* 9 *d.*

Figure 5.2 Detailed breakdown of one example of Hutcheson's analysis of South Sea Stock. This particular calculation assumed the Company priced new Stock sales at £175 per share and chose to absorb the irredeemable annuities.

hypothetical future profits. This was a major milestone in the history of financial thinking. At the time, sophisticated financiers used compound-interest discounting to assess the value of "fixed income" securities like annuities, bonds, and real-estate rents. But Hutcheson's South Sea calculation was one of the earliest documented uses—perhaps *the* earliest documented use—of that technique to evaluate stocks. His crucial idea was that, just like the owner of an annuity or landed estate could expect to earn a steady stream of interest or rents from that property in future years, so should the owner of a stock expect to earn a steady stream of dividends from the Company's profits. Hutcheson's insight remains fundamental to modern financial valuation techniques like "discounted cash flow" (DCF).[45]

In technical terms, Hutcheson's calculations appear almost shockingly modern. Yet the assumptions and motivations behind them were unquestionably early modern. Compared to modern equity valuation techniques, Hutcheson's model appears almost backward. In current practice, a stock's price is generally treated as the output of a valuation model, while information about a company's

[7]

An ABSTRACT *ſhewing the* LOSS *to the New Subſcribers to the* SOUTH-SEA *Stock, at the ſeveral Prices following; and the Yearly Profits on* Trade *neceſſary to make Good the ſaid Loſs, on the Ingraftment of all the Redeemable Debts, which will make the Capital* 28,500,000 l. *and alſo on the Ingraftment of the Irredeemables, which will make the Capital* 43,558,000 l. *And theſe* CALCULATIONS *are made Computing Intereſt at the Rate of* 4 l. per Cent. per Ann.

Nº		Capital of 28,500,000 *l.*			Capital of 43,558,000 *l.*		
		1. The advanced Price, or preſent Loſs, of every 100 *l.* South-Sea Stock ſubſcribed for.	2. The Annual Profits on Trade for Seven Years neceſſary to make good the ſaid Loſs.	3. The Amount of the ſaid Annual Profits computed on the whole Capital.	1. The advanced Price, or preſent Loſs, of every 100 *l.* South-Sea Stock ſubſcribed for.	2. The Annual Profits on Trade for Seven Years neceſſary to make good the ſaid Loſs.	3. The Amount of the ſaid Annual Profits computed on the whole Capital.
		l. s. d.	*l. s. d.*	*l.*	*l. s. d.*	*l. s. d.*	*l.*
1.	at *Par* - -	10 07 6	1 14 6	491,625	19 0 4	3 3 4	1,379,336
2.	at 125 - -	20 13 8	3 8 11	982,062	25 15 0	4 5 9	1,867,549
3.	at 150 - -	30 19 10	5 3 3	1,471,312	32 9 10	5 8 3	2,357,576
4.	at 175 - -	41 6 0	6 17 7	1,960,563	39 4 9	6 10 9	2,847,604
5.	at 200 - -	51 12 0	8 11 11	2,449,813	45 19 7	7 13 2	3,335,816
6.	at 300 - -	92 16 9	15 9 4	4,408,000	72 19 2	12 3 1	5,294,111
7.	at 400 - -	134 1 3	22 6 9	6,366,187	99 18 6	16 13 0	7,252,407
8.	at 500 - -	175 5 10	29 4 2	8,324,375	126 18 1	21 2 10	9,208,887

Figure 5.3   Table summarizing the conclusions of Hutcheson's analyses of South Sea Stock under various conditions. The two panels ("Capital of 28,500,000 *l.*" and "Capital of 43,558,000 *l.*") represent the scenarios in which the Company did not (left) and did (right) absorb the irredeemable debts. From Hutcheson, *Some Calculations Relating to the Proposals*, 7. Courtesy of the Kress Collection of Business and Economics, Baker Library, Harvard Business School. (HOLLIS: 007388124).

commercial prospects are treated as inputs. The goal of valuation is to determine a fair price for the stock, given what is known about a company. Hutcheson did the reverse. He treated the South Sea Company's Stock price as an input, and sought to calculate what those prices implied about the Company's trading ventures. The South Sea Company's Stock price was a transparent number that the public would know with certainty, because the Company had discretion over the initial prices it would offer when selling new Stock to the public. By comparison, only the Company's inside leadership knew about the Company's trading opportunities. In Hutcheson's mind, whatever price the Company assigned to its Stock, it was effectively testifying that it would be able to make the profits needed to justify that price. A stock's price was a company's promise.

Another striking feature of Hutcheson's model was its labor-intensive, spreadsheet architecture. Whereas other calculators generally offered only one or two valuations in their South Sea pamphlets, Hutcheson iterated his valuation procedure over sixteen different scenarios. Hutcheson had a clear reason to do so much computational work. Modeling multiple scenarios was his strategy for dealing with informational uncertainty. In early 1720, much remained unknown about the South Sea Scheme—not only the most opaque and privileged facts, like the Company's trading results, but even more fundamental decisions like how the Company would price its Stock and whether it planned to purchase the "irredeemables." As in his earlier models on the national debt (like Figure 4.2), Hutcheson's South Sea valuation model marshaled all of the facts about the Company that were publicly available and then combined them with selected hypothetical values for certain key unknowns. In doing so, he highlighted how little the public actually knew about the Company and its momentous refinancing Scheme. His political anxieties about secrecy were built into the structure of his calculation.

Because it entertained various different scenarios, Hutcheson's calculation did not definitively say what the intrinsic value of South Sea Stock was. But the warning was clear. As he explained, "If the Computations I have made, be right, it is then evident, That the Gains of the *South-Sea Company*, in Trade, must be immensely Great, to make Good to the New Subscribers, at any high Rates, the Principal Money advanced by them." Hutcheson was skeptical of the Company's prospects in overseas trades, and admitted to believing that "there is no real Foundation for the present, much less for the further expected, high Price of *South-Sea Stock*." More than demonstrating that South Sea Stock had limited value, though, Hutcheson's calculation was intended to represent the political dangers involved in the secretive scheme. He urged the Company to disclose more information and "Explain, from whence their Advantages are to arise," so that "Thousands and Thousands of unwary People may not be Undone."[46] This was a familiar refrain for Hutcheson throughout his time in Parliament. For all its innovation in financial technique, Hutcheson's first South Sea pamphlet was more a polemic about political transparency than a tip-sheet for venturesome investors.

As the Scheme went into action, the Company did little to satisfy Hutcheson's demands for honest disclosure. Instead, South Sea promoters confused matters more, printing what Hutcheson considered misleading calculations that celebrated the Scheme without providing any real information. The most notorious example came in the April 7–9 issue of the *Flying-Post* newspaper. An anonymous calculator began by offering a numerical accounting of the Company's capital resources, the new government debts it stood to absorb, and the cash funds it could theoretically raise through the South Sea Scheme—numbers that were readily available to the public at that point. On the basis of that data, the *Flying-Post* calculator made a series of optimistic claims about South Sea Stock, including that the Stock could pay 9¾ percent annual dividends and that its value could be worth up to £880 per share (at a time when it sold for around £300). In order to justify the latter claim, the *Flying-Post* calculator relied on what Hutcheson called an "extraordinary Paradox, *viz.* That the higher the Price is which is given for *South-Sea* Stock, the greater Benefit will the Purchaser have thereby." Specifically, the *Flying-Post* argued that if South Sea Stock were priced for new subscribers at £300 per share, its "intrinsic value" would actually be £448 (see example in Figure 5.4). Even more striking: if the new stock were priced at £600, that intrinsic value would soar to over £880.[47]

The work of this "ingenious Computer" infuriated Hutcheson, who quickly responded with a pamphlet of his own. He argued that the *Flying-Post* had made a serious error by dividing the immediate profits from the sale of new Company Stock among too small a group of proprietors—including only the original holders of the Stock plus those who had redeemed government debt for Stock, but leaving out those who bought new Stock for cash. The *Flying-Post* calculator effectively suggested that anyone who bought new Stock for cash would receive none of the proceeds of the new Stock sales. (To put it in anachronistic terms, the *Flying-Post* was encouraging investors to buy South Sea Stock by demonstrating that it made an excellent Ponzi scheme!) To Hutcheson, the *Flying-Post*'s shady figures were yet more evidence for why the Company's leadership had to disclose better information to the public. He urged "those who

	(£ in millions, except *share prices*)
South Sea Company Stock:	
"Whole Debt owing to them by the Government"	£42.2
Maximum "United Capital Stock"	42.2
Less: Current "United Capital Stock," inc. Government Debts Exchanged	(21.5)
Remaining "Stock to Sell"	£20.7
Stock Price for Selling New Stock	*£300.*
South Sea Company Capital Assets:	
Cash from Sale of "Stock to Sell"	62.0
Less: Required Payment to the Government	(7.5)
Plus: Debt "which the Government must pay to the Company"	42.2
Total Value of Capital Assets	£96.7
Divided by: Current "United Capital Stock"	21.5
"Intrinsic Value" of £100 share in South Sea Stock	*£448.15s.*

Figure 5.4 Summary of a calculation showing the "intrinsic value" of South Sea Stock, presented in the *Flying-Post, or, Post-master* no. 4,428 (Apr. 7–9, 1720).

have the Honour to serve in the Administration or Direction of the *South-Sea* Company" to do the responsible thing and "disown the aforesaid, and all such Sort of *Calculations*" and "in a plain and easy Method, State the instrinsick Value of their Stock." In the meantime, Hutcheson hoped that his own calculations might help those "unwary People, unskilled in Figures" protect themselves against the Scheme's dangers and help to "explain to them the Mystery which the Advocates for the *South-Sea* Company, with so much Art, endeavour to conceal." That, after all, was the essential power of calculation in political life for Hutcheson: to illuminate mysteries others sought to conceal.[48]

In the summer and fall, as the Bubble began to crest and eventually to fall, Hutcheson's frustrations only worsened. His computations became increasingly judgmental, arguing that the Company's continued secrecy was not just dangerously negligent but actively criminal. On September 10, 1720, as the South Sea Stock price was teetering on the brink of total collapse, Hutcheson raged that he and the rest of the public remained almost "intirely in the Dark" about the South Sea Company's operations, particularly when it came to "the Profits which have arisen, or may arise, from their Home and Foreign Trade." Throughout the summer months, the Company had stoked the public's growing frenzy while failing to provide any of the vital data it had effectively promised. "I was in great hopes . . . to have seen a full and authentick State of the Value of the Stock of this Company," Hutcheson lamented, "by which the fluctuating Price thereof might have been settled, and People might have been at some Certainty."[49]

Instead, the only significant statements the Company had made about its Stock value, as far as Hutcheson could tell, was a series of references to its *market* price. To illustrate this point, Hutcheson transcribed several passages from the *Daily Courant* newspaper (May 19, August 13, August 31, and September 9), in which the Company had sought to demonstrate the Scheme's success by pointing to the high market price of South Sea Stock. The Company was effectively saying: "People thought fit to give 357 *l. per Cent.* for this Stock, and, therefore, it was worth so much." Hutcheson found this market-based logic perverse, comparing it to "Hudibras's Rule"—"What's the Worth of any Thing, But as much Money as 'twill bring?"—a reference to a mock aphorism from Samuel Butler's 1663 satirical poem *Hudibras*. In Hutcheson's mind, for the Company to buy into Hudibras's logic was already to betray its public duty. In selling Stock to the public at high prices, the Company had already made a promise about its Stock's intrinsic worth.[50]

Hutcheson offered the Company's leadership an alternative formula: "Surely, they ought to have set forth, The Money due them from the Publick; The Money gained at those Respective Times by Subscriptions; and, The Profits made by their Home and Foreign Commerce; and to have likewise set forth, The Debts due from them; and, That on the Balance of their Books, their Stock was then worth such a certain Sum." By disclosing that data they could provide a "Proof of the real and intrinsick value of their Stock, as was only fit for them to give." Intrinsic value could be known with high mathematical certainty—proof, even. But, for unfortunate political reasons, access to such proof lay only with "the Directors of this Company, who are in the Secret of this Mystery." The "intrinsick," for Hutcheson, was the privileged sphere of insiders.[51] Ironically, modern scholars have recently cited Hutcheson's 1720 calculations as evidence that contemporary investors were at fault in failing to see through the dangers of the South Sea Scheme. Because Hutcheson "rationally" understood the proper valuation of South Sea Stock, contemporaries ought to have been able to as well. The fact they did not was evidence of their collective "irrationality."[52] But this was not at all how Hutcheson understood the message of his own calculations. In his mind, those numbers showed that the Company, not the investors, was at fault. Only the Company was in a position to have known better.

Mathematical Facts, Notorious to the Whole Nation

During much of 1720, Hutcheson's models analyzing South Sea Stock had been designed to serve a similar purpose to his earlier political calculations on topics like wasteful military expenditure and the national debt. He used calculative models to interrogate suspicious political enterprises, highlight dangerous uncertainties, and pressure insiders into revealing information. After the South Sea Bubble burst in autumn, Hutcheson the calculator took on a new role—from sentinel to prosecutor. In December 1720, as the price of South Sea Stock plunged below £200, Hutcheson was appointed a member of the Commons' committee responsible for investigating the calamitous South Sea Scheme—ironically called the "Committee of Secrecy," as they were meant to carry out their work in secret. In that new position, he found the opportunity to see the inside information he had so long demanded. The investigators requested a massive volume of evidence from the Company, including its account books, transaction records, subscriber lists, and notes on the Directors' deliberations. Hutcheson and his colleagues unearthed a frightening amount of fraud and corruption. Using the ledgers of Company cashier Robert Knight as evidence, the Committee's first published report revealed that the Company had given Stock incentives to many prominent politicians in exchange for their support of the Scheme. Particularly egregious were two large payments totaling £250,000 to Charles Stanhope, a Treasury Secretary and cousin of Whig chief minister Lord James Stanhope, concealed (poorly) through fictitious transactions under the name *Stangape.* Later reports elaborated in shocking detail how the Company used intermediary brokers to hide the ultimate target of stock bribes and how Company employees, specifically a clerk named Christopher Clayton, had ripped pages from critical ledgers and intimidated employees to protect Company secrets.[53]

These particular transgressions were just some components of what Hutcheson and others on the Committee considered a far bigger crime. What bothered Hutcheson most were the string of broken promises and public deceits upon which the Company had built the entire Scheme. In his new prosecutorial role, Hutcheson demanded that the Directors explain the rationale behind the various financial maneuvers they had made throughout the previous year. He particularly wanted Company insiders to account for the implicit promises they had made about the intrinsic value of their Stock. On December 30, the Committee made a formal inquiry to the Company Directors concerning two especially suspicious Company policies: the choice to price new stock offerings during the summer (the third and fourth subscriptions) at £1,000 per share, and the aggressive promise in early September to pay future dividends of 30–50 percent. As the MP Sir James Lowther reported, Parliament and the public were especially aggrieved by those two duplicitous decisions, which many felt had "brought on this calamity and misery upon the nation." The Committee of Secrecy requested that the Directors "do lay before this House the Calculations & Inducements" that led them to implement those two policies.[54]

The Company's Directors would try to explain themselves, including with calculations of their own—on which more in the next section. But Hutcheson remained unconvinced. For him, the evidence was already clear: the Directors could not honestly have believed that the intrinsic value of their Stock merited the outrageous prices they were giving to it, nor could they justify the absurd dividend promises they made. In this context, Hutcheson's calculations found a new purpose. His earlier models had sought to provide "Information and Caution" to the public concerning the South Sea Company's secretive enterprises. After the South Sea Bubble had popped, Hutcheson's calculations became retrospective tools for proving what the Company's leaders must have known. He laid out his accusations in a scathing pamphlet published in early 1721. In the aftermath of the Bubble, the veil of secrecy that had shrouded the Company had been lifted, giving Hutcheson access to crucial insider data. Whereas his earlier South Sea models were necessarily "not exact," because he was "not perfectly informed of the Facts," this new information allowed him to "conclude my writing on this subject with exact Calculations" assessing the value of South Sea Stock in the past—the kind of exact calculations he had so long craved.[55]

From this new position, Hutcheson refashioned his calculations into tools of judgment instead of interrogation. He believed that numbers could prove the Company's worst crime: selling the public a stock whose price they knew, with certainty, could not be justified by its intrinsic value. For example, Hutcheson produced one calculation showing that in April, when the Directors of the Company had taken in the first two money subscriptions (Stock price: £300–£400 per share), and had taken the first exchange of annuities (Stock price: £375), the "Value of 100*l*. Stock" was "not ¼ above 120*l*."[56] (See Figure 5.5.)

This *post facto* computation was substantially simpler than what Hutcheson had offered in early March. For one, it only included one precise value for South Sea Stock, instead of analyzing a range of different possible scenarios. It also assigned no value to the potential future profits from the Company's trading

	(£ in millions, except *share prices*)
Company's Stock Issued:	
Original Stock	£12.9
New Stock Exchanged for Government Debt	3.6
New Stock Sold for Cash	4.1
Total Company Stock	£20.6
Company's "Capital":	
Total Government Debt Absorbed	21.2
Total Cash, including from Stock Sales	12.8
Total Company "Capital"	£34.0
Less: Fee to Government and Other Obligations	(9.2)
"Net Value of Capital"	£24.7
Divided by: Total Company Stock	20.6
Intrinsic Value of South Sea Stock (*per £100 share*)	**£120.**

Figure 5.5 Summary of Archibald Hutcheson's retrospective calculation analyzing the intrinsic value of South Sea Stock as it stood in April 1720, as presented in Hutcheson, *A Computation of the Value of South-Sea Stock, on the Foot of the Scheme as it Now Subsists* (London: s.n., 1720 / 21), 15–16. Reproduced in *The Reports of the Committee of Secrecy to the Honourable House of Commons, Relating to the Late South-Sea Directors, &c. Carefully Corrected*. (London: Printed for Cato, 1721), 16 (first pagination series).

endeavors. While his earlier calculations had mapped out a range of possibilities, reflecting the uncertainties surrounding the Company and its Scheme, his post-Bubble calculation considered exactly one scenario: the truth itself. Once the Company had been forced to reveal its true nature, the uncertain variables that had been built into his earlier calculations were no longer uncertain. As Hutcheson saw it, the evidence uncovered by the Parliamentary investigators confirmed the worst possible scenario: the Company was a vehicle for fraud from the outset, and hence never had any realistic prospects for trading profits. That vital unknown variable was, and always had been, zero.

To Hutcheson, this mathematical "Proof of the real and intrinsick value of their Stock" was proof of the guilt of the Company's Directors. Because Company insiders must have known that the Stock was not worth what they sold it for, they had committed demonstrable fraud in seeking "to raise the *Nominal Value* of South-Sea Stock, to that extravagant Rate to which it was afterwards advanced above the real and intrinsick Value thereof."[57] Hutcheson's pamphlet showed that the South Sea Scheme had in fact been built on a series of systematic deceptions—the initial pricing of the Stock at £300 per share, the increasing price of their subsequent subscriptions (up to £1,000), and the lavish dividend promises. Calculations could demonstrate that each of these choices was "grossly fraudulent." These were mathematical *facts*—both in the modern sense of distinct quanta of true knowledge, and in the older, legalistic sense of things proven *to have been done*. "These Facts are now Notorious to the whole Nation, and have been long so to my self, and many others," Hutcheson wrote.[58]

Hutcheson's dual accounting of the financial value of South Sea Stock and the ethical values (or lack thereof) of the South Sea Directors proved immensely influential in how the British public came to understand the South Sea Bubble. His mathematical analysis became a formal part of the legal case leveled against the Directors by Parliament, written by Hutcheson into the official reports issued by the Committee of Secrecy.[59] Angry critics frequently alluded to Hutcheson's efforts to "to calculate [South Sea] Stock" as key evidence against the Directors, which made it possible to "show how much [they] may have unlawfully cheated the Nation."[60] We have already seen how the polemicists John Trenchard and Thomas Gordon gestured at Hutcheson's calculative work in early December, claiming that the "figures and demonstrations would have told them [the investors], and the directors could have told them," that lofty prices for South Sea Stock had been unjustified. Legal thinkers like the influential Scottish jurist and Whig politician David Dalrymple elaborated upon and formalized Hutcheson's argument about the responsibility the Company's Directors bore for the price of their Stock. As Dalrymple put it, in a pamphlet likely published in the first three months of 1721:

> For the Directors and all those concern'd in the Management of that Stock ought to know the Value of the Stock they have under their Care as well as any Artists ought to know that sort of Work he deals in: If they knew that what they sold was not worth half of what they got for it, they surely were great Cheats, and Guilty of the most heinous Fraud . . . and if they were Ignorant of the Value of that very Stock they had under their Care, it was a supine Ignorance, and does not excuse them.[61]

Through such diverse allusions, the logic of Hutcheson's quantitative argument became a core part of an emerging standard narrative about the Bubble. Many of the key voices who cited or gestured at Hutcheson's calculations—like John Trenchard and Thomas Gordon—were not serious calculators themselves, and may have had little insight into the technical intricacies of his figures. Yet they celebrated Hutcheson's achievement, repeated the logic of his arguments, and treated his calculations as demonstrable facts. In the case of the South Sea Bubble, Britons came to believe that numbers held special and necessary political truths, even if the calculations that produced them remained highly opaque to most. Calculation had come of age in Britons' civic epistemology, and it did so in a moment of crisis.

A Calculated Defense

As an episode in the history of calculative thinking, perhaps the most arresting fact about the South Sea Bubble is that the most intensive, enthusiastic, and inventive calculators were not the financial insiders peddling the South Sea Scheme, but political critics trying to thwart it. This affirms two of this book's key arguments. First, the rising authority of financial calculation in British public life was largely driven by political imperatives as much as economic ones— including political agendas hostile to the growing power of corporations and the new finance. The quantitative age in British civic epistemology thus should not be seen as simply the straightforward triumph of a capitalistic or financial mentality. Second, calculation was frequently an instrument of outsiders, wielded as a tool of interrogation and critique by those who did not have direct access to power and information.

Public calculation, though, was rarely a one-sided affair in eighteenth-century Britain, and 1720 was no exception. During the fall and winter of 1720–1721, the South Sea Company and its allies marshaled an extensive public defense of their own operations. These defenders recognized Hutcheson's calculations as a very serious threat and went to considerable lengths to refute them—one measure of just how influential Hutcheson's calculations were. Some pamphleteers constructed elaborate caricatures of the calculator Hutcheson and his fixation on intrinsic value. Exemplary was a lively text called *Matter of Fact; or, The Arraignment and Tryal of the DI—RS of the S—S—Company,* written under the name "Timothy Telltruth." The pamphlet depicted a fictitious trial, pitting the South Sea Directors against a team of opposing prosecutors representing the angry public. One of them, "Counsellor Query," was an obvious stand-in for Hutcheson. Query leveled an extended technical argument about how the Company's Directors had breached the trust granted them by Parliament by selling their stock at a price they knew to be unjustifiable based on "any reasonable prospect of Profits in Trade, or from any other intrinsick Value therein." Later in the course of the fictional trial, the Directors' imaginary barrister

("Philopatris") deftly outmaneuvers Query by casting doubt on the concept of intrinsic value—"if I know what he means by intrinsick"—and pointing out internal inconsistencies in Query's various computational arguments.[62]

Other propagandists sought to thwart the calculative accusations made by Hutcheson and others by questioning the motives behind them. One vocal respondent was a serial publication with the (perhaps ill-chosen) title *The Director,* which appears to have published at least thirty issues between October 1720 and January 1721. Observing that "we have had a great Cavil among us about the intrinsick Value of *South-Sea* Stock," the *Director* contended that the many calculations purporting to prove the low intrinsic value of South Sea Stock had the destructive effect, and perhaps the intent, of driving down the Stock's market price. This impoverished British investors and made the Stock easier pickings for foreigners. "What then are all our Calculators doing," the *Director* asked, "but endeavouring to run down their own Country, in Favour of Strangers?" The paper cautioned people against being "Calculated out of our Money, and brought to *throw away what we had,* by the Rules of Figures and Fractions."[63]

Though publications like the *Director* urged readers not to put too much faith in accusations based on "Figures and Fractions," many Company supporters—including the *Director*—realized that they could not simply cede the numerical terrain to the critics. Numerous pamphlets, newspaper articles, and manuscripts offered quantitative defenses of the South Sea Scheme and its architects. Some calculators offered figures showing that many people who invested in the Scheme still stood to make solid returns on their investment.[64] Multiple publications tried to defend the efficacy of the refinancing Scheme by calculating how much money it promised to save the government and the British public over time, specifically by reducing interest payments on the national debt. Issue no. 30 of the *Director* included a highly optimistic calculation suggesting the Scheme might have saved the government more than £297 million on interest. (The *Director* prefaced that calculation by explaining that, although the Company's "Opposers are pleas'd to let us see they deal much in Figures and Calculations," they were not the only ones who had numbers on their side.) A similar calculation in a late April 1721 issue of the *Moderator* newspaper determined that those savings reached the only slightly more modest figure of £235 million.[65]

The figures produced by the *Director* and the *Moderator* relied on some rather obvious flights of numerical fancy, and were probably too outlandish to gain much traction with public audiences. A rather more compelling numerical defense was produced by the South Sea Company itself. As mentioned, in late

December, Hutcheson and the Parliamentary Committee of Secrecy had requested that the South Sea Directors "lay before this House the Calculations & Inducements" that had led them to implement several of their dubious policies, specifically the promise made in September to pay 50 percent annual dividends for the following twelve years. On January 3, the Company's Court of Directors assembled to decide what to do about the Commons' demand for such a "Scheme or Calculation relating to the Dividends." After this internal discussion, the Directors instructed a Company employee, Jonathan Webster, to compile such calculations.[66] Webster presented the manuscript calculations he prepared to the Directors on January 11. Those figures were subsequently passed on to Parliament, and a copy came to rest in the Treasury Papers at the National Archives in Kew, hidden among a mostly uncatalogued packet of papers relating to the South Sea Company. (See Figure 5.6.)[67] This remarkable document was obviously an object of concern for Hutcheson and the Committee of Secrecy, and it attracted mention in at least one contemporary account of the Bubble.[68] Yet it seems to have been overlooked in modern Bubble scholarship.

Like many other computations on South Sea Stock, Webster began with a general account of the Company's current Stock structure, computing how much Stock the Company had already issued and how much it had left to issue under the stipulations of the original Parliamentary Act. (See Figures 5.7a and 5.7b for details of Webster's calculation.) Assuming that all of the holders of outstanding public debts decided eventually to buy into the Scheme, Webster concluded that as of late August, the Company still had £14.9 million left in unsold Stock that could have been offered to the public. Assuming that this was all sold at £1,000 per share, the Company could raise £149 million in additional funds. Added to money previously raised from Stock sales, and deducting various debts (like the £7.5 million fee to the government), Webster concluded that the Company effectively had "Money" available totaling £187.9 million. Webster's next computation was the crucial one; Hutcheson would have understood it well. He calculated how much this supposed quantity of cash could be expected to pay out annually over twelve years, assuming it were treated like an annuity at a standard 4 percent compound interest rate: "Now as 9£. 7 [s.] 8½[d.] Imployed at 4 p[er]Cent p[er] Annum Compound Interest will produce an Annuity of 20 shillings p[er] Annum for 12 Years, so 187,854,519.4 being employed at that Rate will produce ye Ann^l. Summ of 20,016,464.9."[69] That is, the "Money" (£187.9 million) held by the Company was mathematically equivalent to an annuity paying out just over £20.0 million per year for twelve years. Adding in roughly £2.1 million in interest that the government could be expected to pay on

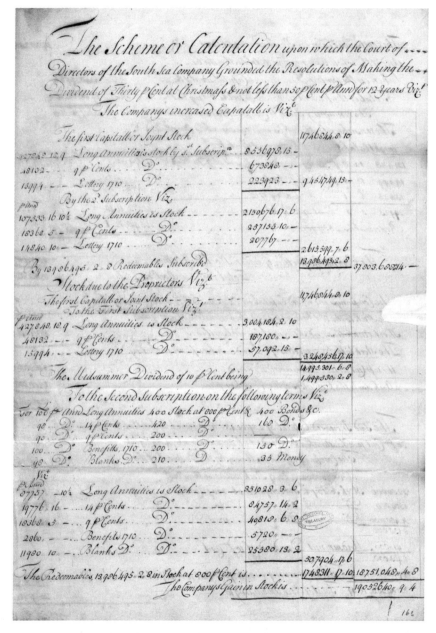

Figure 5.6 Image of Jonathan Webster's calculation, "The Scheme or Calculation upon which the Court of Directors of the South Sea Company Grounded the Resolutions of Making the Dividend," January 1721, Treasury Papers, TNA T1/231, ff. 163–164. Courtesy of the National Archives, Kew, United Kingdom.

government debt held by the Company, plus a modest £400,000 for the "advantages expected upon Trade yearly," he concluded that the Company was in a position to distribute £22.5 million a year in dividends over the following twelve years. That amount came to just over 50 percent of the Company's £43.4 million in nominal Stock outstanding. The result: the South Sea Company could indeed pay 50 percent annual dividends for twelve years.

Many disgruntled contemporaries no doubt would have questioned the rosy conclusions of Webster's calculation. Indeed, one contemporary account called it "absurd."[70] But Webster's calculation was also an elegant computational exercise, whose technical steps were all relatively plausible within the framework of early eighteenth-century thinking about financial mathematics. Two major assumptions made the calculation work. First, it assumed that the Company would be able to convince all of the remaining holders of government debt to buy into the Scheme and then to sell the maximum amount of Stock to the public at £1,000 per share. Second, it assumed the Company could somehow earn 4 percent interest on all of that money and gradually pay it out to its stockholders like an annuity. The first assumption was extremely optimistic, but not entirely unreasonable. Throughout the summer of 1720, public investors had shown themselves more than willing to sign up to buy South Sea Stock priced at £1,000. The second assumption poses an even more interesting set of ambiguities. Mathematically, the calculation was unassailable. Using the mathematics of compound interest, £187.9 million was indeed equivalent to a twelve-year annuity of just over £20 million, at 4 percent interest. What was questionable was how that mathematical relationship would play out in reality. Where was all of this "Money" going to come from exactly? (£187.9 million was an extraordinary sum, and it was optimistic to think there was even that much liquid capital available in Britain, even if investors were willing to entrust it to the Company.) How was the Company going to generate the 4 percent interest it assumed it could earn on those funds? (It obviously could not buy an annuity of that size from a third party.) In what order would the transactions take place? How could the Company guarantee sufficient "cash flow" to make the regular dividend payments?

It seems quite clear: Webster's calculation was a retrospective exercise, created only after the Bubble had burst and Parliamentary investigators had demanded that the Company provide documentation for its earlier policies. The manuscript delivered to Parliament included a "Memorandum" appended to the end, stating that the figures were not an exact rendering of the Company's original computations. It explained that Company Treasurer "M^r. Knight acquainted the

	(£ in millions, except as noted)
"The Companys increased Capitall is":	
"The first Capitall or Joynt Stock"	£11.7
Capital created through the exchange of public debts:	
1st subscription of irredeemables	9.5
2nd subscription of irredeemables	2.6
Redeemables Subscribed	14.0
Total "increased Capitall"	£37.8
"Stock due to the Proprietors":	
"The first Capitall or Joynt Stock"	11.7
Stock owed to those who exchanged public debts:	
To the 1st subscription of irredeemables	3.2
Total owed to "first Capitall" and 1st subscription	£15.0
Plus: "Midsummer Dividend" of 10%	1.5
To the 2nd subscription of irredeemables	0.5
To the redeemables subscribed	1.7
Total "Stock due to the Proprietors"	£18.8
Total "increased Capitall"	37.8
Less: Total "Stock due to the Proprietors"	(18.8)
Company's Current "Gain in Stock"	£19.1
"And if the whole annuities and debts should be taken in the Company's increased Capital would be"	43.4
Less: "Total of the present increased Capital being but"	(37.8)
"The remaining Stock to Compleat the Scheme"	£5.6
Public debts yet unsubscribed:	
Irredeemables	5.2
Redeemables	2.6
Total public debts left to be taken in	£7.8
"The unsubscribed Debts . . . reasonably concluded might be taken in Stock at 1000 p[er] Cent":	
Cost (in Stock) of irredeemables, at £1,000	0.5
Cost (in Stock) of redeemables, at £1,000	0.3
Total cost (in stock) of taking in unsubscribed debts	£0.8
"The remaining Stock to Compleat the Scheme"	5.6
Less: Total cost (in stock) of taking in unsubscribed debts	(0.8)
"The Company's further Gain in Stock will be"	£4.8
Plus: Company's Current "Gain in Stock"	19.1
SSC's "Total Gain in Stock," incl. future subscriptions	£23.9

Figure 5.7a Summary of Jonathan Webster's calculation, "The Scheme or Calculation upon which the Court of Directors of the South Sea Company Grounded the Resolutions of Making the Dividend," January 1721, Treasury Papers, TNA T1/231, ff. 163–164. Courtesy of the National Archives, Kew, United Kingdom.

	Stock Issued	Money Raised
"The Company's increased Capital at Christmas 1720 will Consist of":		
"Stock to the old proprietors and proprietors of . . . Annuities"		18.8
"Stock to the Proprietors of the 1st, 2d, and 3d Subscriptions"		9.0
"Stock due to the Proprietors of the 4th Subscription"		1.2
Total "increased Capital at Christmas 1720":		£28.9
Christmas Dividend of 30% on the "increased Capital"		8.7

"The Company's Gain in Money by the Disposal of the" £23.9mm gain in Stock "will arise thus":	Stock Issued	Money Raised
Money already raised in 1st, 2d, and 3d Subscriptions	9.0	56.8
"The Remaining Stock, if sold at 1000 p[er] Cent"	14.9	149.2
Total "Gain in Money" on sale of Stock, incl. future sales	£23.9	£205.9
"From which gain . . . deduct":		
"The monys to be payd the Government"		(7.7)
"Bonds and money for the annuitants"		(3.0)
"Christmas Dividend" of 30%		(8.7)
Net of: Money owed to the SSC by the Gov't before Christmas 1720		1.4
Total deductions		(£18.0)
Total "Gain in Money," after deductions		**£187.9**

	Present Cost	Annual Payment
An annuity for 12 years, at 4% interest, produces:		
To yield £1 annually	9.42	1.00
Given £187.9mm present cost	£187.9	£20.0
Available for dividends:		
Annual proceeds from "Gain in money," as 12-year annuity at 4%		20.0
Annual interest payments from the government		2.1
"The advantages expected on Trade yearly"		0.4
Total available for dividends, annually		**£22.5**
Total "increased Capitall"		43.4
Proposed 50% per annum dividend		21.7
Total available for dividends, annually		22.5
Proposed 50% per annum dividend		21.7
"Annual Surplus"		**£0.8**

Figure 5.7b   Summary of Jonathan Webster's calculation, continued.

Court that the Paper upon which the Calculation was made is Lost or Mislaid, But he and M^r. Webster declare this"—the January 1721 analysis—"was made agreeable to the same."[71] In other words, Webster's January 1721 calculation was supposedly a faithful effort to re-create the original calculations that Knight and other Company leaders had made in the summer of 1720, which apparently could not be found. Company defenders would later claim in print that their controversial dividend policies had in fact been "grounded upon an exact Calculation that was made some Time before."[72] Perhaps this was true. But it seems far more likely that Webster's calculation was a creative, backdated rendering of what the Company might have been thinking—rather than a true reflection of how the dividend decision had been made.

Webster's calculated defense of the Company dividend policies offers a kind of allegory for the complexities of Britain's new quantitative age. For one thing, it was evidence of the remarkable influence calculative thinking had taken on in British politics. Faced by an onslaught of quantitative criticism from Hutcheson and others, the Company had no choice but to defend itself numerically—even if this meant retroactively concocting a set of figures that the Company had never actually used in the first place. It is especially telling that what drove the Company to produce these elaborate analyses was probably not their own financial designs, but rather the imperatives of public politics.

At the same time, though, Webster's calculations hinted at another, more unsettling feature of the quantitative age: just how flexible calculation could be as a form of political and economic argument. Even in a seemingly cut-and-dried case, like the South Sea Company's "absurd" dividend policy, a skilled calculator like Webster was able to shape the numbers in a way that provided at least plausible support for his side. Arguably the most striking thing about Webster's calculation was that it was *not* entirely ridiculous by the standards of its moment. Some of its key underlying assumptions were highly debatable, but the mathematics was sound and overall it probably fell within the bounds of plausibility. At its core was a particularly ambitious use of the logic of compound interest—largely the same logic that drove the calculation of the Equivalent in 1706, Hutcheson's plans to repay the national debt in 1715, and indeed much of Hutcheson's own analysis of South Sea Stock in March 1720 (as well as Robert Walpole's sinking fund, the subject of Chapter 6). Webster's calculation probably did little to convince the Committee of Secrecy, or the British public, that the South Sea Company's dividend policies had been well-reasoned or justifiable; Hutcheson's argument, that the Company had knowingly deceived the public about the value of its Stock, still carried the day. But perhaps Webster's calculation created just a bit of doubt, or just slightly softened some MPs' retributive feelings toward the Company's Directors.[73] His ambitious numbers showed that, when it came to the most fraught political and economic arguments, there was usually just enough flexibility in the figures to keep the argument going a little bit longer.

Judicious Calculations

The collapse of the South Sea Bubble produced a twofold triumph for Archibald Hutcheson and his calculations. First, his skeptical calculations from early 1720 were seen, after the fact, as prescient warnings that could have spared the nation from ruin. Second, his retrospective calculations after the Bubble burst became a vital political tool for explaining what had happened and, even more importantly, for assigning blame. Economic evaluations of the Company's Stock value were coproduced with ethical evaluations of the Company's leaders and political supporters.[74] Hutcheson's calculations of the intrinsic value of South Sea Stock demonstrated that the Company's Directors had done wrong; conversely, public anger toward the Directors, bolstered by varied evidence of their corruption, reinforced the belief that Hutcheson's calculations were right. (Whether or not the Directors would ultimately receive sufficient punishment in the eyes of the public was another question entirely.)

In the wake of the crash, Hutcheson reaped the rewards of his computational triumph. Fellow Britons lauded him for having "constantly opposed, in Parliament and in Print, the *Fatal South Sea Scheme,* and Early foretold the Ruinous Consequences thereof." Some, like Eustace Budgell, specifically commended his computational achievements. "Your several Calculations of the Value of *South-Sea* Stock have shewn you to be an *honest Man,* and a good *Accomptant,*" Budgell wrote in a post-Bubble pamphlet. "I wish, for the Sake of my Country, I could say half as much of some other people." One poetic epistle to Hutcheson in 1722 pondered how much better off the nation might have been if Hutcheson's fiscal plans had been adopted, instead of the South Sea Company's: "Full happy shou'd we've been, nor e'er disdain'd, If Your Elaborate Schemes had once obtain'd."[75] Applause was not limited to the press. In 1722, Hutcheson was reelected to the Commons, not only for Hastings but also for the prestigious seat representing Westminster. Campaigning largely on his prescience about the South Sea Scheme, he won nearly 4,000 votes and defeated Parliament's long-standing fiscal specialist William Lowndes, who earned fewer than 2,200.

Hutcheson's computational triumphs were not quickly forgotten, either. Throughout the next century, chronicles of the Bubble memorialized Hutcheson and his numbers. Exemplary was Adam Anderson's *Historical and Chronological Deduction of the Origin of Commerce* (1764), a highly influential study republished multiple times in the ensuing century and translated into German in the 1770s. Anderson, a former South Sea Company employee, recalled how at the height of the frenzy in June 1720, "some began to have their Eyes opened by the judicious Calculations of *Archibald Hutcheson,* Esquire, and others." He later called Hutcheson "an ingenious Gentleman," complimented his "fair and candid Calculations," and directly quoted Hutcheson's calculations in explaining the defects of the South Sea Scheme. Later political economists would hold up Hutcheson as a paragon of financial prudence. In an 1816 pamphlet on Britain's ongoing debt problems, Richard Preston recalled how "the able pen of an Hutcheson" had been needed "to open the eyes of the country to the folly" of the South Sea Scheme. Hutcheson's legend lingered on in unexpected places, like in an 1833 travel guide to the Sussex coast. Recalling Hutcheson's generous donations to St. Clement's Church in Hastings, the author described him as "an able financial writer" who "contributed to undeceive persons and save them from the celebrated South Sea bubble."[76]

The personal fame Hutcheson managed to garner from his South Sea calculations was rather rare. As discussed in Chapter 4, most political calculators—like John Crookshanks—worked in relative anonymity and never managed to make any kind of public name for themselves. Hutcheson did. But even after his Bubble triumphs, Hutcheson was not immune to criticism and distrust from his (many) political opponents. He remained a political outsider, struggling against a powerful Whig establishment. His election as MP of Westminster was overturned in December 1722 for election tampering, though he remained in the Commons for Hastings.[77] He continued to fight off accusations of Jacobitism, and was occasionally still called a conjurer.[78] Hutcheson kept up the numerical fight in the next decade, regularly sparring with Robert Walpole and his ministry over questions of public finance. However great his successes in 1720, he never became any kind of authoritative "expert" who was seen to transcend politics.

Hutcheson's reputation dimmed with time. In his 2004 account, financial historian Richard Dale lauded Hutcheson as the "the unsung hero of the South Sea Bubble"—"unsung" only to observers at the turn of the twenty-first century, not to those in the eighteenth.[79] (One culprit for Hutcheson's historical neglect may be Charles Mackay. In his renowned Victorian fable *Memoirs of Extra-*

ordinary Popular Delusions in 1841, Mackay repeated many of Hutcheson's core ideas but mistakenly attributed them to Robert Walpole.)[80] Fittingly, Hutcheson's calculations had a more enduring legacy than he himself did. His numerical analysis laid the foundations of a standard interpretation of what happened during the Bubble—namely, that it was a pathological divergence of the Stock's market price from its true "intrinsic value," engineered by the deceitful Company leaders. Major twentieth-century accounts of the Bubble have downplayed Hutcheson's individual role in the events of 1720 but also largely reiterated his interpretation of what happened in that fateful year. In fact, one recent critic of Hutcheson's, economic historian Helen Paul, has argued that many mistakes in the scholarship on the Bubble can be traced to a persistent overreliance on Hutcheson's interpretations.[81]

Reviving Hutcheson's reputation, and litigating whether it is deserved, is not the chief concern of this study. (It is worth noting that, in many ways, Hutcheson was no hero; his laudable achievements were deeply entwined with highly repugnant personal and political values, including marked anti-Semitism and unabashed enthusiasm for the African slave trade.) Rather, this chapter has sought to use Hutcheson and his calculations as a window into the South Sea Bubble as a political and epistemic event. Britain's experience in 1720—and the role of Hutcheson's numbers therein—offers a vivid example of how financial crises transform public knowledge about finance and, more broadly, about how crises can reconfigure civic epistemologies.

For one, the Bubble marked a significant turning-point in the history of financial thinking in Britain, which reshaped attitudes about the nature of financial value. As we saw, prior to 1720 there was widespread disagreement about what made stocks valuable and how that value might be assessed. That disagreement certainly did not disappear after 1720. But the Crisis did galvanize a considerable amount of public support around one way of thinking, namely the intrinsic value approach expounded by Hutcheson. This vision held that paper financial objects, like shares in a joint-stock company, had a real, intrinsic value that could be calculated. That value derived from both a company's current financial assets and its potential for future profits, quantities that could be related to one another using the mathematics of compound-interest discounting. On the balance, therefore, the South Sea Bubble made the concept of financial value *more* concrete in the minds of Britons, not less. Hutcheson's calculations depicted the stock market as a place where reason and calculation should prevail—even if they failed to do so in 1720 because of the fraudulent actions of some unethical individuals. This story contradicts those narratives of financial history that hold that, prior to the development of modern forms of financial analysis in the twentieth century, stock market investment was generally seen as a form of gambling and not a subject of real knowledge.[82]

More broadly, the South Sea Bubble was a climactic moment in the development of British civic epistemology. It revealed, suddenly and strikingly, the serious dangers brought by the new financial technologies developed in the preceding decades. Like many of those innovations, the South Sea Company's intricate debt restructuring scheme was obscure and untested. Britons were collectively uncertain about how it worked and the risks it posed. What is more, they did not yet have established ways to go about answering those questions on a political level—no clear experts to turn to, guidelines to follow, past experiences to fall back on. The 1720 Bubble was, to a large degree, a consequence of this collective confusion.[83] But the Bubble itself also provided a measure of edification. By offering a painful lesson about the dangers of ignoring mathematical "figures and demonstrations," it helped to consolidate the growing authority of calculation within British civic epistemology. It taught Britons that, in such risky technical matters, where the potential for misunderstanding and deceit was high, numbers were more trustworthy than people.

CHAPTER SIX

Futures Projected:
Robert Walpole's Political Calculations

W HEN CHARLES DAVENANT endeavored to refashion political arithmetic in the 1690s, one particular style of calculation exemplified his
new approach. The formula went as follows: consider a particular tax; using whatever external information available, compute a reasoned estimate of what that tax
ought to yield; compare that calculated expectation to data showing what that
tax *actually* produced; draw a suitable political conclusion. Davenant applied
this algorithm to a new poll tax in his 1695 *Essay upon Ways and Means* and in
his more elaborate analyses of the public revenues in his 1698 *Discourses*. It
embodied Davenant's use of calculation as a tool of critique. Calculation made it
possible to compare the political situation as it was to the political situation as
it ought to be—to measure the distance between ideal and reality.[1]

Davenant was not the only calculator from the era to adopt this style, though.
An especially vivid example, created around 1730, can be found in a short manuscript on customs taxation surviving in the papers of Robert Walpole.[2] The
calculation assessed customs tax revenues raised on the burgeoning Maryland
and Virginia tobacco trade (Figure 6.1). The anonymous calculator surveyed
twelve ships each from Virginia and Maryland to determine the weight of tobacco in an average imported hogshead. Combining this sampled data with information from the Customs administration and Exchequer, he calculated the

Quantity	Value	Unit
"Tobacco . . . brought into Britain from Virginia and Maryland," annually (7-year average)	63,687	Hhds.
Weight of tobacco, per hogshead (based on an average sample of 12 ships)	720	lbs.
Total weight of tobacco imported, annually	45,834,640*	lbs.
"From which . . . being deducted for damage Tobacco cut off before weighed"	(481,301)	lbs.
"Nett" weight of tobacco imported, annually	45,353,339	lbs.
Fraction of "Tobacco used for home consumption," based on the "Lowest Estimate"	1/3	
Quantity of Tobacco used for home consumption	15,117,780	lbs.
Rate of impost on tobacco, "clear of all deduction"	4¾	*d.* / lb.
Imposts on Maryland and Virginia tobacco "should produce annualy"	**£299,206**	£
"The nett money paid into the Exchequer . . . upon the Impost of Tobacco" (7-year average)	160,625	£
Customs paid in Scotland, estimated	4,000	£
Paid out of impost on tobacco, for charge of management	8,000	£
"Nett produce" of imposts on tobacco	**£172,625**	£
"Which is less than it ought to produce upon the foregoing Estimate"	126,581	£
Less: "The price of the Insurance upon the Quantity ship'd may amount"	(6,000)	£
"Sum . . . lost by different frauds committed in the Collection of this Revenue"	**£120,581**	£

*The calculation of total tobacco imported annually (lbs.) is irregular, as 63,687 times 720 equals 45,854,640, not 45,834,640. This is likely a transcription error that compounded—not uncommon in contemporary calculations.

Figure 6.1 Analysis of expected and actual produce of customs taxation on Maryland and Virginia tobacco, c. 1730, adapted from a manuscript calculation in Robert Walpole's papers, Cambridge U.L. Ch(H) Papers 43 / 11 / 2.

net weight of tobacco imported annually and deduced what the government ought to be earning in customs revenue. The conclusion was that the tobacco customs were underproducing by over £120,000 annually. It was a masterful application of Davenant's procedure, combining disparate data to highlight how far fiscal realities fell short of ideal standards. The calculator contended the shortfall stemmed from "different frauds committed in the Collection of this Revenue," like merchants taking excessive "drawbacks" on reexported goods and Customs officials surreptitiously "pilfering" merchandise. Like Davenant, the anonymous calculator concluded that the numbers showed major changes needed to be made in Britain's tax system.

While it followed Davenant's computational procedure impeccably, this later calculation differed from Davenant's in one profound way. Whereas Davenant had critiqued government policy from the outside, this calculation was produced under the ultimate insider: Sir Robert Walpole, often credited with the title of Britain's first "Prime Minister" (1721–1742). The calculation was produced as part of Walpole's effort to overhaul British taxes by instituting new domestic excises on tobacco and wine.[3] This deft manuscript calculation was indicative of a crucial trend in the culture of calculation in British politics: the increasing internalization of calculation as a tool of state. In the first three decades or so after 1688, Britain's new culture of calculation had largely been driven forward by the work of political outsiders like Davenant and Hutcheson, who used computations to question the state and other privileged institutions. Government ministers and administrators had, of course, long kept quantitative records of various kinds. But they often did not value numerical information in the same way these critical outsiders did—recall how the Commissioners of Public Accounts in the early 1690s discovered that the Lords of the Treasury treated certain public accounting data as "wast[e] papers"—or use arithmetical calculation in the same inventive ways those outsiders did as a means of thinking through political problems. As we saw in Chapters 4 and 5, Archibald Hutcheson's calculating career was fueled by a consistent frustration that agents of the British state, and state-sponsored institutions like the South Sea Company, seemed relatively unconcerned about the kind of quantitative information he felt was essential to good government. When it came to calculative thinking, critics of the state were usually one step ahead of the state itself.

At the same time, though, some leading ministers like Lord Godolphin and the former Accounts Commissioner Robert Harley, as well as influential civil servants like William Lowndes, did embrace what Davenant had called the "Computing Faculty" and began incorporating it into aspects of government practice.[4] Political leaders came to appreciate that calculation was a crucial instrument for establishing support within Britain's contentious public sphere, and thus recruited calculators—see Harley and the *Mercator* in 1713, or the Earl of Halifax and John Crookshanks in 1718—to defend the government's position against calculated attacks. This gradual internalization of calculative thinking within state practice reached its culmination in the ministry of Robert Walpole. Under Walpole, calculation—not just numerical record-keeping, but the active and creative use of mathematical reasoning—became a defining feature of the evolving governmentality of the British state.[5] No longer would the British state lag behind its critics in the calculating game.

Since Walpole's own lifetime, many observers have described Britain's first Prime Minister as the embodiment of the "calculating" politician—but in two rather different senses. On the one hand, early chroniclers celebrated Walpole for his skills at numerical calculation and his command of financial technicalities. In the first comprehensive biography of Walpole, published in 1798, William Coxe reported that "it was said of him that he was endowed with a genius for calculation." Another late-eighteenth-century account by Thomas Pownell praised Walpole for having "recovered the administration of the Finances of the Country out of confusion."[6] Twentieth-century financial historians like W. R. Scott and P. G. M. Dickson have confirmed this picture of Walpole as an exceptionally shrewd financial mind.[7] On the other hand, the preponderance of historical accounts depict Walpole as calculating in a different sense: scheming, tactical, duplicitous, a master of manipulating people more than manipulating numbers. To many of his political opponents in his own day, Walpole was corruption personified, whose chief gifts were cronyism, venality, and fraudulence. Later accounts have often drawn attention to Walpole's skills as a calculating political tactician. An 1842 description from the *North American Review* is typical: "He had a clear head, a calculating heart, a mind not overburdened with troublesome scruples of morality, and a thorough contempt for all men."[8] Walpole's definitive modern biographer, J. H. Plumb, depicts Walpole as a complex political figure, who used patronage and proscription to consolidate a Whig "Oligarchy" and establish unprecedented political stability in the nation. Though keenly attentive to Walpole's unique political skills, Plumb has firmly rejected the notion that Sir Robert was some kind of financial or mathematical genius.[9]

Observers have often treated Walpole's two different calculating personae as somehow antagonistic to one another, even mutually exclusive. By contrast, this chapter argues that, for Walpole, these two kinds of calculating skill—the mathematical versus the tactical, his "genius for calculation" versus his "calculating heart"—were essentially intertwined. He was a political calculator in both senses at the same time. Numerical calculation was not a separate, instrumental activity, required for the practicalities of state administration but unrelated to Walpole's political maneuvering. Rather, numerical calculation was at the very center of his political thinking and practice—central to how executive policies were assessed internally, how they were justified publicly, and how political success was defined. Not only was Walpole a highly competent calculator in his own right, but he also assembled a team of quantitative advisers to help him reckon with policy decisions and formulate public arguments. These ranged from a former Governor of the Bank of England, Nathaniel Gould, to an extensive collection of anonymous clerical officials, freelance accountants, and hack writers, including one John Crookshanks.[10]

In addition to using calculation internally to guide policy choices, Walpole and company were also highly skilled in the public art of numerical argumentation. Early in his career, Walpole was disgraced as the result of a ruthless computational attack by Harley and the Tories. It was a lesson he never forgot, as Walpole became keenly aware of calculation's tremendous power as an instrument of public politics. That appreciation was further heightened by the event that brought him to power: the 1720 South Sea Bubble. Hutcheson's seemingly prescient South Sea calculations had shown many Britons the great political utility of numbers, and the dangers of ignoring them. When Walpole became first minister, his immediate challenge was to reorganize the South Sea Company and restore public credit. In doing so, he took a page—often quite literally—out of Hutcheson's book, using similar computational tools to analyze possible financial remedies and to promote them to the public. In the coming years, Walpole's skill at numerical debate proved a major weapon in fighting "Patriot" opponents and in generating confidence in the nation's fiscal health. Walpole's opponents struggled to respond to his numerical tactics. A few tried to dismiss his "multitude of *Figures* and *Calculations*" as a crooked form of political thinking. Others, like Walpole's disgruntled former ally William Pulteney, tried to fight back with numbers of their own. (It was in a fight with Walpole that Pulteney first invoked the phrase "*Facts* and *Figures*" in 1727.) But no one played the numbers game as well as Sir Robert.

Over time, he and his calculating team invented powerful new political uses for calculation. In particular, the Walpolean calculators turned their numbers from the past to the future. Against a backward-looking "politics of nostalgia" often invoked by Patriot opponents like Lord Bolingbroke, Walpole's circle used calculation to develop a distinctly forward-looking political temporality.[11] In doing so, Walpole relied on two calculative tools familiar from earlier chapters. First, Walpolean calculators adopted the multiscenario "spreadsheet" model, so critical to Hutcheson's numerical polemics, as an internal tool for thinking through the future consequences of political decisions. Like many other eighteenth-century thinkers, Walpole turned to numbers as a way to reckon with future uncertainty. But his strategy relied not on probability but *possibility;* he used calculation not to reckon with chance, but to model an array of alternative futures that could be achieved through human action. Second, Walpole's calculators seized upon the exponential mathematics of compound interest as an instrument for projecting present power into future glory. Compound interest was the foundation of one of Walpole's most iconic fiscal projects: his "sinking fund" for defeating the national debt.

Walpolean calculators shifted the battlefield of partisan contestation away from the scorched terrain of the nation's economic past and present, toward the unclaimed spaces of the nation's future. In doing so, they contributed to a reshaping of political time, bringing the political future nearer at hand. Walpole's political calculations were an especially powerful example of the broader ascent of future-oriented thinking in eighteenth-century culture. Historian Edward Jennings argues that England witnessed a distinct "transition, between 1600 and 1800, from an attitude that individual and social destiny is beyond human control to an attitude that the future can be predicted and affected by human behaviour." Scholars have traced the rise of this forward-looking attitude in multiple areas of eighteenth-century endeavor, from law and education to weather forecasting and news reporting.[12] But in the early eighteenth century, when Walpole came of age, this futuristic outlook was still highly controversial. Some people, like the unconventional Tory polemicist John Asgill, embraced prognostication as a virtuous endeavor. "Nothing would do more Service" for his country, he explained in a 1713 pamphlet on political prophecy, "than to give them some *Predictions,* that so being warn'd of their Danger, they may prevent it." Yet many Britons still rejected such forward-looking speculations as fringe, dangerous, even irreligious—reminiscent of the worst excesses of so-called "judicial" astrology. Any claim to use numerical calculations to make assertions about the future risked evoking long-standing associations between mathematics and the occult, between calculation and conjuration. Daniel Defoe lamented in 1704 that even though "the Calculation of Probabilities may give a Man an insight farther into an Affair, and its Consequences, than ever one may think possible," yet "when what was rationally deduc'd comes to pass, the Man is taken for a Conjurer."[13]

By the 1720s—earlier than some historians have thought—this skepticism was definitely subsiding.[14] Forward-looking calculations were going mainstream, and financial politics was vital to that shift. In 1722, for example, Archibald Hutcheson actively trumpeted his foresight about the South Sea Company as a reason for Westminster electors to vote for him. (Still, some critics accused him of being a conjurer.)[15] Walpole took the politics of financial prediction even further than that. He used projective mathematics to claim credit for achievements that had not yet occurred, and would not for decades, like the extinguishment of the national debt by the sinking fund. One especially grandiose pamphlet described in detail how Walpole and his fiscal triumphs would be celebrated by subsequent generations *three hundred years* in the future. That apotheosis of Walpole was, arguably, the apotheosis of Britain's new calculating political culture.

The Affair of the Missing Millions

Born in 1676 to an influential gentry family in Norfolk, Robert Walpole was educated at Eton and King's College, Cambridge. He may have fostered an early interest in a church career, but the deaths of his two older brothers forced him to turn away from his education in 1698 to learn the business of managing the family estates. Walpole's father was an intelligent manager who advanced the family's wealth through agricultural improvements and shrewd investments, and it was likely at the family's Norfolk home that the younger Walpole first learned to calculate. Walpole's father died in 1700, vacating his Parliamentary seat for the "pocket" borough of Castle Rising. Robert was elected shortly thereafter, entering Parliament in 1701. He quickly ascended through a series of technically demanding positions, appointed a member of the Council of the Lord High Admiral in 1705 and Secretary at War in 1708.

In 1710–1711, Walpole enjoyed a brief spell as Treasurer of the Navy, where he suffered one of the era's most violent numerical attacks. Shortly after a landslide victory in the general elections of autumn 1710, the new Tory ministry under Robert Harley undertook an aggressive audit of administrative and accounting practices under previous Whig officeholders—"the Great Plunderers of the Nation," as Defoe had described them. A key target of Harley's interrogation were naval finances, especially the Navy's "imprest" system. The imprest system was akin to a petty-cash system, through which Parliament advanced money upfront to administrators to cover expenses, and then the total funds used were reconciled after expenditures had been made. In April 1711, the new Auditor of the Imprest Edmund Harley, Robert's brother, presented a report to the Commons suggesting that more than £35 million in imprest funds, granted by Parliament before Christmas 1710, had never been accounted for by the appropriate officials. Particularly hard-hit by the accusation was the outgoing Treasurer of the Navy: Robert Walpole.[16]

The figure was shocking, if also aggressively misleading. "In many parts of the Country," Walpole observed with frustration, "our People were taught to think they had been plunder'd of Thirty five Millions." The affair of the disappearing millions gave Walpole a firsthand lesson in numerical combat. Walpole and fellow Whig publicist Arthur Maynwaring published a series of technical pamphlets explaining why that £35 million was not yet unaccounted for, and why it did not mean that £35 million had been stolen.[17] (Walpole might have actually engaged in some fiscal chicanery, but £35 million was an unrealistic amount to embezzle.) They tried to clarify why, due to the vagaries of naval administration, the Navy's spending needed to be reconciled after the fact through the imprest system, rather than "estimated" ahead of time, as had become standard in other areas of government since 1688. Naval affairs were so riddled with the unpredictable that accurately estimating future needs was often futile. "By the nature of these Services it appears impossible, at first sight, to fix and ascertain the exact Expence upon any Head except that of Wages," read one pamphlet. "For how can it be foreseen that the Loss or Damage of Ships by Storms, or by the Enemy, shall be just so much in one Year, and no more?" The most practical way to conduct naval business was to entrust administrators with a fair amount of money up front and have them settle up at the end.[18]

Walpole's expositions of naval accounting procedure were diligent and largely sound, but proved little match for the Tories' numerical attacks. In Plumb's judgment, "Walpole demonstrated forcibly the misrepresentation of the Committee who had used the technical delays of accountancy to try and create an atmosphere of colossal corruption." But Plumb also notes that "there can be little doubt that the accusation of the ministry obtained a wider currency" than did Walpole's defense.[19] Walpole managed to escape any official punishments in the "missing millions" affair, but it cast a massive shadow nonetheless. The Tories did not let up, and in 1712 Walpole served a six-month sentence in the Tower of London on a different set of corruption charges. For Walpole, the £35 million ordeal was an especially brutal introduction to the new culture of computational combat, but it offered lasting lessons about the political power of numbers. It taught Walpole that, in numerical contests, fortune favored the bold.

Luckily for Walpole, the naval scandals did not hinder his political career permanently. During his incarceration, key Whig party personnel regularly visited Walpole, who became a kind of "whig martyr."[20] His administrative skills and his political sacrifices were not forgotten when the Whig party returned to power in 1715. He became First Lord of the Treasury and Chancellor of the Exchequer in 1715, and it was in that position that he engineered the introduction of the sinking fund to the government's finances. But a rift with Whig leaders quickly forced him into opposition in 1717. When many in the core of the Whig ministry became implicated in the South Sea Bubble, Walpole found himself in an opportune position. In 1721, Walpole seized control of the Whig ministry, emboldened by the need to rescue public credit in the wake of the unprecedented crisis. Walpole's efforts to restructure the South Sea Company offered early evidence of the distinctive "computing faculty" he would bring to Britain's chief ministerial post.

Walpole's Computing Faculties

The South Sea Company posed a consuming political problem from fall 1720 through summer 1721. First, there was the matter of making sense of what went wrong, who was to blame, and how they should be punished. Here Hutcheson and the Committee of Secrecy played a leading role. Fury against the Company's Directors and leading executives like cashier Robert Knight was widespread, and some, like the authors of *Cato's Letters,* wished to see the Directors hanged. But there were many borderline cases in which culpability was less clear, particularly for the more peripheral Directors and politicians who had supported the Scheme or profited from it indirectly. Walpole called for leniency, earning the ignominious title of "Screenmaster-General" among critics for shielding many politicians from Bubble-related retribution.[21] Second, there were the thorny questions about what to do for and about the many private investors who had become entangled in the Scheme. Would investors who owed outstanding subscription payments be legally obligated to complete those transactions? Would those who had subscribed to purchase South Sea Stock for cash or government annuities at £1,000 prices be given any retrospective relief? These were hard questions, involving the compensation of some innocent citizens at the expense of others.

Finally, something did have to be done with the ongoing finances of the South Sea Company, and the nation's commerce and credit in general. Though modern scholars argue the overall effects of the Bubble on the British economy were actually relatively mild and short-lived, contemporaries were deeply concerned the crisis would damage public finances and trigger the "Decay and Loss of private Credit."[22] Many factors fueled this fear of a credit crunch. On September 29, 1720, the Bank of England temporarily stopped discounting bills of exchange, closing off a major source of everyday commercial credit. The South Sea Company itself owed money to, or was owed money by, a massive web of counterparties. Britons further worried that foreign profiteers had taken hard currency out of the country through well-timed speculations on South Sea Stock.[23] Figuring out how to resuscitate the nation's credit without further jeopardizing the country's commerce and currency constituted a challenging riddle. There were difficult trade-offs everywhere. For example, any policies that sought to bolster the price of South Sea Stock helped Britons who were left holding Stock, but posed other dangers. As one commentator, perhaps Walpole himself, put it: "An attempt to raise the Stock to a Higher value than it can be supported at, would only involve a new set of persons in the misfortunes . . . & expose the Publick to the great Loss that will be sustain by Foreigners selling out at high prices."[24]

As early as September 1720, business leaders, politicians, and pundits had begun entertaining a variety of treatments for the ailing South Sea Company. In September, the Company sought to prop up its sagging Stock price and appease angry stockholders by retroactively amending the Stock prices given to money subscribers and annuity-holders who bought into the Stock in the summer, dropping the lofty £1,000 price to £400. Around the same time, political leaders led by Walpole began to contemplate more aggressive measures, like the "Bank Contract," a two-part bailout to be executed by the Bank of England. Under the plan, the Bank would underwrite £3 million in South Sea Company bonds and give the Company roughly £3.8 million in government debts held by the Bank; in exchange, the Bank would receive over £15 million in South Sea Stock.[25] The Bank quickly soured on the idea, and the plan died in mid-November. Thereafter, Walpole began to push a new and even more involved solution. This "Ingraftment Scheme," first discussed by Walpole's personal banker and policy adviser Robert Jacombe in October, involved distributing £9 million of South Sea Stock (roughly a quarter of its total stock) each to the Bank of England and the East India Company, in exchange for a comparable amount of stock in those two companies.[26] The Bank and the East India Company entertained the notion around March 1721 but soon abandoned the plan. Between May and July 1721, the House of Commons hashed out a more modest restructuring, ultimately approved in August. The bill excused the Company of most of the £7 million-plus fees it had originally promised to pay the government, but canceled £2 million of the Company's underlying capital (long-term debts owed by the government to the Company). Subscribers who had yet to pay up their full subscriptions were forgiven their outstanding payments and given

Company Stock at £300–400 for what they had already paid, depending on their time of subscription. Remaining South Sea Stock was then to be distributed proportionally to all remaining stockholders.[27]

The South Sea calamity, and Walpole's calculated approach to managing the consequences, was pivotal to his rise to political preeminence. Throughout 1720, Walpole was kept well informed of technical matters on Exchange Alley, largely by his diligent banker Robert Jacombe, who managed Walpole's extensive investments. Walpole tracked the day-to-day movements of important financial indicators, not only the price of South Sea Stock but also of Bank of England stock and South Sea bonds.[28] During the tumultuous autumn, many concerned Britons looked for guidance to Walpole. Correspondents sent Walpole numerous proposals for how to remedy the nation's finances. Some of those projects exhibited a zaniness typical of the era, like one scheme that recommended recapitalizing roughly half of the South Sea Stock into four tiers of annuities to be allocated through a massive lottery.[29] Others were far more substantive, like a letter Walpole received from one Thomas Houghton, proposing that the South Sea Company be granted the right to set up bonded warehouses to rent to merchants looking to reexport goods tax-free—an idea Walpole would later pick up in his own plan to reform the excises in the early 1730s.[30]

Walpole probably paid little attention to some of the more outlandish financial calculations and commentaries that came across his desk, but he read others with great diligence. His manuscripts contain many vivid examples of his own calculative mind in action, including several examples of Walpole reading and responding to other people's financial ideas. One illuminating example, from his working notes on the Ingraftment Scheme, includes bulleted lists summarizing analyses of the Ingraftment Scheme by three different commentators: "Mill^r," "Sloper," and "Hutch.[eson]" (see Figure 6.2).[31]

Walpole reacted to some calculations with particular vigor—including at least one paper by Archibald Hutcheson. The paper in question discussed how beleaguered holders of South Sea Stock would be affected by certain plans for restructuring the Company, like a retrospective reduction in purchase prices for the Stock or Walpole's proposed Ingraftment Scheme. At the time, Hutcheson was concerned that many of the restructuring proposals under discussion led to unfair outcomes, compensating certain investors at the expense of others. In Walpole's manuscript copy, Hutcheson's text was transcribed on one side of each page, leaving the facing side free for comments. Walpole's marginalia reveal a high level of technical command over Hutcheson's figures. The Prime Minister

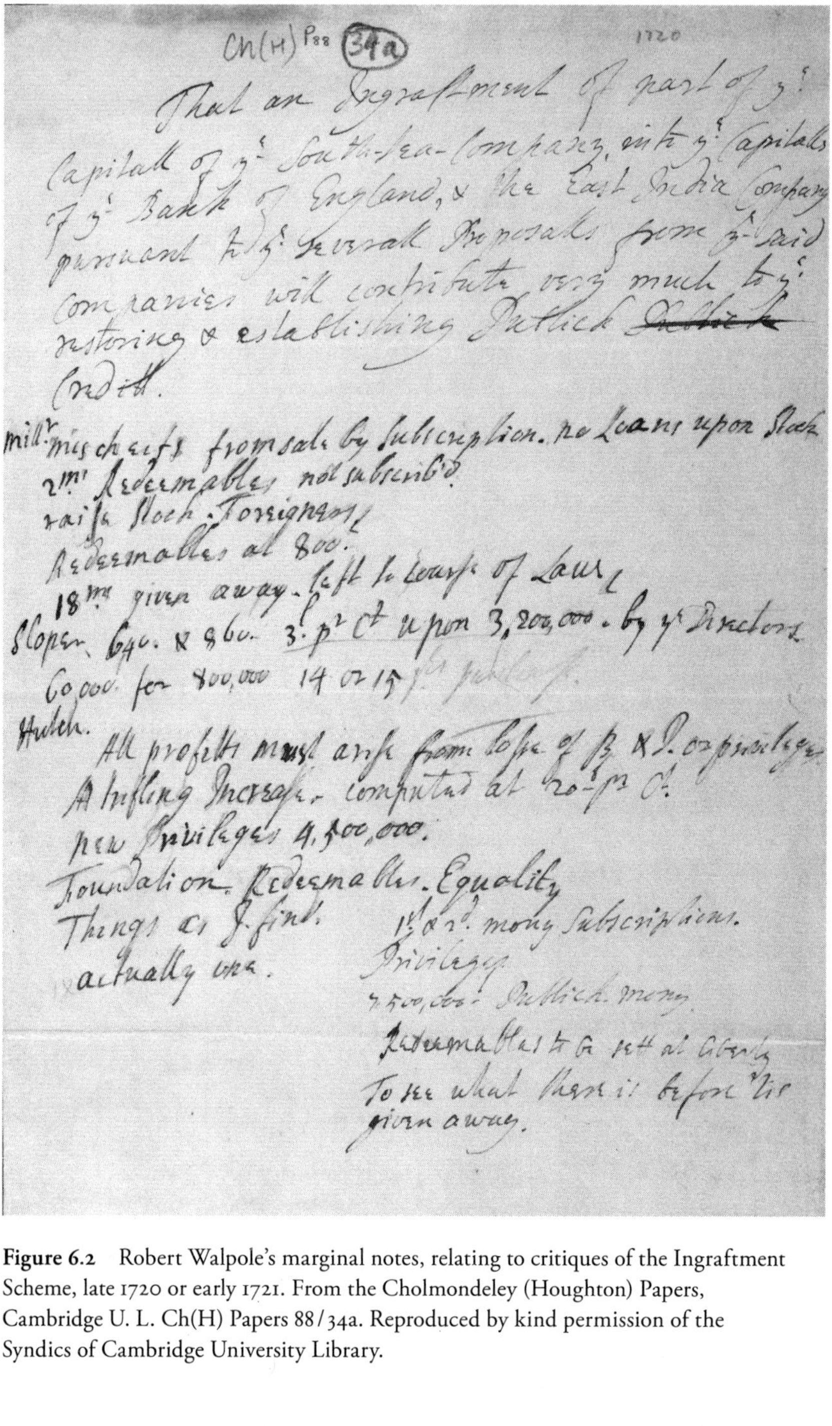

Figure 6.2 Robert Walpole's marginal notes, relating to critiques of the Ingraftment Scheme, late 1720 or early 1721. From the Cholmondeley (Houghton) Papers, Cambridge U. L. Ch(H) Papers 88/34a. Reproduced by kind permission of the Syndics of Cambridge University Library.

showed a particular interest in critiquing his opponent's calculations and figuring out what hidden assumptions lay behind them. For instance, Hutcheson had claimed that, after certain proposed restructuring operations, the value of £100 South Sea Stock would be worth a mere £114.14*s*.2*d*. Walpole pointed out in the margin that: "These Computations are made upon two suppositions 1. that the praemium be paid to the Government 2. that in the present Capital the Subscriptions be Completed." According to Walpole, Hutcheson's conclusion was erroneous, because "both these suppositions are left out by the new s[c]heme." Another marginal comment astutely noted that "in these Computations Mr H makes no allowance for any advantages in trade etc. in the South Sea, nor any advantage from the Bank nor the East India Company. All these several omission[s] will bring the value of £275 present Stock much higher than his imaginary £559.5.10." At times, Walpole's comments revealed an impatient, snarkier side. Referring to one of Hutcheson's analyses about how Stock would be reallocated between the South Sea Company, Bank, and EIC, Walpole noted that the calculation "might have been made by any body of much less capacity and business than Mr. H."[32]

During his manuscript dialogue with Hutcheson, Walpole also used his counterpart's figures to inspire calculative conjectures of his own. At one point, Walpole tried to determine why Hutcheson's computations of the hypothetical value of South Sea Stock under a certain version of the Ingraftment Scheme differed from internal calculations crafted by Walpole's advisers. The difference, he concluded, arose from the fact that his calculators did not assume, as Hutcheson did, that "the second payment of the 2^{d}. subscription [was] pay^{d}." (The question regarded investors who had bought South Sea Stock "on subscription" during the second subscription sale in late April 1720. At issue was whether to assume those investors had paid the second installment payment on their purchases.) Walpole proceeded to recalculate how the prospective value of South Sea Stock would change if he toggled that one assumption: "if both payments are made the Value is £145.08," he wrote, while "if only one payment is made the Value is £146.009." It was only the smallest of monetary changes—just under £1—but this tiny bit of computation reveals the character of Walpole's calculating mind. For one, it leaves little doubt that Walpole had excellent calculating abilities and that he cared about, even relished, numerical details. Even more critically though, it is emblematic of how Walpole used those numerical skills. He used them to imagine the numerical consequences of different decisions or different unknown variables. He used them to model different futures.[33]

Sundry States

In navigating the aftermath of the South Sea Bubble, Robert Walpole relied on forward-looking computational models to think through different policy options. Chapter 5 discussed how Archibald Hutcheson developed the "spreadsheet" model as an instrument of political inquiry. In his models, Hutcheson repeated a single arithmetical procedure multiple times, inputting a range of different values for key unknown variables. Walpole adopted this distinctive style of calculative reasoning as well, but used it as a tool for decision-making instead of critique. While Hutcheson used computational models to make sense of political uncertainties that lay under other people's control, Walpole used them to deal with the various futures he himself might create.[34]

Reorganizing the South Sea Company was an immensely complicated political and economic problem. There were many different people and institutions with a stake in what happened—original shareholders who still held Stock, those who bought into the Scheme early, those who bought in late, those who gave government debts for Stock, those who resold Stock in the secondary markets, plus the government itself, the Bank of England, and so on. Finding some kind of fair solution could seem an insoluble political puzzle. For Walpole, numerical modeling offered a way to make sense of the tangle of different variables, interests, and metrics of success. In May 1721, for example, Walpole received a manuscript calculation from Samuel Hyde, a regular computational adviser. Hyde's dense, three-page manuscript analyzed what would happen if Parliament forgave the roughly £7.1 million in fees the Company had agreed to pay the government back in early 1720. Hyde wanted to determine who would benefit from the £7.1 million refund. One argument against refunding the government fees was that it would benefit the Company's original (pre-1720) stockholders, some of whom had engineered the ill-fated Scheme in the first place. Hyde began by carefully calculating the current value of South Sea Stock, assuming that the government could expect to recover £6.1 million in compensation from the seizure of the estates of the Company's disgraced Directors. As Hyde noted, he drew many of his basic assumptions from "Mr. Hutcheson." Hyde determined the basic value of one share of South Sea Stock to be just over £127.[35]

From this base valuation, Hyde laid out three "suppositions." In each scenario, he calculated how the proprietors of "Old Stock"—those who had bought into the Company in 1711—would ultimately fare. Under the first "supposition," the £7.1 million owed to the government was still demanded. In this base case, South Sea Stock was worth £127 per share. At that rate, he concluded that holders of Old Stock had made a collective profit of over £3.7 million over a ten-year period, even after the Bubble had burst. Given this substantial profit, Hyde concluded, the "Proprietors of Old Stock . . . cannot plead for the Charity of the Legislature." For comparison, Hyde laid out a second "supposition" showing what would happen if the owners of Old Stock were allowed to share equally in Parliament's £7.1 million refund. In that case, the value of South Sea Stock would increase to £151 per share, and the holders of Old Stock would see their profit increase to £6.2 million. Hyde recognized that both of these calculations were unfair in some way. Not everyone who held original South Sea Stock from 1711 necessarily made out well from recent events. They might have bought it in the secondary market well after 1711, or might also have bought "New Stock" at inflated prices later on. Hyde's third scenario explored one rough solution to this quandary: he assumed that half of the owners of Old Stock were "losers" who had been harmed by the South Sea Scheme, while the other half ("not losers") had held the Stock for long enough that they were still net beneficiaries after the Bubble. He calculated what would happen if the "losers" got a share of the government refund, but the original owners of Old Stock—the "not losers"—did not. Hyde did not delve into the logistical challenges of his plan, like how to distinguish "losers" from gainers. But his calculations provided a powerful heuristic nonetheless. By translating economic trade-offs into numerical terms and modeling a set of imaginable political futures, Hyde helped make the enigma of the South Sea restructuring more tractable.[36]

Hyde's three-scenario calculation was hardly the most elaborate example of this kind of projective thinking in Walpole's archive. Especially remarkable was a manuscript paper entitled "A View of the Value of 100 Pounds Capitall in Sundry States of South Sea Stock," also produced in spring 1721, probably slightly earlier than Hyde's calculations (see Figures 6.3a and b). The manuscript showed the output of dozens of repeated calculations assessing various hypothetical values for South Sea Stock, both past and future. The top half of the left-hand page (Figure 6.3a) presented a series of past counterfactuals, showing what the value of South Sea Stock would have been at various moments during 1720, had events in that chaotic year transpired differently. The first line, for example, contended that the Stock would have been worth only £58.19s.10d. per share if the Company had never sold any new Stock to the public. The most optimistic valuation suggested that the Stock could have been worth as much as £542.12s., if

A View of The Value of 100 Pounds

	Value of 100 Capitall Stock

Comp.ª not in Debt, Except to the Publick

If no Subscription had been taken by the Company . . . . 58 : 19 : 10

After the First Money Subscription, and all its payments made — 97 : 0 : 11½

After the Second Subscription, all its payments made . . . . . . 124 : 5 : 0

After the First Annuities Subscribed . . . . 143 : 13 : 0

After the Third Money Subscription, and all its paym.ᵗˢ made . 292 : 6 : 10

After the Second Subscription of Annuities and the Redeemables . . 331 : 9 : 6

After the Fourth Money Subscription, and all its paym.ᵗˢ made . 364 : 13 : 2¼

If The Company had taken all the Debts in, upon the Terms of the Act . . 82 : 19 : 5

If all the Debts had been taken in upon the 1ˢᵗ Scheme, and Capitall compleated at 1000 542 : 12 : 0

If all the Debts subscribed upon 2 Scheme, and Capitall Compleated at 400 . . . . 245 : 10 : 8

If all the Debts had bin taken in 1ˢᵗ Annuities at 375 The Rest at 400, Stock had ⎫
bin made at 300 for money paid on 1ˢᵗ Subscription, and at 400 for money p.ⁿ on 3 last ⎭ 176 : 7 : 6

If Stock at 375, for 1ˢᵗ Subscrib'd Annuities, money p.ⁿ on 1ˢᵗ Subscription at 300. - - ⎫
The money p.ⁿ on, 2. 3. & 4 Subscriptions, The Last Subscrib'd ann and Redeemᵇˡᵉˢ at 400 ⎬ 195 : 19 : 9
no money had bin Lent or Oweing by the Comp.ª and Paym.ᵗ to Publick Remitted ⎭

If the Money Subscribers had p.ᵈ up to 300 & 400 p Cent, The 1ˢᵗ ann'n bin made ⎫
Stock at 375, The Last ann and Redeemables made Stock at 400 p Cent, The ⎪
Comp.ª had bin in Debt 5361182, had Lent no money, and the Publick accepted — ⎬ 204 : 15 : 1½
Stock at 400 p Cent for paym.ᵗ due from the Company . . . . . . ⎭

If The Last Subscrib'd Debts are withdrawn, The Company owe, 5361182, have lent ⎫
11558775 : 11 : 0, And pay the Publick in money . . . . ⎬ 48 : 4 : 10 ⎭

———— and the Publick takes Stock at 400 . . . . 78 : 12 : 6

———— and the Publick Remitts the Payment . . . . 89 : 3 : 5

If The first Subscribed Annuities are made Stock at 375, The Money paid ⎫
upon the four money Subscriptions, The Last Annuities and Redeemables ⎪
made Stock at 300 p. Cent, The Company have Lent 11558775 : 11 : 0, are in ⎬ 166 : 10 : 2½
Debt 5361182, Should Recover 10000000 of the late Directors et al, And — ⎪
The Publick accept Stock at 400 p Cent . . . . ⎭

Figure 6.3a and 6.3b Manuscript calculation from the papers of Robert Walpole (1721), entitled "A View of the Value of 100 Pounds Capitall in Sundry States of South Sea Stock," the Cholmondeley (Houghton) Papers, Cambridge U. L. Ch(H) Papers 88/94a. Reproduced by kind permission of the Syndics of Cambridge University Library.

Capitall in Sundry States of South Sea Stock,

<table>
<tr><td rowspan="2"></td><td rowspan="2"></td><td colspan="3" align="center">Payment to the Publick</td></tr>
<tr><td>Remitted</td><td>in Stock at 400</td><td>In Money</td></tr>
<tr>
<td>Comp.ª have lent
11558778:11:0 upon
Stock &c. and owe
5361182:0:0</td>
<td>2 Subscrbd Ann^d
and Redeemables
at 400 p Cent.</td>
<td>152:8:2¼</td><td>139:3:5</td><td>113:15:5½</td>
</tr>
<tr>
<td></td>
<td>2 Annuities and
Redeemables at
300 p Cent.</td>
<td>136:18:7</td><td>130:17:1</td><td>105:10:1½</td>
</tr>
<tr>
<td>Comp.ª are Rep:
money lent and
owe 5361182:0:0</td>
<td>2 Subscribd Ann^{ties}
and Redeemtes at
400 p Cent.</td>
<td>174:2:1</td><td>165.8.1</td><td>144:19:7</td>
</tr>
<tr>
<td></td>
<td>2 Annuities and
Redeembtes at
300 p Cent,</td>
<td>163:2:4</td><td>155:9:8¼</td><td>135:16:8</td>
</tr>
<tr>
<td>Comp.ᵉ owe 5361182,
have lent 11558778:11:0
and Recover -- -
10000000 of late
Directors et all</td>
<td>2 Subscribd Ann
and Redeembtes
at 400 p Cent</td>
<td>199:16:0</td><td>183:18:7½</td><td>161:5:3</td>
</tr>
<tr>
<td></td>
<td>2 Annuities and
Redeembles at
300 p Cent -</td>
<td>180:1:11</td><td>171:5:7</td><td>149:10:1</td>
</tr>
<tr>
<td>Comp.ᵒ owe 5361182
Rep.ᵈ money lent
and Recover of
late Direct.ᵗˢ et all
10.000.000</td>
<td>2ᵈ Subscribd Ann
and Redeembtes
at 400 p Cent</td>
<td>214:18:5½</td><td>204:4:5</td><td>185:15:11</td>
</tr>
<tr>
<td></td>
<td>2 Annuities
and Redeembtes
at 300 p Cent</td>
<td>201:7:3</td><td>191:18:10</td><td>174:1:6</td>
</tr>
</table>

If 1ˢᵗ Subscribd Ann made Stock at 375; Money p on 1ˢᵗ Subscription at 300, and the money p on 3 last at 400 p Cent.

100 Capit.ˡ Stock

the Company had managed to sell all the new Stock they were permitted at £1,000 per share.[37]

Yet it was the other half of the "Sundry States" manuscript that was especially striking (Figure 6.3b). The figures on that right side amounted to a quantitative decision-tree rendered in tabular form, showing the hypothetical future value of South Sea Stock in twenty-four different states. Along the vertical axis of the table, the calculator laid out eight different scenarios, arising from three different binary variables. The first variable concerned whether or not the Company could expect to recover any money from the Directors' estates. The calculator considered two scenarios: a recovery of £10,000,000 or a recovery of zero. The second variable was whether the Company would recoup the nearly £11.6 million they had made in loans in the previous year. The third variable concerned how the government would retrospectively revise the prices offered to people who had exchanged government annuities for South Sea Stock in summer 1720 at peak prices (£1,000 per share): one option postulated that the price would be dropped to £400, the other that it would be dropped to £300. On the horizontal axis, the calculator considered the effect of one further decision, with three hypothetical outcomes. This concerned the £7 million-plus in outstanding fees the Company owed to the government. Parliament had several options: it could completely cancel the fee, require the Company pay it to the government in Stock, or require the Company pay to it in cash.

Together, the four different unknowns produced twenty-four different scenarios ($2 \times 2 \times 2 \times 3$), arrayed in an eight-by-three grid. Each of the twenty-four entries in the grid represented the value of South Sea Stock under one particular set of choices. The possible valuations ranged from £105.10s.1½d. in the least favorable scenario (the Company recovered nothing from the Directors' estates and none of the money it lent out, reduced the exchange prices all the way to £300, and had to pay the government fee in cash), to £214.18s.5½d. in the most favorable case (when all the assumptions were reversed). By navigating through the "sundry states," a reader could assess the marginal effect that a change in certain assumptions had on the Stock's hypothetical value. The table provided Walpole a quantitative tool to assess the potential impact of his policy decisions—to calculate the political future.

Restructuring the South Sea Company was hardly the last time Walpole and his calculating team would turn to such computational models. For instance, Walpole's calculators crafted a similarly demanding set of calculations to model the consequences of a 1737 proposal by the MP John Barnard to reduce interest rates on public debts to 3 percent.[38] The extraordinary level of computational thinking exhibited by Walpole and his advisers offers vital lessons about both the politics and the epistemology of the eighteenth century. For one, it shows that Walpole's political thinking had a decidedly technical dimension. Trying to reduce Walpolean political practice to the manipulation of patronage and the management of interest groups misses something essential about Walpole's politics, and eighteenth-century politics in general.

Walpole's numerical models, like Hutcheson's, also help reveal a distinctive eighteenth-century attitude toward calculation and the future that has gone unrecognized in scholarship on historical epistemology. Great attention has been paid to the emergence of new ideas about probability in this period, and to how new techniques of probabilistic calculation helped early modern Britons to conceive of the future "as a terrain of calculable risk."[39] Yet "probabilistic" is not the best way to describe forward-looking calculations like the "Sundry States" or Hutcheson's South Sea valuations. Those calculations did not comprehend the different scenarios they calculated as products of chance, and they did not attempt to calculate likelihood. Rather, they were interested in *possibility*, in marking out the space of achievable outcomes and tracking how different political choices might make those possibilities real. This was an active, confident, and largely deterministic imagination of the future, interested in those future possibilities that fell within the bounds of human control. Nowhere was this vision more evident than in Walpole's greatest computational project of all: the sinking fund.

Exponential Aspirations

In the eyes of many contemporaries, Britain's national debt was a cautionary tale about financial invention run amok. The unprecedented debt Britain had amassed since 1688 was contracted in a desultory fashion—an £800,000 loan here, £1 million of annuities there—often on terms that would appear very unfavorable only a few years later. The British government was never at a loss for ideas and projects about how to borrow more money. Much more problematic was how to actually pay that money off, or at least to reorganize those debts into something manageable. The South Sea Scheme was the most notorious attempt to address this latter problem; Walpole's sinking fund was another. While it never produced the drama of the South Sea Scheme, the sinking fund had aspirations that were just as grandiose. Behind the South Sea Scheme had lurked ambitions of conquering global commerce, exploiting the power of public credit,

	Years	Principal (£)	paid from principal (£)	Interest £	s	d
At 10 p Cent	1	50	5	5		
$D[itt]_0$	2	45	5	4	10	
D_0	3	40	5	4		
D_0	4	35	5	3	10	
D_0	5	30	5	3		
At 5 p Cent	6	25	5	1	5	
D_0	7	20	5	1		
D_0	8	15	5		15	
D_0	9	10	5		10	
D_0	10	5	5		5	
			£50	£23	15	

Figure 6.4 A proposed schedule for repaying government tallies, by Wm. Van Laitz. Adapted from Van Laitz, "A Proposal to Ease the Subject," (c. 1696), NLS Adv. MS 31.1.7, f. 44.

colonial trade, and human trafficking. Behind Walpole's sinking fund was a dream of conquering the political future, using the power of mathematical calculation.

Since the late seventeenth century at least, enterprising calculators had tinkered with the idea of using mathematical algorithms to better manage the nation's debt burden. An illustrative early example, from the mid-1690s, was outlined by William Van Laitz in a manuscript proposal surviving in the papers of leading Whig politician and financial policymaker Charles Montagu, later Lord Halifax (see Figure 6.4).[40] Van Laitz explained the vital importance of making the future repayment of government debts transparent and credible. "It is impossible to Restore the paper Credit of the Nation," he explained, "except we can make it plainly appear that a bill which promises to pay 10 pound is Really worth the Value 10 pound or more, and that one Subject is willing to take it from the other in Trade for Debts Due."[41] Van Laitz proposed taking one half of outstanding "tallies" (short-term, unsecured government credit instruments originally denominated with wooden "tally" sticks) and guaranteeing that they be repaid according to a strict ten-year schedule. Instead of paying then-standard interest rates of 8 percent, he suggested the government promise to pay 10 percent interest for the first five years and 5 percent interest for the last five. Van Laitz contended that this would help make those tallies "currant"—equal to their face value—in the marketplace, rather than being traded at a steep discount. Furthermore, the government would end up paying £73.15s., including principal and interest, for every £50 worth of tallies according to Van Laitz's schedule, compared to £90 at the current 8 percent rate. The savings would amount to £325,000 for every £1 million debt outstanding.

Van Laitz's scheme argued that establishing a rigid, transparent, and mathematically grounded procedure for repaying debts was not only sound fiscal policy in the long term, but could actually improve public credit in the near term. Calculating the future could ameliorate the present. The idea of a "sinking fund" took this idea even further. It worked like this: in order to pay off a certain outstanding debt, the debtor—Parliament—would dedicate a certain initial amount of revenue, like some portion of the proceeds of a specific tax, to a "fund" for repaying the debt's principal. ("Fund" here indicated a flow of money over time, not a settled sum or pool of money, the more common modern usage.) Because principal was repaid, the interest in the following year would be lower, and that amount of interest saved would also be contributed to the fund. Each successive year, the principal repayment would increase exponentially. The sinking fund, therefore, harnessed the power of compound interest to decrease debt. As we have seen, the exponential logic that empowered the sinking fund had been at the heart of several of the most provocative calculative projects of the era, from Paterson's Equivalent project to Hutcheson's South Sea valuations to the South Sea Company's own defense of its dividend policies. But while such exponential reckoning was growing in prominence, it still remained unfamiliar and potentially off-putting to many Britons, capable of inspiring both wonder and doubt.

The sinking fund concept had roots in the seventeenth century. One crucial pioneer was Jan de Witt, leading minister in the Dutch Republic (the "Grand Pensionary") from 1652 to 1672 and a key contributor to the mathematical study of annuities. De Witt had instituted a sinking fund mechanism in 1655, as part of a larger overhaul of Dutch public finances. He managed to keep the sinking-fund principle operating for nearly ten years, leading to significant reductions in public debt, until the pressures of war necessitated abandoning the system.[42] In England, the idea of using the exponential power of compound interest to repay debt was contemplated at least as early as 1700, if somewhat indirectly. The conceptual outlines of a sinking fund were evident in a brief, anonymous pamphlet from that year, entitled *A Letter to a Member of the Late Parliament, Concerning the Debts of the Nation,* possibly authored by William Paterson. The pamphlet's key insight came in a brief numerical table projecting how much of the principal of the national debt could be repaid over each of the next six years. The payments grew exponentially larger every year, as if earning compound interest at a rate of 7.1 percent (or £7.2s. "per Cent.").[43] In effect, the author had created a numerical representation of a sinking fund. But that calculator only contemplated one, short example of a sinking fund, and did not explicate the theory behind it in detail. (In 1729, William Pulteney argued that the author of the *Letter*—and not his foe Robert Walpole—was the true, original author of the English sinking fund.)[44] The first British calculator to explore the exponential principles behind a sinking fund extensively was possibly Archibald Hutcheson, in his 1715 *Proposal for the Payment of the Publick Debts*. As discussed in Chapter 4, one of Hutcheson's first "spreadsheet" models was a five-by-four chart showing the number of years it would take to repay a debt of £45 million assuming different initial annual payments and different interest rates, given that payments were made according to sinking fund rules (see Figure 4.1).

Ultimately it was Walpole, though, who made the sinking fund a political reality. He introduced it as part of a larger parcel of fiscal reforms in 1716–1717, during his first stint as Chancellor of the Exchequer. One of his goals was to consolidate many of the government's disorderly income streams into a few specific revenue "funds," based on their fiscal function (the "General" fund was for general use, the "Bank" fund for servicing Bank of England debt, and so on).[45] An even more immediate concern was to restructure the nation's mess of outstanding debts to achieve more flexible terms and lower interest rates. Walpole, with the help of the ever-reliable William Lowndes and other computational assistants, carefully analyzed how much money would be saved to the public by refinancing all outstanding debts to interest rates no higher than 5 percent. Estimates suggested that the total savings to the public could equal £500,000–700,000 annually, just on redeemable debts alone. More complicated was whether it might be possible, with sufficient incentive, to get holders of long, irredeemable annuities to exchange them for new securities at lower interest rates. (The South Sea Company would, of course, propose a particularly enticing solution to this problem a couple years later.) Walpole calculated that, if the irredeemable holders could be convinced to exchange, a reduction of interest rates to 4 percent would save up to another £250,000 annually.[46]

Walpole wanted to put those interest savings to work. In particular, he wanted them "strictly applied towards the discharge of yᵉ National Debt."[47] In a series of manuscript calculations, Walpole speculated about the potent benefits that could be reaped by using saved interest to pay down debt:

250,000$^£$. p^r Annum, w[hich] added to y^e Savings upon y^e redeemable Fonds makes a Fond of above 700,000$^£$ p^r. Annum to be strictly applied towards reducing y^e National Debt, & it is to be observ'd that a Fond of 700.000$^£$ p^r. Annum, will by compound interest reckon'd at 4$^£$. p^r.C^t. & annually only, pay off the present Debt of y^e nation in less than 30 years.[48]

This was the essence of the sinking fund: allowing interest savings to compound year over year. Harnessing the power of compound interest, such a fund could repay Britain's "present Debt" in thirty years. Modern historians have often downplayed the role that calculative thinking played in Walpole's sinking fund plan, depicting it as a "simple, practical and sound" device rather than a truly mathematical idea.[49] But the intricate and imaginative calculations in Walpole's papers show that was not the case. (See Figure 6.5.) Walpole was speculating about the fund's exponential possibilities from the beginning. The sinking fund was an experimental financial technology, a calculative *project,* like Paterson's Equivalent. And like the Equivalent, it required asking Britons to put their political faith in the numbers.

The Politics of "Facts and Figures"

Walpole's sinking fund went into operation in 1717, but neither Walpole nor the public finances reaped much benefit from it at first. Shortly after Walpole instituted his fiscal reforms, he fell out with the Whig party leadership of James Stanhope and Lord Sunderland, resigning his position as Chancellor of the Exchequer in April 1717. With their engineer now on the sidelines, Walpole's plans for conquering the national debt languished, and Whig leaders entertained new fiscal solutions, like the South Sea Scheme. Yet the principle of the sinking fund remained in place as a government policy through the 1720s. In the aftermath of the South Sea Bubble, Walpole and his supporters recast the sinking fund as a reminder of what might have been had Walpole never lost control of Britain's finances.

Between 1721 and 1724, Walpole made great strides toward consolidating his power as Britain's leading minister: he had brought a measure of resolution to the South Sea affair, neutralized key political rivals, appeased the often unruly political interests of the City of London, and instituted new tax policies amenable to a broad range of constituents in city and country.[50] In the process, Walpole and his publicists developed a distinctive style of public argumentation and

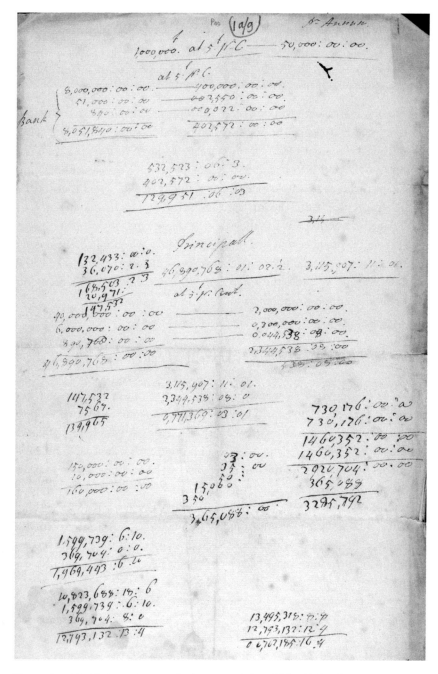

Figure 6.5 Untitled manuscript calculations in Robert Walpole's hand, concerning interest savings on the national debt, c. 1716. The Cholmondeley (Houghton) Papers, Cambridge U. L. Ch(H) Papers 49 / 1a / 9. Reproduced by kind permission of the Syndics of Cambridge University Library.

political epistemology. As historian Simon Targett has argued, the Walpolean Whig political thinkers emphasized the fundamentally prudential character of government. While they acknowledged that human self-interest and corruptibility posed threats to the public welfare, they did not believe that the solution lay in the ceaseless defense of public virtue against moral corruption, as their Country-minded opponents did. Rather, these Court Whig writers emphasized the positive role of government in maintaining political balance and preserving life, liberty, and (especially) property. It was the job of a forward-looking minister to stand above shortsighted partisan squabbling, reconcile competing interests, and seek feasible solutions to advance the public good. While it emphasized prudence and practicality, Walpolean political thinking did not simply amount to "unprincipled" or "empirical" pragmatism.[51] In the 1720s, several Walpolean Whig writers contended that the *practice* of government itself ought to be subjected to principled, theoretical analysis, especially in economic and financial matters. Walpole and his supporters stressed the importance of abstract reasoning—especially mathematical calculation—in elevating policymaking beyond a focus on short-term results and putting it on a more solid, enduring foundation.[52]

The prominent *London Journal,* a key organ of Walpolean propaganda, made a case for this epistemological style in May 1724. Promoting the ministry's recent act for an excise on tea, chocolate, and coconuts, the *Journal* began by bemoaning the corrupt tendency of private interests to derail policies that had been calculated for the public good. It then moved on to explicate the broad fiscal benefits of the new excises.[53] The *Journal* admitted there was no way to know for certain that the unprecedented taxes would produce the revenue that the ministry projected it would. The ultimate produce of those taxes was a *"Matter of Fact,* which *Time* and *Experience* alone can shew to the Conviction of *All."* But the *Journal* explained that this lack of certainty did not discredit the arguments in favor of the new tax scheme, as opponents contended it did. Of course the effects of new and innovative policies could never be known with certainty ahead of time; to insist on such impossible certainty was a cynical partisan maneuver. Policies did not have to be time-tested already to be wise and justifiable, the *Journal* argued. "In the mean while, *Theory* in these Cases, built upon former and long Experiences, and form'd upon just and reasonable *Calculations,* is no very Uncertain or Hollow Ground to go upon." When it came to the new excise taxes, there were very solid grounds for moving forward. *"They* who are known to be the *Best Judges,* and most *Exact Calculators* of such Affairs as concern both the *Merchant* and the *Government,"* the author explained, do "believe and affirm" that the policy was sound and stood to contribute at least £100,000 per year to reducing the national debt."[54]

The epistemological stance taken by Walpole and his propagandists was evidence of the evolving place of quantitative thinking in British politics since 1688. In one way, the *London Journal*'s argument—that speculative calculations could be the basis of sound political knowledge, even if they were not entirely certain—was a familiar one. This was very similar to Charles Davenant's defense of political arithmetic in the 1690s. Davenant, though, had made that argument as a relative political outsider and government critic. He had to justify the value of imperfect calculations because, without access to inside government data, imperfect calculations were all he could muster. Yet Walpole and his supporters were the consummate government insiders, using numerical projections and estimates not to critique state action but to justify it. They still had to work to establish the credibility of their calculations, as there were always political opponents waiting to undercut them. But the fact that Prime Minister Walpole and his supporters so enthusiastically embraced calculation was a mark of how much authority numbers had gained in British political culture. Once a tool primarily for critiquing political power, public calculation was becoming an instrument for managing public sentiment.

For Walpole, it was the sinking fund that demonstrated most clearly how "just and reasonable *Calculations*" could point the way to prosperity and civil concord. In the mid-1720s, Walpole's administration began to promote the value of that forward-looking technology aggressively. At first, the sinking fund seems to have been broadly popular—an easy sell compared to the disastrous South Sea Scheme. In a speech at the opening of Parliament in 1724, King George lauded Walpole's innovative device, proclaiming that "it must be a very great Satisfaction to all my faithful Subjects, to see the sinking fund improv'd and augmented, and the Debt of the Nation thereby put into a Method of being so much the sooner gradually reduc'd and paid off."[55] Soon, though, the sinking fund became entangled with other, more contentious parts of Walpole's fiscal agenda. In spring 1725, Walpole asked the Commons to assume responsibility for debts incurred on the Civil List, the dedicated revenues given annually for the use of the Crown. It was a controversial move, which some worried would undermine Parliament's independence from the monarchy. One such opponent was Walpole's former political ally William Pulteney. Beginning in early April, Pulteney made a series of heated speeches in the Commons, accusing Walpole and his government of waste, incompetence, and corruption, and demanding to see comprehensive accounts on pensions and Secret Service payments made by the Crown. Walpole soon ejected him from positions of executive responsibility. In response, Pulteney joined with other disaffected political figures like Lord Bolingbroke in a growing "Patriot" faction opposed to the Walpole ministry.[56] In biting pamphlets and the famed *Craftsman* newspaper (founded 1726), these Patriot opponents questioned the honesty, ingenuity, and success of Walpole's crafty fiscal management, positioning themselves as the defenders of Country, republican principles against the scourge of Walpole's Court.

In the face of this emerging opposition, Walpole's supporters mounted an aggressive defense of the minister's economic record. One tactic was to stress the theoretical—and quantitative—soundness of Walpole's fiscal policies, especially the sinking fund. The decisive statement came in a pamphlet entitled *An Essay on the Publick Debts of This Kingdom,* published in spring 1726 and widely cited in Whig newspapers.[57] The pamphlet was likely written by MP Nathaniel Gould, a former Bank of England Governor, opponent of the South Sea Scheme, and Walpole surrogate. The avowed objective of Gould's *Essay* was a familiar one: to demonstrate that Britain's national debt had improved over the recent past, thereby showing the effectiveness of the current ministry.[58] (Recall that this was precisely the question Hutcheson and Crookshanks debated in 1718.) In particular, Gould wanted to dispel a "general Suspicion of the Inefficacy" of the sinking fund, no doubt being fomented by Pulteney and his associates.

In order to achieve this, Gould offered calculations—both backward-looking and forward-looking. One section of the pamphlet used past and present public accounting data to show that the national debt had declined £2.1 million since 1717. Gould did not just account for the past, though. Another set of calculations projected the future power of the sinking fund, showing that £50 million worth of principal debt would be repaid by 1756, assuming 4 percent interest. These projective calculations exemplified the Walpolean style of political reasoning. Gould acknowledged that certain key pieces of information, like the yearly tax revenues available to devote to the sinking fund in future years, were necessarily estimates. Yet a well-calculated policy did not require a perfect accounting of every detail. What mattered were the guiding principles that would direct that policy in the future. And, he stressed, the mathematical principles behind the sinking fund worked. In order to drive this argument home, Gould took a didactic approach, carefully instructing readers about the mathematical rules by which the sinking fund operated. By mobilizing the "common Rules for calculating the Increase of Principal Sums continued at Compound interest," Gould explained, one could easily project the progress of any sinking fund at various interest rates and payment levels. Gould even presented a hypothetical calculation showing that, if the nation were to take on £15 million in *new debt* every year for over a century, the sinking fund would assure that the entire balance was still repaid within 105 years.[59]

In the face of such triumphant computational claims, Walpole's opposition took up two different argumentative strategies. One was to dismiss the very nature of the mathematical arguments behind the sinking fund. This was the approach taken by one of the earliest and most inflammatory critiques, *Remarks on a Late Book, Intitled,* An Essay on the Publick Debts of This Kingdom, *in Which the Evil Tendency of That Book, and the Design of its Author, Are Fully Detected and Exposed.* The *Remarks* began by calling Gould's *Essay* "one of the most pernicious Books, that has been published for several years." It went on to explain that whatever success the *Essay* had earned "depends chiefly on the ignorance of his reader," who was exploited by the *Essay*'s "prolix, diffused and abstruse manner" and particularly its "multitude of *Figures* and *Calculations.*" By contrast, the *Remarks* resolved only to argue in plain language, "unless it be in such small sums as are intelligible to the meanest capacity."[60] The critic conceded that the mathematics behind the sinking fund might work as Gould suggested, under ideal circumstances. But the national debt was no mathematical recreation: "This sort of reasoning may serve well enough to try a man's talent at *figures,* and shew to what height interest upon interest may be carried in *Theory;* but I hope that we shall never see it put into practice." The *Essay*'s dizzying figures distracted the public from the gravity of the nation's debt problem, and perhaps even hid a more nefarious design to increase public borrowing.[61]

Other opponents chose not to dismiss the calculated arguments of the Walpoleans, but to respond with numbers of their own. Foremost among these counter-calculators was William Pulteney, who hit back in a series of Parliamentary speeches and technical pamphlets. As Plumb notes, "Pulteney had an excellent head for figures and loved using it." He was especially inclined to use his calculating skills to find faults with others' policies and other politicians.[62] In this sense, he was very much an heir to the Country tradition of calculative criticism fostered by Davenant and Hutcheson. Pulteney felt that Gould's *Essay,* and Walpole's sinking fund more generally, amounted to dangerous projects in computational sophistry, a "confused Jumble of Figures" designed to trick the public. By fixating so heavily on the mathematical properties of the sinking fund, rather than its empirical effects, Gould and company had actually served "to pervert the great and good Ends proposed by it, to the most pernicious Purposes . . . *viz. An Encouragement and Foundation of a Succession of endless Debts and Taxes.*"[63]

In response to the speculative and duplicitous mathematics of the sinking fund, Pulteney offered a different calculative style—based not on mathematical principles, but on steadfast numerical facts and the rigor of accounting. On February 23, 1727, he stated his main conclusion in a heated Commons debate. "Notwithstanding the great Merit that some had built on the sinking fund," he declared, "it appear'd that the National Debt had been increased since the setting up of that pompous Project."[64] Pulteney elaborated on this point in his 1727 pamphlet, *A State of the National Debt.* In order to drive the point home, he invoked double-entry bookkeeping, a "most plain and intelligible" technique. He crafted a series of meticulous "*Debtor* and *Creditor*" accounts that compared the national debt as it actually stood in September 1725 to what that debt would have been had no new debts been contracted since Walpole first took control of the public finances in 1716. The implication was that Walpole, rather than reducing the national debt through his sinking fund, had actually been responsible for adding £7,764,037.15s.0¼d in new debts. For Pulteney, the painstaking numerical precision of his accounts—down to the quarter-penny—provided a defense against the sophistic generality that had come to characterize discussions of the nation's finances in the age of Walpole. "*Facts* and *Figures* are the most stubborn Evidences," he appealed. "They neither yield to the most persuasive *Eloquence,* nor bend to the most imperious *Authority.*" As discussed in the Introduction, Pulteney's spirited comment on the stubbornness of numbers was perhaps the first ever printed use of the phrase "facts and figures"—soon to become an icon of the quantitative age.[65]

Despite his appeal to the plain intelligibility of numbers, Pulteney's actual calculations were not especially clear or easy to follow. His double-entry accounts, presented in a series of appendices, were dense and convoluted. (See, for example, Figure 0.1.) His explanations of those accounts in the body of the text were even more trying, in part because Pulteney clung to an old-fashioned habit of writing out all of the text's numbers in words ("*Seven Millions, seven hundred sixty-four Thousand, and thirty-seven Pounds,*" and so forth). The substance of Pulteney's argument was often highly technical. Because the South Sea Scheme and its clean-up had reshuffled so many of the nation's debts, much of Pulteney's computational work went toward making the 1716 accounts comparable to those for 1725. Many of the criticisms he leveled against Gould's *Essay on the Publick Debts*— like the critiques Hutcheson and Crookshanks had leveled against one another a decade earlier—came down to subtle disputes about accounting method, related to how the *Essay* classified various past debts. Pulteney's unequivocal

"Facts and Figures" were, ultimately, laden with theoretical assumptions and disputable interpretations.[66]

Walpole's calculators were predictably quick to denounce Pulteney's haughty claims to empirical rigor. One manuscript commentator, for example, offered a brief exposition of the "many gross mistakes and impositions" within *A State of the National Debt,* citing double-counted payments and inconsistent valuations as indicative of the "disingenuity, or ignorance of the Author." Pulteney's numerical assertions met with fiery opposition in Parliament, particularly from Gould. During the February 23 debate, that "eminent Merchant" verbally impugned Pulteney's printed calculations, stating that "if he [Gould] understood any Thing, it was Numbers, and he durst pawn his Credit and Reputation to prove that Author's Calculations and Inferences to be false and erroneous." (Pulteney responded in kind, claiming that "he would likewise pawn his Credit and Reputation to make good his Assertion.") The author of the *Essay,* presumably Gould, published his own *Defense of* An Essay on the Publick Debts later in 1727. In order to undercut Pulteney's claims to the numerical high ground, Gould presented a sly series of calculations that showed just how much he and Pulteney actually agreed upon. To his Walpolean opponents, Pulteney's inflexibility in interpreting numbers was a sign that he cared—to mix metaphors—about the letter of the mathematical law and not its spirit. In a particularly memorable barb, Gould lamented Pulteney's state of confusion: "What a Misfortune is it, to understand Arithmetick with no better luck in the Application of it."[67]

The debate between Pulteney and the Walpolean Whigs over the sinking fund offers yet another example of the polyvalent authority of numbers. As in arguments about the Equivalent in 1706, or disputes over the Anglo-French trade balance in 1713–1714, Britons saw different virtues in numbers and had different ideas of what made a good calculation. Pulteney celebrated his "Facts and Figures" for their stubbornness and exactitude; Walpole's supporters celebrated their calculations for the theoretical principles and mathematical imagination underlying them. The fact that the two sides fostered these different epistemological attitudes about calculation meant they calculated in different ways, which further fueled the numerical dispute between them. In this case, Walpole's team generally had the better of things. Through the end of the decade, Pulteney's dogged legislative efforts to rein in Walpole's fiscal policy were routinely defeated.[68] Given the extent of Walpole's political majority, Pulteney was fighting an uphill battle to be sure. But it also seems that Pulteney's punctilious, bookkeeping style of public calculation (not to mention his ponderous speeches) was just not quite as effective as the bolder, more evocative, and more forward-looking style being fashioned by his Walpolean opponents.

"Freer and Happier Every Year"

In the late 1720s, Walpole and his publicists fashioned the sinking fund into a political icon, symbolic of Walpole's genius and prowess as a leader. Against claims by Pulteney, they stressed that Walpole himself was the fund's sole inventor, and that it was "unthought of, unheard of" before Walpole implemented it in 1716.[69] The talismanic status of the sinking fund gained official sanction from Walpole's Parliament. Shortly after King George I's death in 1727, the House of Commons presented their *Humble Representation of the House of Commons to the King* to George II, representing Parliament's ostensibly official view on the nation's finances. The sinking fund earned a starring role. In fact, the Parliamentary authors claimed that "had this Method . . . been further pursued and without Interruption, the Dangerous and Mischievous part of the late *South-Sea Scheme* might have been avoided." The *Representation* even entertained some aggressive, futuristic mathematics of its own. It argued that the £800,000 that had been added to the annual sinking fund since 1717, if treated like a perpetual annuity valued at twenty-five years purchase, "Makes a real Profit to the Publick Amounting to Twenty Millions." The sinking fund became such a potent symbol of the efficacy of the Walpolean state that writers sought to justify other, unrelated public policies by drawing analogies to Walpole's great fiscal invention. A 1727 pamphlet advocating a hawkish military policy toward Spain chose the telling title: *Great Britain's Speediest Sinking Fund is a Powerful Maritime War, Rightly Manag'd, and Especially in the West Indies.*[70]

Empowered by the sinking fund, Walpolean propagandists made ever bolder proclamations about the future benefits to be wrought by Walpole's brilliant leadership. One of the most extraordinary political pamphlets of the era was a 1728 offering entitled *Remarks on the R—p—n [Representation] of the H—of C—ns to the K—g; and His M—y's A—s—r. Address'd to All True Britons.* The author described Walpole's policies in exceptionally glowing terms, even by the hyperbolic standards of the day. "The sinking fund is deservedly a darling Project," the author wrote, "glorious for the Contriver, and happy for his country; on which all our other National Happiness chiefly depends." That praise was little compared to what was to come, as the author pivoted from the present to the future.[71] "Permit me, for once," the author wrote, "to personate a future Historian, who, in the advanced Ages of Time, some three hundred Years hence . . . he'll be obliged to give the World the honest, candid Account following." The hypothetical future historian of 2028 described how Walpole, "one of the greatest Men that ever *Britain* bred," had met the financial challenges of his era. He "found Means to discharge the [nation's debts] with Honour, and put them into such an annual Course of Payment, that they have been all paid off above two Centuries ago." Walpole's (future) retirement was described as a moment of great joy for the Prime Minister, "having had the satisfaction to see his Country grow freer and happier every Year by the sinking fund, the Establishment of which was wholly owing to himself."[72]

The hagiographic future historian in the *Remarks on the R—p—n* conjured a remarkable vision of political time. It imagined a grand and distant future, but also a precise and measurable one—three centuries in the future, more than several lifetimes but not yet an eternity. And it did not just envision a future, but a *future past;* it traveled far into the distant future to look back at the near future. This was an orderly, predictable, calculable view of future time— exponential discounting in narrative form. It was a vision of the future that was made imaginable by the mathematical armature of the sinking fund. The optimistic, even arrogant, futurism expressed by Walpole and his publicists in the late 1720s marked a new departure in the history of political thinking. Intellectual historians have observed that the financial transformations of the seventeenth and eighteenth centuries, especially the growth of public debt, had a dramatic effect on contemporary conceptions of temporality. J. G. A. Pocock has observed how some Britons during the financial revolution understood themselves to be living in a time when "war and money have speeded up the operations of society" and thus "political behavior [was] based upon opinion concerning a future rather than memory of a past." In particular, Pocock notes that "the growth of public credit obliged capitalist society to develop as an ideology something society had never possessed before, the image of a secular and historical future." In Pocock's account, the novel future wrought by the politics of British debt was often an anxious one, in which future generations were sold out for the benefit of present interests and left to reckon with the catastrophe destined to arise when the debts came due.[73] But there was another side to the proverbial futuristic coin. The obverse of such negative prognostications of debt-fueled political doom was the positive mathematical foresight of Walpole and his sinking fund. The Walpolean Whigs looked forward to a controllable future that could be made orderly through calculated policies and strong leadership.

By defining the government's theoretical ability to repay its debts, the sinking fund quantified and concretized the elusive notion of "public credit."

In the 1720s, calculation entered government. The state had not originally been the prime mover in the emergence of Britain's quantitative age, which had been driven by the work of calculating outsiders trying to call the government itself to account. Walpole learned from this combative calculating culture through his own harrowing political experiences, like the "Missing Millions" affair, and through engaging with the work of critical calculators like Hutcheson in the public sphere. He subsequently brought those quantitative lessons into the internal routines of the state. In doing so, Walpole and his calculators crafted new techniques of policy analysis within the state and a bold, forward-looking style of public rhetoric. At the same time, the future entered government as well. The calculations upon which Walpole relied were not simply notional calculations, crafting a gestural future. The calculated future that Walpole and his advisers constructed in their internal deliberations—the future exemplified by the "Sundry States" model—was labor-intensive, carefully measured, and technically intricate. It relied on mathematical reason to connect current government action to future political outcomes. The complementary future that Walpole and his propagandists projected in the public sphere—the future of the sinking fund—was more imaginative. But it, too, relied on a belief that the state could control the nation's political and economic trajectory, with the help of mathematics.

Robert Walpole's political calculations challenge us to rethink our understanding of the history of the future and how it has been governed. As noted at the beginning of this chapter, scholars have identified many ways in which forward-looking thinking was on the rise in eighteenth-century culture. Yet others have suggested that the future did not become an object of direct *political* control, computation, and contestation for centuries. Timothy Mitchell has recently argued that this kind of calculated future entered government only in the aftermath of World War II. That new future was organized around a new conceptual object, "the economy," understood not just as a static aggregate but as a "dynamic set of forces" that was "capable of unlimited growth." This new economic future was made thinkable and tractable by new calculating technologies, like the price deflator and the logarithmic plot, which became embedded in the reasoning practices of Western states.[74] The future calculated by Walpole and his colleagues prefigured the economic future Mitchell describes in important ways. It was an exponential future, but one in which the acceleration of future time was understood to be manageable and predictable. It was a future built out of specific, iterative computational technologies—the repeated reckonings of the "spreadsheet" model and the projective mathematics of compound interest—that were directly employed in governmental practice. It was also a future that emerged alongside a newly confident, expansive, and assertive vision of the power of the state. There were of course myriad differences between those futures as well, particularly the radically different nature of "government" and extent of governmental power in the two moments. The purpose of pointing out the homology is not to assign Walpole points for his novel contribution to modernity. Rather, it is to suggest that the calculated future has a longer and richer history. Understanding how the future was governed in an earlier past may help us better understand its problematic governance in later times.

The Future of Walpole's Sinking Fund

The story of Robert Walpole's calculations has a telling epilogue. The future he calculated did not, for the most part, come to pass. His sinking fund did not conquer the national debt. Almost three hundred years later—approaching the projected moment when the Walpolean "future Historian" was to write—the British national debt stands around £1.8 *trillion*. The immediate reason Walpole's darling project failed was Walpole himself. He simply was not very interested in following through with the strict discipline his fund demanded. Beginning in 1727, Walpole rerouted around £100,000 per year designated for the sinking fund to increase the Civil List granted to the new King George II, despite vehement criticisms that such an action was a sacrilegious encroachment on the blessed fund. In 1729, he raised a new debt of £1.25 million, dedicating the sinking fund revenues to cover the interest. He repeated that maneuver multiple times in the ensuing years. By 1735, the fund had been mortgaged entirely, all of its income used to cover interest on new debts. The fund survived only as an administrative device, an "accidental Appendix only to the Office of the Treasury." So a mere seven years after Walpole's advocates were predicting three hundred years of national prosperity on the back of the sinking fund, that fund was defunct, destroyed by its own inventor.[75]

Retrospective observers have told the story of Walpole's misbegotten sinking fund in two ways. On the one hand, many eighteenth-century economic commentators argued that Walpole had in fact come up with a brilliant way to deal with the nation's debt, and lamented that he did not have the political virtue to carry it through. Ironically, this actually became Pulteney's position by the mid-1730s.[76] Later advocates for Walpole's abandoned scheme included the century's great commercial historian Adam Anderson and, most famously, Richard Price (on whom more to come).[77] Writing in the 1760s and 1770s, both Anderson and Price campaigned for the fund to be revived and lamented how much sounder the nation's finances would have been had the fund been allowed to work its exponential magic. Walpole's betrayal of his sinking fund hardly dampened long-term enthusiasm for that financial technology, which became highly popular with late eighteenth-century finance ministers including Jacques Necker in France, William Pitt the Younger in Britain, and Alexander Hamilton in the United States.[78]

The second way observers recount the story of Walpole's sinking fund is as a kind of calculated bluff, aimed at calming creditors and generating confidence in the public's finances. The fund's underlying mathematical principles and long-term projected effects, some suggested, were a matter of marketing. Indeed, this seems to have been the opinion of David Hume—who, like Price, will get much more attention in Chapter 7—and his friend Adam Smith. "Sir Robert Walpole endeavoured to shew that the public debt was no inconvenience," Smith explained in one of his University of Glasgow lectures in 1763, "though it is to be supposed that a man of his abilities saw the contrary himself."[79] Modern historians have reiterated the notion that Walpole "never expected the whole National Debt to be redeemed" and instead "persevered with the sinking fund for political as much as for financial reasons."[80]

It is hard to know exactly what to make of Walpole's betrayal of the sinking fund—whether it was an unplanned or opportunistic lapse in discipline, or rather part of some longer-standing plan to manipulate the public by promising one thing and doing another. Whether or not Walpole ever intended to follow through with the sinking fund, one thing is clear: he did believe, in some way, in the possibility of the calculated future that the sinking fund evoked. Adam Smith was wrong to assume that Walpole could not have put any real faith in his extravagant exponential calculations. As Sir Robert's remarkable numerical archive shows, Walpole invested great authority in such numbers. But Walpole understood that it took action to make a calculated ambition into a reality, and whether he chose to take such actions came down to a different political calculus.

Walpole also understood that he was living in a quantitative age. He knew that numbers had power, not just to model his own political options but to shape how his fellow Britons imagined their collective future. Not only did Walpole believe in the calculated possibility of the sinking fund, he believed that the public would believe in that future as well. This is a remarkable testament to how much authority calculation had gained in British public life in the ensuing decades. By the 1720s, the forward-looking computations associated with Walpole's sinking fund became a foundation not only for a battery of fiscal policies, but for an entire political mythology, three centuries of imagined prosperity. (Indeed, if Walpole somehow meant the whole thing as a trick, that would only be more telling—the sinking fund was a political gambit based not just on Walpole's own faith in calculation, but on his faith in the *public's* faith in calculation.) For some observers, though, the fact that Walpole's fiscal future did not come to pass as he had calculated was a disconcerting sign that numbers could lie, that perhaps they had been lying all along. If Walpole's sinking fund marked the apotheosis of Britain's new quantitative age, the failure of that project marked the beginning of a new chapter in that story as well. The eighteenth century was not just an age of quantification, but also of its discontents.

Figures, Which They Thought Could Not Lie: The Problem with Calculation in the Eighteenth Century

ONE OF THE MOST striking testaments to the authority numerical calculation had attained in British politics appeared in 1735. The source was a typically fiery pamphlet with a typically bland name: *Some Considerations Concerning the Publick Funds, the Publick Revenues, and the Annual Supplies, Granted by Parliament*. Amid a series of numerical arguments about the state of public finance, the author excoriated his opponents for their low and libelous tactics. In particular, he accused them of propagating fraudulent numerical claims:

> I have learn'd that the Poison has spread itself through the Nation, and that honest and very well-meaning Persons, when they saw a Representation of Facts so call'd, cloath'd in the Dress and Appearance of Calculations and Figures, which they thought could not lie, have been stagger'd.[1]

There was nothing new about political actors seemingly stretching the numbers to fit their political interests. But the exasperated author was not simply complaining about being the victim of computational calumny. What concerned this author most was that his opponents' numerical lies seemed to be widely believed. Many "honest and very well-meaning Persons" had somehow come to think that "Calculations and Figures . . . could not lie."

The author of this statement was no innumerate reactionary. It was Robert Walpole, the most calculating of all politicians, writing in response to his nemesis

William Pulteney. Walpole had not given up on calculation altogether by 1735, though recent political frustrations like the embarrassing defeat of his "Excise Scheme" may have made him somewhat weary of the political numbers game.[2] Rather, Walpole's decades of experience had made him a keen observer of his nation's civic epistemology. In 1735, he observed something that others would soon recognize as well: the endless succession of calculations inundating British public politics had steadily changed how citizens reacted to numerical arguments. Many Britons seemed to trust numbers simply because they were numbers.

The first six chapters of this book have traced the proliferation of creative quantitative thinking in British politics in the half-century after the Revolution of 1688, of which the inventive calculations by Hutcheson and Walpole in the 1720s were outstanding examples. This efflorescence of political numbers only continued through the remainder of the eighteenth century. Enterprising calculators extended political calculations into new social and geographic settings—city and country, metropolitan Britain and its overseas colonies—and into new public problems—agriculture, disease, drunkenness. Showdowns over financial politics, like the disputes over the management of the Seven Years' War (1756–1763) and subsequent clashes over taxation in the American colonies, continued to generate a substantial volume of printed numbers. Leading statesmen followed Walpole in relying heavily on numerical calculations to analyze policies and shape public perceptions.

The story of how calculation attained such a privileged place in British civic epistemology is as much about numerical consumption as production, though. By the middle decades of the eighteenth century, reverence for "facts and figures" had spread far beyond the few individuals who did the calculating. It extended to a much broader audience of politically engaged, literate Britons, who did not have great computational skills themselves but who might glance at a table of calculations in a newspaper or hear fiscal figures cited in conversation. Understanding how cultural products, including calculations, were received by the "broader public" is a notoriously difficult challenge for historians. Nonspecialists who might have nodded approvingly at political calculations about trade balances or the national debt left relatively few traces of that activity.[3] Yet varied evidence justifies the conclusion that, on the balance, the British public had come to grant special authority to numerical calculation by the second half of the eighteenth century. Number-laden political texts, like David Hartley's *The Budget* and Richard Price's writings on American independence, became legitimate bestsellers. The calculative practices honed in the public-financial controversies of the post-1688 era also began to appear in a broader range of cultural spaces. Quantitative culture went local as various calculators, from concerned clergymen to curious physicians, turned numerical tools upon their own communities. Those new calculative practices also began to leave their mark on a wider range of public media, from introductory textbooks and reference works to popular literature and graphic satire.

At the same time, Britons increasingly observed and commented on their fellow citizens' esteem for quantitative arguments. Perhaps the clearest evidence that Britons had come to invest new authority in numerical calculation was that some of their countrymen began to worry that they trusted in numbers too much. Among the first to voice this concern about the public's numerical gullibility were political calculators like Walpole, frustrated by how easily the public seemed to buy into the (inferior) numbers produced by their political opponents. This was a powerful sign that mid-century calculators envisioned their audience far differently from calculators a half-century earlier. Turn-of-the-century arithmeticians like Davenant, Arbuthnot, and Paterson hustled to convince readers to care about numbers at all. By the 1750s, calculators could take for granted that their public audience not only cared about the numbers, but found something about numbers immediately persuasive.

This fear of the public's numerical gullibility was just one of several new concerns about calculative thinking that arose in mid-century. Did the new appetite of statesmen for numerical information threaten the liberties of citizens? Did the proliferation of numerical texts produce "information overload" that hindered real commercial understanding? Was calculation becoming an especially pernicious form of political deceit? As we saw in the Introduction, early modern Britons had long held suspicions about calculation, seeing it as pedantic, ungentlemanly, even occult. These traditional criticisms did not disappear in the eighteenth century (or today). But in the mid-eighteenth century, a second generation of numerical anxieties arose alongside them—that there were too many numbers, that they had become too powerful, that the public put too much faith in them. These were fears that only made sense once numerical thinking had *already* become deeply entrenched in political practice and public culture. Collectively, these new mid-century concerns constituted a new, modern critique of calculation.

This new wariness about numbers sometimes manifested itself in passing quips and impatient asides. In 1769, for example, an anonymous writer in the *Critical Review*, a popular periodical with Tory and Anglican leanings, offered a telling remark while reviewing a recent pamphlet by William Knox. Frustrated by what seemed like an endless and inconclusive exchange of figures about colonial fiscal policy, the author concluded that Knox's text "furnishes us with fresh reasons for observing a political scepticism in all finance-matters that are determined by facts and figures."[4] Others, like David Hume, felt that the nation's problem with numbers merited more than just a passing comment. For Hume, Britain's numerical fixation was not just a bad habit, but a deep failure of the nation's collective political thinking. In the course of his varied political and economic essays, as well as his mammoth *History of England,* Hume laid out an extended critique of political calculation. He argued that his fellow citizens failed to appreciate the severe limitations of numerical reasoning: that it was often inaccurate, usually inconclusive, and easily manipulated. He coined a memorable motto for this numerical skepticism—and one of the epigraphs for this book—in a 1752 essay critiquing the balance of trade: "Every man, who has ever reason'd on this subject, has always prov'd his theory, whatever it was, by facts and calculations."[5] Hume felt that Britons' reckless reverence for numbers helped perpetuate dangerous generalizations about political economy, including an obsession with trade balances and complacency about national debt. It also left Britons susceptible to calculated political deceptions, like Walpole's sinking fund.

As the second half of the eighteenth century saw new concerns about the peril of political calculation, it also gave rise to new, and even more profound, beliefs about calculation's political possibilities. No one spoke more zealously for the transformative power of numbers than the "good" Dr. Richard Price, a Welsh minister, moralist, and mathematician whose remarkable résumé included authoring the foundational text of modern actuarial science and "the most famous British tract" on the war with the American colonies.[6] If Hume offered the era's most incisive attack on calculation, Price offered its most ambitious defense. Price, like Hume, was worried about the state of contemporary political thinking. But he looked back on the rise of political calculation, particularly the Country calculating tradition that flourished in the decades after 1688, not as a problem but as an unfulfilled promise. In the 1770s, Price made the promises of quantitative thinking a cornerstone in his ambitious system of political philosophy. A consummate rationalist, Price believed that the ability to reason, unrestrained by passion or ignorance, was fundamental to liberty, whether in individual morals or national politics. In politics, he felt that mathematical calculation, fueled by accurate fiscal accounts, was the purest form of public reason. The practice of political calculation was, in Price's mind, essential to civil liberty.

The contrast between Hume and Price serves as a fitting place to end this history of quantification and its discontents. The two were both deeply concerned with problems of civic epistemology and were astute witnesses of how political knowledge was made in the eighteenth century. Each commented extensively on the computational conflicts that make up this book. Hume reflected on the balance of trade debates, Price collected a history of Parliamentary public accounting, and both mused about debt and sinking funds. Each, in a way, had his own story about Britain's escalating love affair with numbers: for Hume, it was a story of deception and decay; for Price, a story of a dream unrealized. The two positions they staked out would come to represent a defining tension of the quantitative age.

Calculating, through Most of the Considerable Parts of the Kingdom

During the period from the end of Walpole's ministry in 1742 to the 1770s, the culture of calculation forged in the uncertainty of post-1688 politics settled into a permanent feature of British political life. Using numbers to make political arguments and to evaluate collective problems became normal, pervasive, and expected. There is evidence that Britain's culture of political calculation was growing along various dimensions: the size of its audience, the extent of its presence in public media, the range of communities who engaged in it, the diversity of problems to which it was applied. Most broadly, it seems likely that across this period, the overall number of Britons able and interested to read public calculations increased. In her study of the social uses of political arithmetic, Joanna Innes writes that "the number of those competent to grasp the point of such enquiries probably grew" over the eighteenth century.[7] By the middle of the century, calculators came to directly comment on the existence of a numerate audience interested in political calculation. Take, for instance, William Allen, who drafted a 1736 pamphlet on *Ways and Means to Raise the Value of Land* (reprinted in 1742 as *The Landlord's Companion*). Allen included an estimate of the annual produce (£9.5 million) and total value (£13.4 million) of the nation's livestock, a calculation he included "for the Satisfaction of my Readers who delight in Political Arithmetick."[8]

Calculative reasoning was also becoming more prevalent across various media. Around 1700, political calculation had often been its own best advertisement; the growing pile of numerical pamphlets about taxation, trade, and debt signaled to casual readers that numbers mattered. Over the course of the eighteenth century, political calculation began to be discussed and celebrated in an array of other genres and cultural contexts. One such context was education. The booming mathematical textbook industry gave increasing attention to the political uses of numbers. Edward Hatton's 1721 *An Intire System of Arithmetic,* reissued in 1731 and 1753, included an entire chapter on "Political" arithmetic, between sections on "Sexagesimal" and "Logarithmical" arithmetic. Works of political calculation also took up positions of prominence in reference works and libraries. Articles defining political arithmetic and summarizing key findings were included in new compendia of knowledge, like John Harris's *Lexicon Technicum* of 1710 and Ephraim Chambers's famed *Cyclopedia* of 1728. Publishers reissued pioneering computational works, most notably the five-volume *Political and Commercial Works of That Celebrated Writer Charles D'Avenant* (1771). The editor, MP Charles Whitworth, credited Davenant's writings with forming the "foundation of our political establishment." Quantitative writings became a common part of private collections. William Allen noted, for example, that the calculations of Davenant and Gregory King were "in the hands of many political readers."[9]

Creative writers and artists also began to integrate political numbers into new forms of literary expression. Exemplary was Daniel Defoe's 1722 novelistic history *A Journal of the Plague Year,* which used numerical tables of mortality data to structure a fictionalized account of the devastating 1665 epidemic. Literary scholars debate whether the *Journal* reflected Defoe's optimism or pessimism about the capacity of numbers to make sense of pressing human problems. Indeed, not all of the new attention given to quantification was positive. Most famously, Jonathan Swift satirized calculative reckoning in his 1729 *A Modest Proposal,* which outlined a mock project—bolstered by arithmetical arguments—to solve childhood hunger in Ireland by encouraging poor Irish parents to sell their children as food to the wealthy. That such calculations became the subject of parody was a sure sign of their cultural influence.[10] As will be discussed later, the rage for calculation in political life even became a target for one of the era's quintessential cultural media: the graphic satire.

At the same time, political calculation was being practiced in an increasingly wide range of geographic locations. Around 1700, political calculation had been predominantly a metropolitan activity, centered on London and, to a lesser degree, Edinburgh. (This fact did not escape William Allen, who suggested that predecessors like Davenant and King might have avoided major mistakes "if they had travelled through most of the considerable Parts of the Kingdom.") But over the course of the eighteenth century, calculative culture also took hold in smaller communities. Calculation went local. Physicians, clergymen, antiquarians, and agricultural reformers began to gather new data on the people, prosperity, and produce of the countryside. These local calculators introduced new granularity to quantitative inquiry, using numbers to explore the differences between parts of the nation and to map what Innes calls "the distribution of happiness and pain across the social body."[11]

Local calculators also played a key role in two of the most significant computational pursuits of the later eighteenth century: the dispute over the nation's demographic trajectory and the development of life-expectancy tables for use in insurance. The question of whether or not Britain's national population was expanding or contracting was among the most heated calculative controversies of the century. Many saw population as a key metric of national strength and governmental efficacy. In 1721, the French cultural commentator Montesquieu touched off an international debate on depopulation when he argued, in his *Persian Letters,* that European populations were one-tenth what they had been in ancient times. For Montesquieu and like-minded observers, depopulation testified to numerous ills in modern life—militarism, arbitrary government, religious repression, luxury, urbanization, and colonialism. Whereas Montesquieu's *Letters* contained only the most haphazard numerical evidence, subsequent debate became intensively quantitative, especially in Britain. Many philosophical and computational luminaries would weigh in, including Hume, Price, and Arthur Young.[12]

In Britain, the depopulation debate was energized by new practices of local quantification. In lieu of a national census, population debaters had to do what many calculators before them did: use creative calculations to make a case from patchy data. Calculators fortified well-worn data sources, like the London Bills of Mortality, with new and creative forms of evidence, like local surveys and window tax returns. In the 1750s, Reverend Richard Forster surveyed his and eight neighboring parishes in Berkshire in order to count what proportion of houses paid the window tax. He then applied that benchmark ratio to scale up national window tax figures and form an estimate for national population. Around 1780, John Howlett, a vicar from Essex, gathered parish register data from local clergy to formulate his population analyses, while William Wales, an astronomer and mathematics instructor in London, created a printed demographic questionnaire that he circulated as far away as Carlisle in northern England.[13]

Both Howlett and Wales were on the optimistic side of the population debate, and they crafted their own calculations to respond to pessimistic ones offered by Richard Price. Price was not only an influential combatant in the depopulation controversy, but also authored one of the era's most enduring pieces of local political calculation: an updated set of life-expectancy tables based on 4,689 deaths recorded in the town of Northampton. He developed the table for the Society for Equitable Assurances on Lives, the first company to apply probability mathematics to the life insurance business. Price's Northampton table would remain the single definitive actuarial table available for decades, until Sun Life's actuary composed a new one in 1812—based on data that had been meticulously collected in the town of Carlisle by a local physician, John Heysham, between 1779 and 1787.[14]

The individual who best exemplified this circulation of computational thinking between city and countryside was Arthur Young. He quite literally followed Allen's directive to travel "through most of the considerable Parts of the Kingdom." Born in London in 1741, Young spent some of his precocious early years trying his hand as a political pamphleteer and novelist, before embarking on a three-year experiment managing a family farm in Berkshire, where he took a keen interest in agricultural improvement. In 1767, he set off on the first of a series of "Farmer's Tours," collecting unprecedented numerical data on the land, livestock, and people of rural Britain. This numerical information became the foundation for novel analyses of farm productivity, local variations in rents and wages, and national income. Young kept up extensive correspondence with international agricultural reformers, carried out his own experiments on agricultural techniques, and extended his agrarian fact-gathering trips to Ireland, France, Spain, and Italy. Over his career, he published dozens of pamphlets and helped found an influential new journal, *The Annals of Agriculture* (1784–1815). In 1793, he was appointed the first secretary of the new Board of Agriculture.[15]

Computational culture not only took root in the countryside in England, but in more distant regions as well. In 1729, for example, Arthur Dobbs's *Essay on the Trade and Improvement of Ireland* offered quantitative analysis of Irish commerce based on customs data and his own local surveying. Notable numerical advances came in Scotland as well. In the mid-1750s, the first systematic estimate of the Scottish population was calculated, an innovative exercise designed by clergyman Alexander Webster, using questionnaires and life-expectancy tables. In 1761, Scotland received its own series of commercial statistics, as Scottish trade data began to be separately reported in the Customs House ledgers.[16] Even more striking was the pivotal role that calculation came to play in eighteenth-century reasoning about Britain's overseas colonies. Britain's makeshift imperial administration developed a steady interest in gathering demographic, commercial, and geographic data about the colonies (though it never employed political arithmetic as a tool of colonial engineering in quite the way William Petty had envisioned in the seventeenth century). Shortly after it was founded in 1696, for example, the Board of Trade set off on "huge burst of census-taking" in the American colonies. The Board attempted demographic surveys encompassing all Britain's colonies in 1721, 1731, 1755, and 1773, though with mixed success. Beyond this administrative brand of colonial quantification, Britons increasingly turned to calculation to conceptualize, evaluate, and critique Britain's inchoate empire. A 1721 report composed by the Board of Trade used a variety of creative calculations to assess the value of various colonies to British commerce, emphasizing that colonial trade accounted for one-third of Britain's shipping trade.[17]

Over the century, colonial questions also attracted keen attention from public calculators in the print media. For example, the mid-century period saw a slew of computational pamphlets on the costs and benefits of the West Indian sugar plantations. A 1738 pamphlet by John Bennet, lobbying on behalf of the Caribbean enslavers, included "several calculations" showing the supposedly desperate state of the sugar trade, including a pessimistic calculation of "the real intrinsick Value of the Land and Stock of Barbadoes." Roughly twenty years later, the prolific calculator Joseph Massie argued against the sugar planters' interests by calculating the exorbitant profits made by the West Indian sugar monopoly (£8 million over thirty years, he estimated).[18] Exercises in political arithmetic appeared in Britain's overseas colonies themselves, including a famous 1755 exercise entitled "Observations Concerning the Increase of Mankind," which made provocative projections about the exponential population growth of North America. Its author was Benjamin Franklin.[19]

As calculators extended numerical techniques to new geographic spaces, they also extended them to new questions. One such area, exemplified by Arthur Young's arithmetic, was agriculture. Another was health. Mortality and morbidity were among the oldest objects of political calculation. John Graunt had organized evidence on fatal illnesses in the 1660s, while the publication of mortality figures during seventeenth-century plague epidemics was one of the earliest examples of public calculation about a social phenomenon. In the eighteenth century, "medical arithmetic" began to flourish as a specialized practice, as calculators sought to quantify the frequency and danger of diseases, to investigate the efficacy of certain remedies, and to interrogate new models of disease causation, like the neo-Hippocratic correlation of health and environment. One especially prodigious medical calculator was James Jurin, a secretary to the Royal Society and ardent Newtonian, who published a series of quantitative investigations about smallpox inoculation in the 1720s. Many readers reported to Jurin that they were convinced of the value of inoculation by his diligent figures. Jurin's other pursuits included gathering data on mortality in provincial areas, developing a network of observers to collect local weather data, and treating Robert Walpole with an experimental remedy for bladder stones (one that may have proved lethal).[20]

Calculation also took an increasingly sociological and, indeed, moral turn in the middle and later eighteenth century. As Joanna Innes has masterfully shown, calculators increasingly sought to quantify "social 'happiness'" by turning their calculations upon poverty, employment, education, hunger, crime, and vice. One particularly striking example of such social-moral calculation was the analysis of alcohol abuse. Two flurries of such calculation arose in 1736 and 1751, triggered by legislative proposals (ultimately successful) to place discouraging taxes and regulations on the sale of gin. Moral reformers, many of whom were clergymen, crafted quantitative pamphlets to support these anti-vice policies. Citing Davenant and Petty as inspiration, they attempted to demonstrate that gin use decreased population and to quantify the total costs of drunkenness on the nation.[21]

Over the course of the eighteenth century, political calculation became more dispersed and more diverse: geographically, socially, topically. It would, of course, be wrong to suggest that this was simply a process of diffusion, from the "center" of political and intellectual power outward to more distant sites. Different communities fostered their own reasons to calculate, and made their own, often innovative contributions to quantitative practice. Scholars have been particularly attentive to the ways that colonial settings and the imperatives of colonialism fueled calculative effort.[22] It would also be wrong to suggest that the forms of public quantitative thinking discussed in this book—that is, political and predominantly public calculations focused on economic and financial problems—were the sole driver of quantitative enthusiasm and creativity in the eighteenth century. The mid-century period gave rise to new kinds of calculating personalities, like doctors, clergymen, and agricultural reformers, who drew ideas and inspiration from their own fields. There were certainly a variety of historical forces at play driving the rise of quantitative activity in these different domains, not just public politics.

Nonetheless, the quantitative civic epistemology that arose earlier in the century—driven largely by partisan concerns and organized largely around questions of public finance—did energize, inform, and condition this more diverse and expansive calculative culture in important ways. Innes argues that arithmetical innovations in the later part of the century moved beyond older questions of "national power." True, but these new calculations were still discernibly *po-*

litical, and the influence of the kinds of calculative practice discussed in this book can be seen in several ways. Many of the influential calculators who introduced quantitative thinking into new domains in the mid-century period had direct connections to the adversarial world of Parliament, parties, and the pamphlet press: James Jurin worked for Robert Walpole; Arthur Young dabbled in party polemics early in his career; Richard Price was a famed publicist for radical political causes as well as a demographer. Many new calculative projects developed in the period, like the attempts to put a value on the sugar trade or to calculate the public costs of drunkenness, were directly aimed at influencing Parliamentary policies. Stylistically, computational "representations" and "examinations" analyzing population growth in the Americas or smallpox inoculation maintained the antagonistic tone characteristic of earlier pamphlets on excise taxation and the national debt.[23] Calculation remained an instrument of dispute. When thinkers later in the century turned to calculation to advance an argument about a social question, and did so in the belief that such calculative arguments bore certain virtues and would attract a certain response from public audiences, they were building upon decades of civic-epistemological groundwork. The post-1688 tradition of using numbers to win public arguments remained alive and well—now in churches and hospitals, towns and villages, near and far.

The State of State Calculation in the Mid-Eighteenth Century

Calculation remained alive and well within the centers of state power, too— though not in ways always obvious to subsequent generations. As retrospective observers have noted since the early nineteenth century, the mid-eighteenth century yielded few durable, institutional changes in how the British state managed and used quantitative information. By comparison, the first two decades after 1688 had produced the Commission of Public Accounts, the first regular series of Treasury accounts, and the office of the Inspector-General of Imports and Exports; the turn of the nineteenth century would prove even more fruitful, yielding the Board of Agriculture (1794), the first national census (1801), standardized government accounting procedures (1820s–1830s), and the General Register Office (1837), among others. Despite the surge in enthusiasm for reforming public accounting after 1688, relatively little progress was made in increasing the rigor and professionalism of state accounting procedures through much of the eighteenth century. Administrative historians have largely echoed the verdict of John Sinclair, who complained in 1790 that "since the reign of Queen Anne the national accounts are far from being distinguished for their regularity and precision." Balanced annual accounts of government finances were not consistently available for public view until 1823.[24]

Yet practices of numerical calculation did play an indispensable role in British statecraft through the eighteenth century—if not in the disciplined, institutionalized form that would come to be characteristic of modern bureaucracies. This was certainly the case in Walpole's administration, as we saw in Chapter 6, and would continue to be so for his successors. Leading ministers throughout the period sought the counsel of skilled calculators for their internal deliberations and their public engagement. William Pitt the Elder relied on Joseph Massie, George Grenville on Thomas Whately, and the Earl of Shelburne on Richard Price. Some prominent ministers in the era were exceptionally well-known for their calculating spirit. Henry Pelham, Prime Minister from 1743 to 1754, was known as an uninspiring orator in large part because his speeches were, in one scholar's words, "filled with facts and figures relevant to the subject under debate." The divisive and disagreeable George Grenville, Prime Minister from 1763 to 1765, was derided by some contemporaries as "the financier." According to one (perhaps apocryphal) story, he was once at a concert and became so fixated on convincing a neighbor about the merits of "some grand fiscal scheme" that he began writing out details on paper atop the piano.[25]

While Grenville's fondness for numbers may have been especially zealous, many mid-century statesmen accumulated numerical data and commissioned calculations to help analyze policy questions. The surviving archives of politicians like Charles Townshend and the Earl of Shelburne are riddled with quantitative material. The precocious and volatile Townshend (1725–1767) was among the most influential political figures of the era of the Seven Years' War. An MP for two decades, he held positions on the Boards of Trade and of Admiralty and finally as Chancellor of the Exchequer (1766–1767). Shortly before his untimely death he effected his most (in)famous achievement: the so-called "Townshend Acts," which levied taxes on a series of goods, like paper and tea, imported into Britain's American colonies. Townshend was a prototypically calculating statesman. He collected numerical data about the national debt, the sinking fund, historical budgetary surpluses, and customs and excise taxation, as well as various technical reports on topics ranging from the history of Parliamentary supply proceedings to barrel gauging to "The French Debt in 1762." Like Walpole, Townshend accumulated this material both "for his own use" in internal deliberations and also use in public numerical contests in the pamphlet press.[26] For example, he collected several different manuscripts commenting on and critiquing Thomas Whately's technical 1767 pamphlet *Considerations on the Trade and Finances of This Kingdom,* written in defense of the previous Grenville administration. At least one of the surviving manuscripts includes Townshend's own running commentary on the pamphlet.[27]

The kinds of numerical information statesmen collected reflected broader trends in numerical culture at large. Townshend's papers include ample numerical information about food, reflecting new political attention to questions of agriculture and hunger. This alimentary information included a manuscript entitled "General Accounts of Corn Consumed," showing the amounts of barley, oats, rye, and wheat grown, consumed, exported, and imported, as well as a printed text showing the prices of wheat and malt at Windsor dating from 1646 to 1745, a project started by William Fleetwood. Such time series data was representative of a mounting interest in historical data and in quantifying change over time. Townshend's papers include numerous examples of such chronological data series, including lengthy series on government budgetary surpluses (1718–1760) and the produce of the Customs (1711–1765). The latter paper on customs concluded with a page of manuscript comments identifying trends in the data and offering key historical context, an indication that Townshend and his advisers were actively thinking through customs data.[28] An especially remarkable example of such chronological quantification survives in the archives of the 2nd Earl of Shelburne, an influential Whig politician who also happened to be a grandson of the political arithmetician William Petty (and was, in fact, named William Petty himself). Over his lengthy career, Shelburne held positions as Secretary of State (1766–1768), Home Secretary (1782), and Prime Minister (1782–1783). In his library, Shelburne possessed a massive, elaborately bound folio volume containing balance-of-trade accounts for dozens of Britain's trading partners. Countries were arranged alphabetically, from Antigua to Virginia. For every region, data was given for each of nine different historical years, from 1716–1717 to 1759–1760, showing import-export figures, key articles traded, bullion movements, and other details.[29]

Calculation undoubtedly remained central to the reasoning practices of Britain's political leaders. As noted, though, the numerical practices of statesmen did not necessarily translate into enduring institutional changes. Several factors were at play. For one, ministerial leadership during the second half of the century was quite volatile, particularly between the death of Pelham (1754) and the rise of Lord North (1770–1782). That period witnessed eight different ministries led by seven different individuals. Such turnover was not conducive to building durable institutions. Yet there were also deeper, political-cultural reasons why the British state did not—or, rather, could not—undertake more formal and visible quantitative projects during this period. As we have seen throughout this book, reverence for numerical information and calculative thinking within British politics had often been driven by suspicions about centralized governmental power. Skeptical citizens wanted numerical information about their government; but they did not necessarily want their government to have extensive numerical information about them. As ministers like Walpole came to appreciate the political and administrative value of numbers, and internalized calculation as a state tool, some citizens came to worry that a state too well-informed might be a danger to civil liberty.[30]

In some critical ways, then, the political dynamics that produced Britain's quantitative age actually worked against the expansion of state quantitative power. This helps to explain one of the most oft-cited quantitative failings of the eighteenth-century British state: the abortive efforts to establish a national census. In the early 1750s, a coterie of interested doctors and calculators began to agitate for a general reform of the procedures used in registering births and deaths. Leaders included John Fothergill, an Edinburgh-trained physician and writer on epidemic diseases, and James Dodson, a master at Christ's Mathematical School and prominent financial mathematician. Desirous for better data on which to ground life insurance and annuity projects, these reformers sought to standardize how births, marriages, and deaths were recorded at the parish level. The cause attracted support from key politicians, including the Duke of Newcastle, then a Secretary of State, and the MP Thomas Potter, who pushed to extend the project into a more radical national census scheme. In fall 1753, Potter introduced a Parliamentary bill drafted by Customs official Corbyn Morris, formerly one of Walpole's propagandists and a prolific political calculator. Under the proposal, population figures would be gathered by an annual household survey taken by local overseers of the poor, who would pass their findings to parish clergy and other local notables, who would collate and deposit them with the Board of Trade. Supporters boasted that a national census would aid in the more equitable imposition of taxes, the raising of armies, the administration of poor relief, and the regulation of immigration.[31]

The Census Bill provoked a spirited debate in Parliament beginning in late 1753, reproduced for a wide readership in the *Gentleman's Magazine*. The bill provoked various concerns. Most dramatic was a religious fear that the project would violate the Biblical prohibition against King David's sin of "numbering the people." More consequential, though, were political arguments against the census, inspired in large part by Country sentiments. Many Britons feared that the census, like Walpole's ill-fated Excise Scheme in the early 1730s, would foster an army of prying government officials who would violate citizens' privacy and upend the social deference to which upper-class Britons felt themselves entitled. Some also feared that such mass enumeration would reveal vital or embarrassing information to Britain's foreign adversaries, including the nation's susceptibility to smallpox. The most vocal spokesman for the opposition was MP William Thornton. A national census would certainly not advance the national interest, he exclaimed, but rather the interests of various national enemies—"the *Spaniards* and the *French*" abroad, "Place men and Taxmasters" at home.[32] While supporters tried to downplay the Census Bill's ambitions, suggesting that it was simply intended "to gratify . . . curiosity," Thornton remained skeptical. Curiosity was no reason to jeopardize the nation's liberty and security. As Thornton asked: "We are to entrust petty tyrants with the power of oppression . . . to subject every house to a search; to register every name, age, sex, and state, upon oath; record the pox as a national distemper, and spend annually 50,000 pounds of the public money—for what?—to decide a wager at *White's!*" Potter's Census Bill ultimately passed in the Commons but was defeated in the Lords, not to be revived for fifty years.[33]

"Ye Best Things, When Corrupted, Become Ye Worst"

The failure of the national census project did not reflect public indifference to "facts and calculations," but rather a mounting public concern about them. Britons recognized that numbers *were powerful*—and by the mid-eighteenth century, they were beginning to realize that this power brought problems. The threat of state surveillance by "petty tyrants" was only one such danger. Britons also began to worry about the overload of numbers being produced, about the new kinds of numerical tricks being played by calculating politicians, and especially about the quantitative gullibility of the public.

Often these critiques were voiced by calculators themselves. By the 1750s, some of the most devoted political calculators were grumbling about how the public's numerical credulity had made honest calculation harder. One of those frustrated calculators was Richard Forster, a participant in the population debates. Forster was optimistic that Britain's population was growing and published two papers in the Royal Society's *Philosophical Transactions* to that effect in 1757. He aimed to rebut William Brackenridge's recent claims that the London Bills of Mortality showed Britain's population plateauing in 1728 and falling after 1743. Forster questioned Brackenridge's unpatriotic conclusion and his methodology, particularly his exclusive reliance on data about births and deaths. In contrast, Forster tried to develop a more accurate estimate of living Englishmen based on local sampling and data about window tax receipts. As the dispute progressed, Forster grew impatient, unable to silence an opponent who misused data and abused the public's quantitative credulity. "No Man Living has a deeper Sense of ye Merit of Mathematical Argumentation than myself," he wrote in a 1760 manuscript. "But then we must not go too far; well knowing, that ye best things, when corrupted, become ye worst. And it appears to all considering People, that this sort of Reasoning has been abused." It was not just Brackenridge he was worried about, but a broader pattern of numerical recklessness. Too often, it seemed, numbers were "applied to improper Subjects" or "to proper Subjects indeed, but upon wrong or scanty Principles."[34]

At the same moment, the era's licentious quantitative practices were also beginning to wear upon Joseph Massie. Based in London, Massie served as an economic adviser to the elder William Pitt, who was effective head of government from 1756 to 1761. During that period, Massie produced over twenty pamphlets, many defending Pitt's fiscal policies and his management of the Seven Years' War. Massie seized upon public calculation as a way to disavow the public of dangerous misunderstandings fostered by Pitt's opponents. For example, Pitt's adversaries had spread the pernicious rumor that Britain's amalgam of land and consumption taxes had become so arduous under Pitt's watch that some wealthy Britons were paying "Twelve, or Fourteen Shillings in the Pound" (60–70 percent) of their incomes in taxation. Massie responded to this charge in a 1756 pamphlet, *Calculations of Taxes for a Family of Each Rank, Degree or Class.* Through a careful calculation, he showed that effective tax burdens were merely half those rumored: roughly £334 for a wealthy landowner earning £1,000 yearly. For Massie, this delusion about tax burdens was symptomatic of a mounting problem. The public was taken in by unsubstantiated numerical claims far too easily. "Though asserting one or other of these Things doth not prove them to be true," he wrote, "yet . . . such Assertions have the same Weight as Proof, with all such Persons as believe them; and the Effects wrought thereby, in the Minds of such Persons, will be the same in Quality and Degree, as if they were produced by Fact instead of Fiction." To a sympathetic audience, baseless assertions cast in quantitative form could appear as established facts, and function like facts once they began to circulate in public conversation.[35]

Through his pamphlets, Massie hoped to raise the level of computational discourse. He published data-rich pamphlets and urged his opponents to engage in fair, open, and constructive numerical fights, rather than resort to selective numerical sniping. (One economic historian calls Massie a notably "responsible" pamphleteer who "stands out from the general press of pamphleteers both in ability and assiduity.")[36] He went even further in his 1760 treatise *A Representation Concerning the Knowledge of Commerce as a National Concern*. Noting that commerce as a "Branch of Knowledge, still continues at a very low Ebb in this Kingdom," Massie laid out a broad program for reforming how the nation thought through commercial questions. He was particularly concerned with how Britons made use of empirical facts and figures. The nation did not just need more commercial data, Massie felt, but better data. In fact, the frenzy of political calculation in preceding decades had already produced an overwhelming supply of material. As it stood, "acquiring a Knowledge of Trade as a national Concern" was forbiddingly laborious, "for a Man must first collect Fifteen Hundred, or more, commercial Books and Pamphlets." But most of this overload of information was nothing but noise. "I much doubt," Massie wrote, "whether the national and valuable Part of the commercial Matter in them, will more than fill One Folio Volume of the larger Sort." Instead, he proposed a precise catalogue of sixteen particular facts he considered essential to understanding each branch of Britain's commerce and industry, including eight key numerical metrics: materials costs, export figures, the average time taken in various steps of production, and so on. Such a systematic database of commercial numbers would help to identify critical unknowns about the nation's trade and serve to discipline government record-keeping. Crucially, by providing authoritative numerical benchmarks, Massie's database would help resist the proliferation of "false Facts" and "erroneous Calculations."[37]

Forster and Massie were primarily concerned about the public abuse of numbers by their calculating colleagues. For other critics, the most troubling problem came when state authorities took advantage of their privileged access to information and citizens' numerical reverence to manipulate public opinion. Those concerns were voiced in one of the most successful political pamphlets of the period: David Hartley's *The Budget* (1764), an attack on the authoritarian fiscal policies implemented by Prime Minister George Grenville in the wake of the costly Seven Years' War. Grenville had imposed a host of new taxes and strict commercial regulations on Britain and its colonies (including, in 1765, the notorious "Stamp Act" imposed on the North American colonies). The budget figures Grenville published in 1764 suggested this austerity was working: by cutting wasteful expenses and increasing tax yields through anti-smuggling and other measures, the government had brought revenues above expenses and repaid £2.8 million of debt.[38]

A tenuous coalition of Tories and centrist Whigs supported Grenville's authoritarian approach. Many others fiercely opposed it. For some, Grenville's attempt at "oeconomy" was penny-wise but pound-foolish. A contemporary graphic satire lampooned Grenville, "the Great Financier," for imagining that such fractional savings could balance out the nation's £140 million in debts. The accompanying verses quipped: "Such wonders our Grand Financier can dispense, That he'll pay off ten millions by saving ten pence."[39] More radical Whig critics warned that Grenville's oppressive policies threatened civil liberties. David Hartley, the polymathic son of an influential philosopher (also David), became a leading voice of this radical opposition. In *The Budget*, Hartley cited several metrics showing the nation was in a graver economic condition than when Grenville had arrived: stock prices had fallen 10 percent, interest rates had risen, and the annual revenues directed to the sinking fund had declined £250,000. In the tradition of calculating outsiders like Davenant and Hutcheson, Hartley used deft arithmetical analyses to show that the administration's boasts—like its claim that raising customs on tea had improved revenues by £391,000 annually—were implausible.[40]

Hartley was not just bothered by Grenville's fiscal policies, but also by his use of "the budget" as an actual numerical object. Hartley rebuked the haughty administration for having "condescended, by an advertisement in the public papers, to explain the *Budget* to the meanest capacity," and for having assumed that "the wonders of the *Budget* must need make the ignorant start, and admire the transcendent talents of the ministry, who have advertised such miracles."[41] Hartley resented how the Grenville administration tried to exploit its privileged control over fiscal data to manipulate a seemingly "ignorant" public into believing it had performed fiscal miracles. While Hartley explicitly chided Grenville and his publicists for thinking so little of the British people, his pamphlet betrayed an implicit anxiety that, in fact, many were likely to be taken in by Grenville's numerical tricks.

Hartley's *Budget* proved a great success, a "runaway bestseller" that went through seven editions—another clear indication of the British public's appetite for quantitative fodder. The pamphlet's publisher, the radical Whig John Almon, remarked in August 1764 that "it is amazing how it has opened the eyes of the public."[42] One critic was distressed to admit that Hartley's text had become so popular "that in almost every Company the first Question is—*Have you seen the* BUDGET?" One measure of *The Budget*'s influence was the fevered reaction it generated from Grenville's supporters. An anonymous pamphlet entitled *An Answer to* The Budget derided Hartley's offering as "absurd and ill-written," and disrespectful to "his Majesty's faithful Servants." What bothered the author of *An Answer* most was the fact that the public welcomed Hartley's technical arguments so "*mechanically*," even though "not above one Man in a hundred understands what is meant by the Word BUDGET." *An Answer* actually called such lay readers "Ignoramus's," suggesting that such "People much better mind their Business, and let *State Affairs* alone." Other critics expressed a similar sentiment. *The Wallet: A Supplementary Exposition of the Budget* spent twelve pages trying to dismantle Hartley's calculations, noting that "there may be many too weak to see through the flimsiness of the author's reasonings." The most robust counterattack came from Thomas Whately, a secretary to the Treasury and one of Grenville's most influential economic advisers (as well as an amateur Shakespeare scholar and a prominent commentator on gardening). His masterful pamphlet, *Remarks on* The Budget, deployed a battery of creative computational tools against Hartley. Whately, too, worried about Britons' blind numerical faith. In fact, he suggested that one of the key reasons he needed to publish his own calculations was to help enlighten those members of the public who simply "admit the Truth of Calculations, and Conclusions, rather than bear the Trouble of examining them." Hartley's foes were afraid of the same thing Hartley was—namely, that the general public might be too willing to believe the numbers thrust upon them by crafty calculators.[43]

The most striking thing about the dispute over *The Budget* was what it revealed about how the calculating combatants had come to imagine their public audience. By the 1760s, calculators were no longer especially concerned that readers might simply dismiss or distrust their own calculations, as calculators in Davenant's generations had been. Rather, calculators were newly worried that those readers might be too inclined to trust the numbers wielded by their opponents. This was remarkable testimony to just how much authority numbers had come to garner in British civic epistemology. For many political calculators, of course, lamenting the public's numerical gullibility was just another part of the evolving numbers game. Hartley and his critics did not wish for the public to dismiss all numbers, just those inferior, unsophisticated, and deceptive ones peddled by their adversaries. But at least one eighteenth-century observer thought the problem with Britons' rage for calculation went far deeper.

Heliogabalus's Cobwebs; or, David Hume's Problem with Calculation

David Hume needs little introduction. Born 1711 in Edinburgh, Hume was the greatest luminary of Scotland's Enlightenment, renowned for his transformative contributions to numerous areas of intellectual inquiry, including epistemology, ethics, political economy, aesthetics, and history. Hume was also a keen observer of his own contemporary moment, particularly of the habits of thinking that prevailed in eighteenth-century public life. Hume felt that Britons' automatic reverence for numbers indicated the troubled state of British political thinking. His contemporaries failed to appreciate the limitations of calculation, leading them to put too much faith in numerical generalizations and leaving them vulnerable to quantitative casuistry. His critique of calculation, developed across a series of political essays and his famed *History of England*, constituted one of the sharpest accounts of both the power and the pitfalls of numerical reasoning in the eighteenth century.

Hume's skepticism toward calculation combined his theoretical commitments concerning epistemology, political philosophy, and political economy with his own personal experiences and observations of British politics. Hume maintained close ties to the worlds of commerce, finance, and government throughout his life. In 1734, he served as an apprentice to a sugar merchant in Bristol. In the late 1730s, he befriended a circle of Scottish political and economic leaders in London, including Archibald Campbell, founder of the Royal Bank of Scotland. His close friends and correspondents included leading Edinburgh merchants and financiers, economically minded politicians like James Oswald of Dunnikier, and influential commercial thinkers like Josiah Tucker and Adam Smith. Hume relished what he learned from such connections. In 1744, for example, he complimented Oswald for his genius in explaining "the whole Oeconomy of the Navy, the Source of the Navy Debt; [and] many other branches of public Business." He even made his own forays into public administration and accounting. From 1746 to 1748, after being rejected for a professorship in Edinburgh, he took a position as personal secretary to Lieutenant General James St. Clair. He traveled with St. Clair on diplomatic missions around continental Europe, offered observations about other nations' fiscal policies, and even kept the army's financial books.[44]

Beginning in the early 1740s, after publishing his *Treatise of Human Nature* (1738) and at the end of Walpole's ministry, Hume began a sustained investigation into the politics and culture—indeed, the civic epistemology—of his own time. The first products of that endeavor were a series of essays, the earliest published in 1741. Many of the most influential, particularly on political economy, appeared in his well-received *Political Discourses* in 1752. The second key output of that project was his *History of England*, a massive, six-volume enterprise completed between 1754 and 1761, which, even more than his philosophy, made Hume famous in his own day. While he observed British politics with a critical eye, he was no radical reformer. He appreciated the relative stability of British political life since the Hanoverian Succession and especially since the rise of Walpole. In fact, scholars have long debated whether Hume was essentially a political conservative (and indeed a Tory), cynical about the need or possibility for real political change. But there was one area of political life where Hume definitely thought improvement could be made: the way his contemporaries thought through political problems.[45]

In Hume's mind, the nation suffered from a collective failure of political reasoning. Unable to properly evaluate knowledge-claims, Britons had become reckless in their political thinking. Hume was frustrated, for instance, by his fellow citizens' failure to understand the relationship between "*particular* deliberations and *general* reasonings." Contemporaries were often far too quick to extrapolate general conclusions from particular instances. Such carelessness was evident in the hyperbolic tone that dominated partisan polemics. Britons of all parties tended to interpret specific policy errors or the failings of individual politicians as grand constitutional trials. "Those who either attack or defend a Minister . . . always carry Matters to Extremes," Hume explained in his 1741 essay "That Politics May be Reduc'd to a Science." "His pernicious Conduct, it is said, will extend its baneful Influence even to Posterity, by undermining the best Constitution in the World." An exemplary offender was Bolingbroke, whose screeds in the *Craftsman* claimed Walpole was destroying Britain's hallowed constitution. Such myopic thinking was especially evident in matters of "*commerce, luxury, money, interest,* &c.," as Hume explained in the opening essay of his 1752 *Political Discourses,* "Of Commerce," which began with an extended discussion of the shoddy reasoning that dominated discourse on commercial and economic affairs.[46]

To a degree, these failures of political and economic thinking reflected more general tendencies in human understanding. As he explained in his epistemological writings, Hume believed that humans had a natural capacity for generating probable knowledge in the face of uncertainty, but that they regularly ran into trouble in doing so—misevaluating the quality of empirical information, failing properly to "proportion [their] belief to that evidence," and forming unjustified and prejudiced rules on the basis of specious analogies. They were thus far too willing to accept extraordinary claims without adequate justification—whether reports of religious miracles or politicians' grandiose proclamations about the nation's commerce, credit, or constitution. But while such erroneous thinking was common, it was not incurable. Through his writings, Hume hoped to foster a more circumspect and epistemologically modest style of political deliberation that avoided overgeneralization, shortsightedness, and hyperbole.[47]

Britons' troubled patterns of political thinking brought real dangers, Hume believed. Because politicians and the public had become so rash in their reckoning, British politics became a histrionic game of accusation and adulation played for the affections of ill-informed citizens. Citizens became increasingly vulnerable to the manipulations of skilled performers, of whom Walpole was the most skilled of all. (Hume viewed the guileful Walpole with a mix of fear and admiration: "As I am a Man, I love him; as I am a Scholar, I hate him; as I am a Briton, I calmly wish his fall.") Ambitious political leaders like Walpole were best served by stretching the facts and making utopian promises, rather than engaging in a sound and measured discussion of "the general course of things." This was particularly true in matters of finance and commerce, where highly misleading ideas—like the primacy of the "balance of trade" and Walpole's sinking fund—had taken hold because they played well with public audiences, not because they explained what was truly in the public interest.[48]

The contemporary fondness for calculation was a key indication of the poor state of the nation's collective political thinking. Rather than serving as a useful counterweight to unsubstantiated speculations, Hume argued, numbers fueled Britons' tendency to make dangerous generalizations. Throughout his political essays, Hume showed a persistent skepticism toward quantitative thinking. He resisted placing undue weight on numerical data in his own arguments. When he did cite a number, he often rounded it off in a deliberate performance of epistemological modesty. In his essay "Of Luxury," for example, he claimed that Louis XIV kept in his pay "above 400,000" men and then added a footnote stating an inscription in the Place-de-Vendome gave the number at 440,000— as if to point out how little credit he gave to that number's precise value. At other times, he relished identifying contradictory numerical evidence and calling out writers who fell prey to numerical absurdity. In the opening of his essay "Of Public Credit," while talking about the tendency of ancient governments to store up treasure in prosperous times, he offered an extended note explaining a discrepancy between Plutarch and Quintus Curtius over whether Alexander the Great had seized 80,000 talents or 50,000 talents when he captured the city of Susa.[49]

At other times, Hume was more deliberate about his intolerance for numerical enthusiasm. This was most evident in his lengthy "Of the Populousness of Antient Nations," his salvo in the population debates. Hume's goal in that text was to discredit the notion, propounded by Montesquieu and Scottish clergyman Robert Wallace among others, that the world had been more abundantly populated in ancient times. His method was telling. Unlike many combatants in the population debates, he did not resort to direct enumeration. He felt that numerical data available from ancient sources were too "uncertain" and "imperfect"; so, too, were recent exercises "in computing the greatness of modern states." Numbers simply were not going to settle the matter: "Many grounds of calculation, proceeded on by celebrated writers, are little better than those of the emperor *Heliogabalus,* who form'd an estimate of the immense greatness of *Rome,* from ten thousand pound weight of cobwebs, which had been found in that city."[50]

Instead of venturing his own numerical argument, Hume chose to cast doubt on all the bombastic demographic data commonly cited from classical sources. As he bluntly put it: "With regard to remote times, the numbers of people assigned are often ridiculous." Much of the second half of the essay was filled with snarky rejoinders about ancient calculations, like one particularly egregious claim from Appian: "Julius Cæsar according to Appian, encountered four millions of Gauls, killed one million, and made another million prisoners. . . . No attention ought ever to be given to such loose, exaggerated calculations." Hume admitted that there was a certain "temerity" in this "critical art" of debunking ancient calculators without offering anything constructive in return. But he argued it was a necessary defense, because "the license of authors upon all subjects, particularly with regard to numbers, is so great." He noted at least one inherent weakness that made quantitative evidence especially unreliable. "All kinds of numbers are uncertain in antient manuscripts, and have been subject to much greater corruptions, than any other part of the text," he explained, "and that for a very obvious reason. Any alteration in other places, commonly affects the sense or grammar, and is more readily perceiv'd by the reader and transcriber." Numbers were too prone to errors of transmission. While context helped to identify misspelled or mistranslated words, miscopied numbers easily went unnoticed. (Recall the many miscopied accounts of the Equivalent in Chapter 2!) This charge could be leveled against modern numbers as well as "antient manuscripts." It was an observation that ran directly counter to Crookshanks's claim that numbers "never lose their Denomination and true Value."[51]

Hume did not limit his critiques to ancient numbers. More recent generations also attracted his ire, as evident in his *History of England*. The nation's political history was a heated issue in mid-century politics, as contemporary politicians relied upon retrospective judgments about past leaders in building their own political narratives. One particularly contentious figure was King Charles II. Whig-party mythology, built around an image of 1688 as a triumphant salvation of the nation's constitution, required that Charles and his brother, James II, be represented as corrupt, wasteful tyrants. But Hume felt that such grandiose accounts had led to Charles II being unfairly treated in the historical record. Hume observed that Charles's later Whig critics often relied on flawed quantitative arguments. For example, a common argument was that Charles had been flush with revenues and had only run into fiscal problems through his own mismanagement. The year 1675 offered a good illustration. "Several historians have affirmed, that the commons found . . . the king's revenue was 1,600,000 pounds a year, and that the necessary expence was but 700,000 pounds." But, Hume continued, there was no documentary evidence for this claim, "and the fact is impossible." Hume's *History* did not just call out the numerical excesses of fellow historians. He called out historical politicians, too—like the Brooke House Commission. In the late 1660s, that short-lived body had reported that Charles II had let £1.5 million of public money go unaccounted for. Hume countered with a lengthy footnote showing £1.5 million worth of errors and uncertainties in the Brooke House report.[52]

The abuse of numbers had only gotten worse in Hume's own day. He felt that reckless, inconclusive, and deceptive calculations had generated misunderstanding about key political-economic issues. For example, in his gloomy essay "Of Public Credit," which warned of the dire, long-term consequences of public indebtedness, Hume chided the notion that the national debt could be managed—let alone harnessed for the public good—by mathematical fixes. Hume singled out Archibald "Hutchinson" and his mid-1710s "scheme" to "make a proportional distribution of the debt amongst us" as one indication of that quixotic attitude. Even more problematic was Robert Walpole and his calculated sinking fund, which perpetuated the dangerous "paradox" that public debts were not only sustainable but beneficial to the nation over time—a deceptive notion that, Hume speculated, Walpole used as a political ploy and did not even believe himself.[53]

Fondness for (mis)calculation also fueled his contemporaries' preoccupation with trade balances. In his influential "Of the Balance of Trade," Hume tackled the common fear that a nation with a negative trade balance would ultimately see "all their gold and silver . . . leaving them." He offered a novel explanation for why this "very groundless apprehension" did not materialize: in a nation that exported more than it imported, the specie flowing inward would raise prices at home, including for exports; foreign demand for those expensive goods would wane, reducing exports, while domestic appetites for cheap imports would increase, correcting the balance. Hume's "specie-flow" model has been credited as a foundation of modern monetary theory and one of the first conclusive challenges to "mercantilist" theories.[54]

But another feature of Hume's famed "Balance of Trade" essay has often gone overlooked: it was a stern warning about the problem with calculation. As he explained in a November 1750 letter to James Oswald, a key goal of his essay was "to remove people's errors, who are apt, from chimerical calculations, to imagine they are losing their specie." Britons' mistaken faith in "chimerical calculations" about the balance of trade led to dogmatic arguments, destructive trade and currency policies that hurt national industry, and fearful economic sentiments that inhibited men of commerce from taking productive economic risks.[55] And, what was worse, those calculations could be generated to support almost any political position. And here we return to the third epigraph that began this book:

> 'Tis easy to observe, that all calculations concerning the balance of trade are founded on very uncertain facts and suppositions. The custom-house books are own'd to be an insufficient ground of reasoning. . . . Every man, who has ever reason'd on this subject, has always prov'd his theory, whatever it was, by facts and calculations.[56]

Notably, as Emily Nacol observes, "Hume's solution to this problem [was] not to promote a more careful, detailed, and secure system of recording the to and fro of commerce," but was rather to be found in a more holistic and less "particularistic" understanding of the system of international trade. Better numerical data and more precise calculations were not the answer.[57]

For David Hume, after all, calculation itself was a problem, and a big problem indeed. When it came to questions of population, trade, and debt, calculation was never up to the task. Available numerical evidence proved unstable, unreliable, and inconclusive. The public's failure to recognize the limitations of quantification fueled hyperbole and hubris in political discussion, as people believed that complicated problems could be settled by a few accounting figures. Numerical reckoning promoted the consolidation of dogmatic economic principles that were highly misleading, like the obsession with the balance of trade, or downright dangerous, like the idea that the fate of the national debt could be predictably calculated. It also left Britons susceptible to the calculated deceptions of politicians like Walpole. Hume offered perhaps the first thorough diagnosis of a fundamental dilemma of the quantitative age: the more people believe numbers cannot lie, the easier it is to lie with numbers. The more we trust in numbers, the more they let us down.

The Calculating Divine

Hume's gloomy attitude about calculation was probably more the exception than the rule in the mid-eighteenth century, though. That period also yielded some remarkably hopeful statements about the potential for numbers to advance the public good—none more powerful than that of the "good" Dr. Richard Price. Like Hume, Price had a long and eclectic intellectual career, ranging across moral philosophy, mathematical probability, demography, public finance, and political theory. Drawing upon all of these elements, Price formulated a distinctive, programmatic vision for British civic epistemology, which sought to put numbers at the very heart of the nation's political thinking. Mathematics was the ultimate expression of rational thought, Price believed, and the greatest weapon against uncertainty and doubt; correspondingly, mathematical calculation about public accounts and other political-economic data was the ultimate form of *public* reason.

Price was born in 1723 in Glamorganshire in southern Wales.[58] He was educated as a preacher in dissenting academies, first in Talgarth, Wales, and subsequently in Moorfields, London, where he absorbed a Unitarian, philosophical strain of dissenting Protestantism. Ordained in 1744, he became a family chaplain to a wealthy merchant, a position that provided a convenient platform from which to pursue a budding interest in moral philosophy. He became minister of the famed dissenting meetinghouse at Newington Green in 1758, and that same year earned acclaim for his *Review of the Principal Questions and Difficulties in Morals,* where he laid out his "rational intuitionist" ethical theory. His *Review* argued that people are born with fundamentally correct intuitions about moral "right" and "wrong," which are irreducible to baser instincts like happiness or self-interest. Determining moral duty requires reasoning freely upon those intuitions to reconcile moral pressures. Price's theory challenged Francis Hutcheson, Hume, and others who contended that morality derived from non-rational sentiments or was cultivated by social interaction. Price's ethics drew heavily upon Platonist influences and especially his dissenting Christianity. Price believed the universe was providentially ordered and that God had created humans as autonomous, reasoning agents as part of his divine plan.[59]

Price soon extended his defense of rationality, and his struggle against Humean skepticism, beyond ethics into metaphysics and epistemology. He believed that reason allowed access to moral truths, which ultimately had a divine origin; similarly, he believed that reason allowed people to access fundamental truths about nature, which ultimately had a divine order. Price stridently rejected Hume's claim that people can never fully comprehend the "secret powers" behind natural phenomena and that our perception of cause-and-effect was a product of habit. "Surely," Price wrote, "never before were such pains taken to produce darkness and perplexity on a point so plain."[60]

In the 1760s, Price's concerns about natural order, causality, and the possibility of certain knowledge found an intriguing outlet: the mathematics of probability. In 1763, Price edited and published a mathematical essay "on the doctrine of chances," which he found amid the papers of his late acquaintance Reverend Thomas Bayes, a fellow dissenter. Bayes's essay asked: Given a number of observations about some random event—say, drawing 1,000 balls from an urn containing an unknown number of black and white balls—can you estimate the probability of the next trial—for example, the likelihood the 1,001st ball is white? More precisely, can you quantify how certain you are that your estimate is correct?[61] As Price explained in his introductory remarks to the essay, this was not "merely a curious speculation in the doctrine of chances, but necessary . . . [for] a sure foundation for all our reasonings concerning past facts, and what is likely to be hereafter." In other words, Bayes had provided a mathematical procedure for evaluating the strength of our knowledge about causes and effects. As Lorraine Daston has explained, Price saw Bayes's calculations as an answer to Humean skepticism. As Price wrote, the point was "to shew what reason we have for believing that there are in the constitution of things fixt laws . . . [and] therefore, the frame of the world must be the effect of the wisdom and power of an intelligent cause."[62] His work on Bayes's papers marked Price's coming-of-age into a calculating life. In 1764, he published a complicated, technical supplement to Bayes's first essay, also in the *Philosophical*

Transactions. The following year, he was elected a Fellow of the Royal Society in his own right, where he would continue to present and publish mathematical papers, most notably on Britain's demography.[63]

Through his work on probability, Price learned that mathematics could be a powerful defense against the specter of uncertainty. This was not purely a metaphysical problem for Price; uncertainty could arise from social as much as natural causes. In time, Price became increasingly interested in using mathematical calculation to solve practical questions and help citizens live more certain and stable lives. In 1771, he published his *Observations on Reversionary Payments,* nearly 350 pages of intensive, technical discussion on the mathematics of annuities, insurance, and public finance. Price's *Observations* was inspired by a recent spate of insurance "projects" in London and Edinburgh that promised to provide financial annuities to widows in case of their husbands' deaths. Finding many of these schemes dangerously flawed, Price undertook to prove "the inadequateness of their plans, by undeniable facts and mathematical demonstration." One recent project met with Price's approval, though: the "justly stiled" Equitable Assurances on Lives and Survivorships, a London company founded by mathematician James Dodson and "guided, in every instance, by strict calculation." Price had advised the Equitable since 1768, and his *Observations* offered further advice on how to improve that enterprise. He especially stressed that it was essential for the company to "be under the inspection of able mathematicians." Price's work was foundational to what would become known as "actuarial" science. His nephew, William Morgan, became the Equitable's first "actuary" in 1775.[64]

Price's mathematical analysis of "reversionary payments" constituted a two-fold assault on the specter of uncertainty. First, Price's calculations provided a defense against a very human kind of uncertainty: the deceptions of the marketplace, particularly unscrupulous insurance salesmen peddling the wares of "bubble" companies. In a personal letter, Scottish philosopher Thomas Reid praised Price for having "raised an allarm, and shaken the foundation of many Visionary Schemes which would have produced Ruin to many innocent families." Reid lauded Price for his facility in the "Science of Numbers," and for having shown that "even the most abstract parts of it . . . [have] great utility in the affairs of Life." Second, Price's calculations protected against uncertainties that lay outside human control. Mathematically sound insurance promised a defense against the whims of future economic fortune, particularly for vulnerable individuals like widows. Historian Peter Buck argues Price's actuarial mathematics was a consciously democratizing political project. Insurance and pensions constituted a stable form of property, offering diverse Britons the economic security needed to be republican citizens, regardless of social class.[65] Price's pioneering actuarial investigations would soon become crucial instruments of an emerging logic of "risk" central to modern capitalism. Yet, in the context of the eighteenth century, they offer even more evidence that the calculating culture of that era was driven as much by political imperatives as by profit motives.[66]

Motivating Price's inquiries into insurance was a sense that uncertainty threatened the polity, whether generated by the vicissitudes of the market, the vices of men, or the vagaries of nature. In the 1770s, the conquest of political uncertainty became ever more central to Price. His political commitments had been refined through his connection to the "Honest Whigs," a circle that included Joseph Priestley, James Burgh, and Benjamin Franklin. Around 1771, he also earned the attention of the Earl of Shelburne, a leading figure in the opposition Whig party. Shelburne would soon become Price's foremost patron, and Price Shelburne's go-to calculative adviser.[67] In many ways, Price and his fellow Honest Whigs carried on the earlier Country tradition in British politics. They shared the Country fixation on civic virtue, the fear of corruption, and the belief that calculation could be a virtuous instrument for political reform. Price celebrated earlier Country calculating projects, like the Commission of Public Accounts in the 1690s. He also lamented how such efforts at governmental accountability and rationality had fallen by the wayside during the age of Walpole, leaving an obscure and uncertain system of government.

For Price, nothing demonstrated the mounting uncertainty that faced the polity more than the national debt, which had grown from under £55 million when Walpole entered office to roughly £130 million in 1770. The third of three major chapters in his *Observations on Reversionary Payments* included a lengthy discussion "Of Public Credit, and the National Debt." Price ran off a familiar Country list of "evils and dangers" produced by national indebtedness— "increasing the dependence on the crown, by jobs and places without number; occasioning execrable practices of [Exchange] Alley; rendering us tributary to foreigners," and so on. But Price felt the menace of debt went deeper. Indebtedness rendered a polity more subject to the vagaries of future chance, including wars, trade downturns, and political crises. This collective vulnerability threatened to "check the exertions of the spirit of liberty." As citizens became more entangled with the public funds, willingly or otherwise, they would hesitate to jeopardize precarious public credit by voicing opinions that might "throw things into confusion." This would silence the "jealous and watchful" questioning necessary for free government.[68]

Price's anti-debt message (and his title "Of Public Credit") echoed his rival David Hume in several ways, but his solution to the problem did not. Price felt the answer lay in calculation—in fact, in a sinking fund. Price saw Walpole's original 1716 sinking fund project as a great missed opportunity, a "sacred deposit" betrayed by Walpole himself. In Price's mathematical reckoning, Walpole's alienation of the sinking fund to cover day-to-day governmental expenses effectively sacrificed a compound-interest return for a simple-interest one, an infinitely costly proposition over time. Price advocated reinstituting Walpole's fund and placing it under the oversight of a politically independent commission in order to prevent future encroachments. For Price, a mathematically driven system of fiscal administration could bring new safety and stability to the polity. Price's *Observations* included extensive calculations showing the remarkable speed with which the sinking fund could retire massive debts, along with intricate analyses showing why the fund was a more effective method for debt retirement than alternative strategies, such as exchanging the government's perpetual bonds for fixed-term annuities.[69] Just as Bayes's calculations restored faith that nature was ordered by "fixt laws," so could the exponential mathematics of a sinking fund restore Britons' confidence that their polity was on an orderly course. It could also free the nation from the uncertainties of political life—the erratic passions of wasteful politicians, or the unpredictable tyranny of future wars. It could give the nation what Peter Buck has termed "mathematical autonomy."[70]

Price also saw the emancipatory power of calculation as a critical part of the solution to another great political problem: the American question. Price's writings on the conflict with the American colonies brought him great public renown. His February 1776 *Observations on the Nature of Civil Liberty, the Principles of Government, and the Justice and Policy of the War with America* sold some 60,000 copies in its first six months. In those writings, Price brought together his moral philosophy and his Country politics in forming a refined, and distinctly quantitative, vision for what Britain's civic epistemology ought to look like. At the heart of Price's thinking was a belief that reason was essential to liberty, whether for individuals or for nations. In his moral philosophy, Price had argued that rational deliberation was fundamental to an individual's "moral liberty." In his American writings he argued that politics was analogous: the ability of a nation to deliberate rationally on its laws was essential for "civil liberty." Price believed that the essence of all forms of liberty—moral, civil, as well as "physical" and "religious"—was what he called *self-government.* And self-government required reason.[71]

Price contended that, in order for any individual (or nation) to govern his own actions in a truly autonomous way, he had to be able to make sound, unbiased decisions. Therefore, incomplete rationality was a form of servitude. An irrational individual was stripped of his moral liberty when his passions came to rule over his reason; an irrational *nation* might find its civil liberty in peril any time it lost its ability to reason collectively—for example, if the legislative process was overwhelmed by a tyrant or demagogue, or if the public became disengaged, uninformed, or corrupted. Without reason, liberty risked becoming licentiousness. "Reason in man," Price explained in his subsequent *Additional Observations on Civil Liberty*, "like the will of the community in the political world, was intended to give law to his whole conduct and to be the supreme controlling power within him."[72]

The problem was that "civil government, as it actually exists in the world," fell short of such rational standards. Like Hume, Price was profoundly distressed by the state of collective political thinking. As he explained in *Observations on Civil Liberty*, nothing revealed this civic stupor more clearly than Britain's disastrous approach to the American colonies, which showed a consistent pattern of "mistake, weakness, and inconsistency." Price demonstrated this by recounting Britain's American tax policies dating back to the 1733 Molasses Act, showing how they had been dictated by pettiness, greed, and a creeping "spirit of despotism." His *Additional Observations* went even further, providing a lengthy catalogue of flaws in Britain's political faculties: a "general indifference" born of luxurious living; mass cynicism resulting from "having been often duped by false patriots"; the silencing of the critical energies of Parliament; the infrequency of elections; and of course the national debt.[73]

Price was disappointed by the common methods Britons used in thinking through the American question. For example, many seemed unable to understand Britain's relationship to its colonies as anything but a narrow legal matter. In *Observations on Civil Liberty*, Price encouraged readers instead to "try this question by the general principles of civil liberty" and move beyond this fixation on "*Precedents, Statutes,* and *Charters*." Focusing on pedantic issues, Britons overlooked much more essential ones—the kinds of issues that became much clearer by looking at the numbers. For example, traditional arguments in favor of Britain's right to rule in the Americas took no account of the numerical fact, exposed by Benjamin Franklin decades ago, that "in 50 or 60 years [the Americans] will be double our number . . . and form a mighty empire." Similarly ignorant was the common claim that the American colonists somehow owed metropolitan Britain economically for its military-fiscal support. Price argued that the colonists had actually supported Britain's public finances in myriad direct and indirect ways. "Were an accurate account stated," Price explained, "it is by no means certain which side would appear to be most indebted." Such an "accurate account" had never been seriously attempted. Britons were not nearly so reasonable.[74]

Price believed that better numbers would make for more rational political deliberation, thus advancing liberty. He attempted to enact this kind of calculative rationality in his own writings. Of the nearly two hundred pages in his *Additional Observations on Civil Liberty,* nearly half contained technical material, on topics ranging from public credit schemes to historical British trade data to the public finances of France. (Those pages are mostly omitted from the leading modern edition of Price's American writings, an indication of how scholars have neglected the centrality of calculation to Price's political thought.)[75] To Price, those "various and extensive" calculations were hardly arcana; they performed the very kind of political reasoning he hoped would govern British politics. In fact, he claimed that those were the most important parts of his *Additional Observations,* "so important that it is probable I should *not* have resolved on the present publication had it not been for the opportunity . . . to lay [those] observations before the public."[76]

Price suggested that calculation could even be used to help solve the American question, not just to understand it better. He regularly touted the power of the sinking fund to ameliorate colonial relations by stabilizing uncertain fiscal relations between Britain and the colonies. His 1776 *Observations* ended with a proposal, advanced by Shelburne, that would facilitate reconciliation by requiring the colonies to assume a calculated proportion of Britain's national debt and make regular payments into a sinking fund. In both his 1776 and 1777 pamphlets, Price quoted a June 1775 resolution of the Continental Congress stating a similar plan. In Price's mind, a problem that was leading to war might well have been solved by calculation.[77]

Of course, British fiscal and imperial policy was never so well-calculated. One striking, and little noticed, aspect of Price's American writings was the keen attention he paid within them to questions of public accounting. He argued that a key reason for the stupefied state of Britain's economic and colonial policy was the fact that Parliament had become woefully negligent in keeping numerical account of the nation's finances. "Nothing is more the duty of the representatives of a nation than to keep a strict eye over . . . the money granted for public services," he explained in his *Additional Observations.* Over time, Price became ever more convinced of the necessity of transparent public numbers for civil liberty. In January 1778, Price published a combined version of his *Two Tracts* on the American question, including a new "General Introduction." It contained an unusual exercise: a timeline of Parliamentary financial oversight, meticulously compiled by Price from historical Parliamentary records. He concluded that, beginning in the late 1710s, "a great change has taken place" in Parliament's concern with fiscal numbers. Despite the fact that "the public accounts have been growing more complicated; and the temptations to profusion and embezzlement have been increasing," MPs could not even be bothered to attend on days when accounts were passed. This change amounted to "little less than the total ruin of the constitution."[78]

Through his American writings, Price sketched a model for civil government built on a rational armature of mathematical thinking. Diligent and transparent numerical accounts would help identify administrative corruption, allocate resources and responsibilities between different constituencies (like Britain and its colonies), and provide the material for reasoned economic policies. Mathematically engineered fiscal policies like the sinking fund would protect the public's resources against the whims of fortune and of ministers. In many ways, Price continued the project of earlier Country calculators, especially Archibald Hutcheson, whose quantitative analyses aimed to shed light on the crippling uncertainties that plagued British politics. But he went beyond these predecessors, making calculation not just a tool of political critique but the linchpin of a political philosophy.

Price also drew a new and powerful connection between calculation as an act of both individual and collective reasoning. Despite his interest in using a kind of "social science" for governmental reform, he was no "technocrat." In fact, Yiftah Elazar's recent study contends that Price's originality as a political thinker lay in seeing "the right to participate in politics as part of the liberty and dignity of the individual." Price looked forward to a polity in which mathematical reforms could make both more rational government *and* more rational citizens. He made this point especially clearly after the American war, in his 1784 *Observations on the Importance of the American Revolution*. Price had been a strong supporter of the American colonists, and he subsequently took the time to offer the newly independent United States some advice. "Nothing is more necessary than the establishment of a wise and liberal plan of education," he wrote. Discouraging the teaching of dogmas or rigid systems, Price urged that the goal of instruction ought to be: "to induce, as far as possible, a habit of believing only on an overbalance of evidence, and of proportioning assent in every case to the degree of that overbalance, without regarding authority, antiquity, singularity, novelty, or any of the prejudices which too commonly influence assent." And, he went on, "nothing is so well fitted to produce this habit as the study of *mathematics*."[79]

Plus or Minus

David Hume and Richard Price laid out two radically different visions of the place numerical calculation ought to hold in Britain's civic epistemology. For Hume, calculation was an unstable, inconclusive, often corrupted way of thinking that fueled carelessness, dogmatism, and deception. For Price, calculation was the purest form of political reason, which had a unique power to guide virtuous action, conquer uncertainty, and promote civil liberty. For Hume, numbers and calculations exacerbated peoples' chronic inability to properly "proportion" their belief to evidence; for Price, numbers and calculations were the best tool for doing exactly that, "proportioning assent." Perhaps more remarkable than their differences, though, is what Hume and Price had in common: they were responding to essentially the same problem. They both diagnosed a fundamental deficiency in how the British public thought through political problems, especially in public finance and political economy. Both saw this as a fundamentally historical problem and, in their own writings, traversed much of the historical terrain covered in this book. Most critically, both identified the public use of numbers as central to the problem, whether because Britons relied on numbers too much or because they had yet to capitalize on their true potential.

Hume and Price serve as fitting figures to tell the final chapter of our story. Each formulated a view of numbers that would become archetypal: we can't live with numbers, but we can't live without them. Each had an enduring influence on later political thought and culture, embedding their numerical sentiments in various ways. Of the two, Hume's legacy is the better documented (and more debated). But Price's contributions to future practices of political and economic thinking were arguably just as influential and perhaps even more wideranging. Price's calculated conception of political rationality left its mark in the widespread fashion for sinking funds in late-eighteenth-century public finance, exemplified by the younger William Pitt in Britain and Alexander Hamilton in the young American republic, and in the surging popularity of life insurance in both countries in the early nineteenth century. Indeed, Price's legacy was almost certainly greatest in the United States, where he was actually invited to come and live by a 1778 motion of Congress, likely in hopes he would become the new nation's finance minister.[80]

The historical significance of Hume and Price as individuals, though, is only a minor part of this broader story. Their debate serves a more important heuristic purpose: it frames the essential tension of the quantitative age. Hume and Price each embraced, with particular force and clarity, one opposing position regarding the vices and virtues of political calculation. But modern quantitative life would not resolve into a Manichaean showdown between Humean skeptics and Pricean believers. Most people who confronted the complexities of calculation would find themselves caught anxiously in between, sympathetic to both Hume's position and Price's, doubting numbers one minute and placing great trust in them the next.

This kind of internal conflict over calculation was already evident among Hume and Price's contemporaries, including some of their closest contacts and rivals. One example of such a numerically conflicted individual was Adam Smith. From one perspective, it seems easy to place Smith as an heir to the numerical skepticism of his friend, Hume. In Book IV of his 1776 *Wealth of Nations,* amid his "digression on the corn trade," Smith famously wrote that "I have no great faith in political arithmetick." Throughout that text, Smith made only sparing reference to previous calculators and was careful to explain that he meant "not to warrant the exactness" of any computations he did include. But in many ways, Smith stopped short of Hume's more thorough skepticism about quantification. Smith relied considerably on numerical data in building his analysis, notably in his historical discussions of money and prices, which drew upon work by Gregory King, William Fleetwood, and Arthur Young. Mary Poovey suggests that Smith had his own, distinctive ambivalence about numbers. While he had serious misgivings about most existing numerical data, he was optimistic about the illuminating potential of numerical facts in general. In Smith's mind, existing numbers, like trade balances, measured the wrong things; once the operations of the market were truly understood, the correct variables could be identified and true numerical guidance obtained.[81]

Another example was Edmund Burke, whose relationship with numbers was even more complicated than Smith's. On the one hand, Burke produced one of the most memorable screeds against the calculating habits of the eighteenth century: his November 1790 *Reflections on the Revolution in France,* written in response to a notorious pro-Revolutionary speech by none other than Richard

Figure 7.1   James Gillray, "Smelling out a Rat; or the Atheistical-Revolutionist Disturbed in his Midnight 'Calculations'" ([London]: Published by H. Humphrey, Dec. 3, 1790). The seated figure in black is Richard Price; the disembodied nose depicts Edmund Burke. Image reproduction © The Trustees of the British Museum.

Price. Burke's famed treatise, long cited as a founding text of modern conservatism, presented a merciless attack against "the calculating divine" Price and his insinuation that the "constitution of a kingdom be a problem of arithmetic." Price was but one of many offenders, in Britain and France, who had replaced wisdom with calculation and had tried to overhaul constitutions "with no better apparatus than the metaphysics of an under-graduate, and the mathematics and arithmetic of an exciseman." As Burke famously lamented: "The age of chivalry is gone.—That of sophisters, oeconomists, and calculators, has succeeded; and the glory of Europe is extinguished for ever." Yet Burke was not always a foe to numerical thinking. In fact, he was a longtime veteran of Britain's political numbers game. Throughout his early career, he had relied on combative calculations to make arguments: in his 1769 *Observations on a Late State of the Nation,* written against a recent pamphlet by William Knox, in conjunction with former Prime Minister George Grenville; in a 1775 speech "on conciliation with the colonies," in which he formulated key quantitative arguments about colonial population and the importance of colonial trade; and in his prosecution of Warren Hastings, former Governor-General of Bengal, for corruption, beginning in 1786. Burke had done more than his part to make the late eighteenth century an age of calculators.[82]

The quantitative age was not just a phenomenon of high intellectual history, of course, and the characteristic ambivalence about numbers was felt far beyond the rarified reflections of Smith or Burke. The earlier sections of this chapter have shown how, in the mid-eighteenth century, both quantification and its discontents became increasingly visible in a wide array of cultural outlets—in math textbooks and encyclopedias, in the *Gentleman's Magazine* and *Critical Review*, in the writings of small-town clergymen and anonymous political hacks. The problem with numbers even came to be visible in that most vivid of eighteenth-century genres: the graphic satire. One of the most memorable renderings of the Price-Burke debate over the French Revolution came in a December 1790 image, attributed to the famed caricaturist James Gillray, entitled "Smelling out a Rat; or the Atheistical-Revolutionist Disturbed in his Midnight 'Calculations.'" (See Figure 7.1.) The image depicts a frightened Price, dressed in the dour black of a Dissenting divine, looking over his shoulder at Burke, depicted as a disembodied nose bearing the totems of Church and King. The image was no doubt meant to poke fun at both the bombastic Burke and the pedantic Price. The extraordinarily rich iconography merits endless analysis, but there is one aspect of the image that stands out in this story, a single word at the end of the caption: " 'Calculations,' " presented in eighteenth-century scare quotes. It was, of course, a pun, easily understood by contemporary viewers, evoking two different meanings of the word: *calculation* as mathematical reasoning and as devious plotting. Were Price's calculations honest or deceptive, numerical or conspiratorial? Or, even more to the point, was the very act of calculation itself an act of intrigue? Those were the kinds of questions that would long trouble the quantitative age.

Conclusion

GILLRAY'S PUN on Richard Price's "Calculations" brings us back once again to the observation laid out at the very beginning of the book: that modern quantitative culture is fundamentally two-sided. On the one hand is the belief that numbers and calculations constitute a distinctly powerful, incisive, and trustworthy form of knowledge—that "*Facts* and *Figures* are the most stubborn Evidences." On the other is the belief that numbers and calculations often confuse, disappoint, and mislead—that they are the damnedest form of lies. What made Gillray's pun effective was that it was able to evoke both of these sentiments at once. The viewer was meant to look upon Price with a measure of sympathy, to cheer on his tireless efforts to pursue the political good through the honest power of numerical "Calculations," in the face of overbearing verbal assaults from Burke. The viewer was also meant to look upon Price with a measure of suspicion, to see his late-night numerical efforts as part of a more sinister program of political "Calculations" that threatened the nation's traditional values. Gillray's image captured a constitutive tension at the heart of the quantitative age.

Calculated Values has sought to shed light on this tension by reconstructing how numerical calculation first attained the authority it has come to hold in many modern civic epistemologies. This book has argued that, in Anglophone political culture, this new quantitative authority arose during a decisive period of less than a century following Britain's 1688 Revolution. Driving this transformation was public political *dispute*. Numerical calculations came to garner new esteem because they proved incredibly useful for interrogating policies, critiquing opponents, and advancing arguments within the highly adversarial political culture that arose after 1688. Britons' emerging belief that numbers and calculations were a distinctly stubborn, honest, and incisive way of making public knowledge about the economy and the polity grew out of their use of numbers in instrumental ways to advance political interests.

This is, I hope, an unexpected story for many readers, which challenges common intuitions about political and quantitative knowledge. In contemporary life, the authority of calculation derives in large part from the belief, or aspiration, that numerical knowledge can transcend politics, creating knowledge that is *im*personal, *non*partisan, and *a*political. Numerical facts and figures are imagined to resist the pressure of political values, objectives, and interests. There are countless times, of course, when numbers and calculations get caught up in political controversy, of course. But these are explained away as instances of the quantitative becoming corrupted by the political, in which unscrupulous individuals seek to pervert true facts and honest calculations in the service of political ends. Those who manipulate the numerical facts are exploiting people's trust in numbers.

This image of the conflict between the quantitative and the political simply did not obtain in the eighteenth century, though. Political calculators in that period were not exploiting some preexisting authority that numbers held in the minds of the British public. Rather, it was the use of numerical calculations to pursue political ends that generated their power and prestige within British civic epistemology. To put it bluntly: Britons did not come to fight with numbers in the eighteenth century because numbers were already believed to be trustworthy and authoritative; numbers came to be seen as trustworthy and authoritative because Britons fought with them. Early political calculators like Charles Davenant, William Paterson, and Archibald Hutcheson could not take for granted that their intricate arithmetical calculations would bear special weight with political leaders or the public. Rather, they turned to calculation because it proved vitally useful for formulating certain kinds of political arguments, like measuring the efficacy (or inefficacy) of practices and policies, demonstrating the need for more intelligent leadership, and highlighting dangerous political secrets. In the fiercely combative political environment in which they operated, their numerical salvos tended to evoke responses from their adversaries. Increasingly, calculations were answered with more calculations, creating a pattern of quantitative controversy.

With each new round of numerical combat, calculation became more embedded in the practice of politics and more familiar to the British public. Calculators fought mercilessly over what the "right" numbers were. They also presented an array of different ideas about what made numerical calculation a virtuous way of thinking: its "common sense" clarity, its mathematical rigor, its association with the shrewd know-how of merchants, its ability to "undeceive." For all they disagreed about, though, calculators mutually affirmed the sense that numbers were indispensable to the political process and that somewhere, amidst all the contested digits, there existed essential numerical truths. These true numbers were imagined to lie beyond the flawed efforts of individual calculators, who were seen as far less trustworthy than the numbers over which they fought. Rarely did any calculator seem to attain such irrefutable numerical truth, of course. No political calculation was ever completely certain or secure from political critique. But this did not stop calculators from trying to get there, and to produce a calculation that was more correct—closer to the truth—than their opponents' calculations.

Is this, then, ultimately a story about *objectivity*—about how numbers came to be seen as a distinctly objective form of knowledge? In the course of researching and writing this book, I have gone back and forth on this question. The best answer I have come to is: "Sort of." On the one hand, eighteenth-century political actors did not think about the virtues of numerical calculation in terms of its "objectivity" at all. They certainly did not use the language of objectivity, either literally or in some embryonic, "proto-" form. As Daston and Galison have shown, the modern concept of objectivity only began to appear in the early nineteenth century, conjoined to a new concept of "subjectivity."[1] Even more, what made numbers appealing as a medium of public knowledge to eighteenth-century Britons were not really the features generally associated with modern visions of quantitative objectivity—the fact that quantitative techniques yield knowledge that appears rule-bound, mechanical, and free of subjective bias. That is part of what makes the history told here intriguing: it reveals that the authority of numbers in public life, at least historically, cannot be entirely explained in terms of objectivity.

Yet, on the other hand, this story does offer something to our understanding of the history of objectivity and the role of numbers within it. It helps to explain how, once the concept of objectivity was formulated and identified as a preeminent epistemic virtue in modern public cultures, numerical calculation came to be seen as an excellent mechanism for achieving it. Long before the modern concept of objectivity, many politically engaged English speakers would have already absorbed a set of powerful expectations about the virtues of quantitative "facts and figures" and their particular benefits for making public knowledge. Numerical claims were already seen to be especially incisive, honest, and stubborn, with a particular proximity to political truth—though not yet "objective." Numerical claims were already likely to evoke certain sentiments, and a certain credence, from many members of the political public—though not because people imagined them to be produced through mechanical means that limited the incursions of subjectivity. It is not difficult to imagine, though, that once objectivity became a political and epistemic virtue worth pursuing, numerical calculation, already carrying certain authority for other reasons, was seen to be a good tool for doing so. Reconstructing exactly how the authority of numbers became yoked to a new concept of objectivity must remain, though, a project for another time.

The frantic numbers game that developed in eighteenth-century British politics created enduring attitudes about the specialness of numerical knowledge as a tool of public reasoning; but that numbers game also produced new frustrations, doubts, and fears about numerical calculation. Some began to worry that the nation had become numerically gullible. Calculators were vexed by how easily ignorant public readers were taken in by the lesser numerical efforts of their political opponents. More searching critics like David Hume feared that the rising rage for numbers indicated a deeper pathology in British civic epistemology. In Hume's mind, Britons grossly overestimated what numbers could do. Whereas many looked to numbers as clear guides to the political truth, numbers proved consistently inadequate. It was possible to justify almost any political position with facts and calculations. At best, the nation's computational fervor produced an overload of numerical noise that distracted from real issues. At worst, calculation constituted a new form of political casuistry and promoted dangerous new political superstitions. Savvy state actors like Robert Walpole realized that, because quantitative thinking had come to bear such authority with many in the public, calculations offered an especially useful tool for molding public opinion and consolidating power. The very same practices of political calculation that produced Britons' belief that "Truth and Numbers are always the same" could, when viewed from a slightly different perspective, reveal numbers' distinctive capacity to deceive.

But even Walpole was not using calculations in a way that was fundamentally more cynical or dishonest than other calculators of his era. He had been a victim of vicious numerical attacks himself early in his career, and he was trying to outcalculate opponents like Hutcheson and Pulteney, just as they were trying to outcalculate him. The story of Britain's early quantitative age cannot be broken down into a conflict between good calculators (and calculations)—seeking to analyze numbers in honest ways to obtain objective political truths—and bad calculators (and calculations)—seeking to manipulate the numbers in dishonest ways to advance subjective political interests. All calculators were interested in some vision of the numerical truth, and all of them were in pursuit of some political end. The calculations they produced were searching, insightful, and often highly credible; they were also opportunistic, biased, and invariably imperfect. All the calculations in the period were, in some sense, like Price's midnight "Calculations" in Gillray's punning image.

The story of Britain's early quantitative age does more than just corroborate a point that critical scholars of quantification have made many times, though: namely, that quantitative ways of knowing, like all ways of knowing, have politics. Rather, what is particularly illuminating about this story is that, in the eighteenth century, those politics were what generated numbers' authority. Calculation attained an elevated position in Britons' civic epistemology *because of,* not *in spite of,* its overt political affordances and applications.

Some may find this a somewhat unsettling, even disheartening, tale. It may suggest that the origins of modern enthusiasm for quantitative knowledge in public life is somehow ignoble, more about pursuing political gain than solving social problems. So what can be said of the effects of Britons' new rage for calculation in the eighteenth century? Did it help the polity to better address collective challenges? Did it advance knowledge and understanding? Admittedly, the evidence of how numbers were used in eighteenth-century Britain offers little reason to believe that numbers might *solve* the hardest political questions. In many ways, David Hume was right. Numbers rarely provided conclusive political answers. Rare was a case in which political calculations told a clear and decisive story that was widely accepted by the public. The exceptional case was Archibald Hutcheson's piercing critique of the South Sea Scheme, and even those calculations were always highly contested and only attained widespread assent after the calamitous bursting of the Bubble appeared to confirm their accuracy. Far more often, the most pressing numerical questions—about the fair size of the Equivalent, about the Anglo-French trade balance, about the ebbs and flows of the national debt—never quite met with satisfying answers.

Critically, these frustrations were not simply the result of *technical* failings in computational methods. The very structures of political antagonism that made calculation such an attractive instrument of political thinking also made it extremely difficult, often impossible, for calculators ever to achieve conclusion. Eighteenth-century calculators strove constantly to make their calculations more accurate, to improve their calculative techniques, and to find better sources of data. But this tended simply to provoke their opponents to do the same, producing copious numbers but little closure.

If viewed primarily as a means to achieve political answers, calculations were often disappointing in eighteenth-century Britain. But that did not mean they did not advance understanding. The ugly, cynical, and frequently inconclusive calculative battles that made Britain's quantitative age were often highly productive, even revelatory. They were a tremendous engine for the identification, accumulation, and organization of new data. Because public calculators craved numerical data for partisan ends—because it revealed a political secret, showed the incompetence of a rival, helped rebut an opponent's argument—data itself took on new value. Those conflicts helped stimulate the creation of new analytical techniques, like the proto-regression analysis Davenant pioneered in his analysis of excise yields, or the multi-scenario spreadsheet models fashioned by Hutcheson. They also helped to formulate, refine, and publicize powerful (if problematic) economic concepts. Some, like the balance of trade, have since waned in prominence, but others, like the use of exponential discounting to value stocks, remain central to modern thinking.

Today, a common criticism leveled against the use of quantitative techniques as instruments of political evaluation, accountability, and deliberation is that quantification tends to flatten out the nuances of complex, human problems, producing what Porter has called "thin description." In a recent study of the use of quantitative indicators in global governance, anthropologist Sally Engle Merry argues that "despite the value of numbers for exposing problems and tracking their distribution, they provide knowledge that is decontextualized, homogenized, and remote from local systems of meaning."[2] Perhaps the most refreshing feature of eighteenth-century political calculations is that, whatever their shortcomings, they were rarely guilty of this kind of thinness and superficiality. The work of calculators frequently made political conversations deeper, richer, and more subtle, shedding light on complexities that otherwise might not have garnered political attention. This came with pitfalls of its own, of course. Calculative controversy ran the risk of making collective inquiry deeper, but also narrower. Numerical contests like that over the balance of trade could be sticky, trapping thinkers within a confined set of technical questions. This could have the effect of shutting out alternative arguments and modes of reasoning that did not fit within the shared calculative framework.

The beginnings of the quantitative age in Britain offer a remarkable archive of historical experience for thinking about calculation and the position it holds in modern civic epistemologies—the sources of its authority, its (perhaps obscured) political functions, its promises and pitfalls. The remainder of this Conclusion looks forward from the eighteenth century. The next two sections look at these questions genealogically. What enduring legacies were left by the calculative politics of the eighteenth century? How does the eighteenth-century history told in this book help make sense of later developments? The final section examines these questions analogically. In what ways might the calculative politics of the eighteenth century illuminate the calculative politics of the twenty-first century? And what edification, or inspiration, can we draw therefrom?

The Quantitative Age, Beyond the Eighteenth Century

The quantitative age that arose after 1688 was, I contend, in many ways the beginning of the quantitative age that persists today in the political cultures of Britain, the United States, and elsewhere. Of course, much has changed in the interim: state powers of quantification have expanded dramatically, new statistical techniques and computational technologies have been invented, new quantitative disciplines and expert identities defined. In the course of time, attitudes and expectations about quantitative thinking evolved as well, in ways that built upon, but also obscured, their historical precursors. The line between the political arithmetic of the 1690s and twenty-first-century practices of political calculation is certainly not a straight one. But there are vital continuities nonetheless—chief among these, the authority of calculation. The primary legacy of the calculating conflicts of the eighteenth century for Anglophone political culture was a robust esteem for numerical calculation as a way of political thinking and a medium of public knowledge. This essential confidence in calculation has remained basically intact, though never universal or uncontested, at the core of Anglo-American civic epistemology for three centuries. To demonstrate that historical continuity in a rigorous way would, of course, be a far greater task than can be achieved in these few remaining pages. Yet it is possible to sketch a map of some promising historical connections that merit further archival exploration.

Most immediately, it is possible to follow the culture of calculation that began in Britain in the eighteenth century forward as it continued to grow into the next century. Comparatively, the place of numbers in nineteenth-century British culture is vastly better known by scholars. This was the century of the first official census and the founding of the General Register Office, of *Facts and Figures* magazine, of statistical innovators like Edwin Chadwick, Florence Nightingale, and Francis Galton. It was also the century when Charles Dickens railed against contemporaries' obsession with "facts and figures" in *The Chimes* (1845) and when Benjamin Disraeli (perhaps) quipped about "lies, damned lies, and statistics."[3] It was an age when quantification—and its discontents—was an almost inescapable part of British life. There can be no doubt that the calculative culture of Nightingale and Dickens's Britain differed in important ways from Davenant and Hutcheson's. The volume of numerical data on political, economic, and social questions ballooned. Powerful new mathematical techniques emerged, most notably a new probabilistic understanding of normality and variation within populations. Calculation became the basis of well-defined professions, like the actuaries that powered the booming Victorian life insurance industry. New objects of knowledge were created, none more important than "society" itself, understood as an entity distinct from politics or commerce and governed by its own statistical laws.[4]

Attitudes toward numbers and calculation changed in some important ways as well. Most importantly, calculative thinking seemed to lose much of its emotional energy. Calculation came to seem rather boring. As noted in the Introduction, the "facts and figures" of Dickens and *Facts and Figures* magazine were seen as "dry food," cold, spiritless, and mechanical—a far cry from the heated, passionate, imaginative "facts and figures" deployed by William Pulteney and his contemporaries. How and when did this change? The period around 1800 appears to have been a key turning point. With the formation of new statistical bureaucracies, political calculation increasingly became an activity carried out by disciplined functionaries in organizational settings, as well as by enterprising individuals fighting personal and partisan battles. This no doubt contributed to a cultural sense that working with political numbers was not an especially adventurous affair. (Though it should be noted that the contentious, calculative politics of the eighteenth century are part of this story as well. As discussed in Chapter 6, the development of state calculating capacities under Walpole and successors was closely tied to the perceived numerical demands of public politics.) Public perceptions about calculation were also shaped by changing attitudes about the nature of the human mind and the value of different kinds of mental labor. Lorraine Daston has argued that the turn of the nineteenth century witnessed a marked shift in European attitudes toward mathematical calculation as a cognitive activity. Whereas eighteenth-century thinkers had celebrated the art of calculating as a manifestation of intelligence, "allied with the higher mental faculties of speculative reason and moral judgment," by the beginning of the next century it had "become mechanical, the paradigmatic example of processes that were mental but not intelligent."[5]

For all these ways that nineteenth-century quantitative culture differed from that of the preceding century, there were crucial continuities nonetheless. The nineteenth-century statistical age built upon eighteenth-century materials in myriad ways: upon a long tradition of using numbers to formulate political arguments in the public sphere; upon widespread civic expectations that numerical information was a public good and that government ought to be accountable in numerical terms; and especially upon fundamental epistemological beliefs about the virtues of quantitative thinking. These foundational elements of Britain's quantitative civic epistemology preceded—and, I suspect, enabled—the census, the General Register Office, and what Ian Hacking famously called the "avalanche of printed numbers" in the early nineteenth century. Recent scholarship on nineteenth-century Britain has revealed one clear continuity with the calculating culture of the eighteenth century: in the nineteenth century, as before, statistical thinking about Britain's people and prosperity was often motivated by the public, not the state. Recent studies by Oz Frankel, Kathrin Levitan, and Tom Crook and Glen O'Hara have deemphasized the importance of the British state as the primary creator and controller of numerical knowledge in the nineteenth century. Instead, those scholars highlight the importance of the public sphere as a critical site of quantitative engagement and innovation. In her recent cultural history of the census, for example, Levitan observes that in nineteenth-century Britain, "statistics was both the domain of government and the domain of civil society." By examining these long-term continuities, it becomes possible to see other, related developments in a richer historical perspective, including the nineteenth-century invention of "society" as a distinctive object of knowledge. Frankel writes that, in the nineteenth century, the "discovery" of the concept of society "took place in the public sphere and in the realm of politics"—precisely where the most creative quantitative thinking had transpired in the eighteenth century, as well.[6]

The legacies of Britain's eighteenth-century culture of political calculation were not limited to Britain, however, but left impressions wherever British political culture and governmental practice were enacted. In particular, I suspect that the events described in this book may offer new insight into the historical foundations of the quantitative practices that were such a crucial instrument of British colonialism. The role of numbering and calculation in colonial government is a far bigger topic than can be done justice here. But it is worth pointing out two important points of connection between eighteenth-century calculating culture and later colonial quantification, in order to suggest further avenues of inquiry. First, one (though certainly not the only) reason quantification became such a crucial instrument for British colonial agents may have been that Britons were so accustomed to using numerical calculation as an instrument of dispute in pursuing political agendas, deliberating on policy, and contesting institutional power struggles. Arjun Appadurai suggests as much in his study of the role of "number in the colonial imagination" in nineteenth-century India. "Numbers regarding castes, villages, religious groups, yields, distances, and wells were part of a language of policy debate," Appadurai observes, "in which their referential status quickly became far less important than their discursive importance in supporting or subverting various classificatory moves and the policy arguments based on them." When it came to numbers, "their referential purpose was often not so important as their rhetorical purpose."[7]

A second connection concerns the particular expectations that Britons had regarding the virtues of quantification as a way of knowing. In her pathbreaking study of numbers and numbering in Nigeria, anthropologist Helen Verran offers a telling description of the British colonial government's attempt to institute a new census in Ibadan in 1921. "The British see a census as a *revealing* of resources," Verran writes. "What motivates them is a concern to see how many colonial subjects they have. They want an *unhiding,* a bringing forth of all from the hidden and inaccessible quarters of Ibadan."[8] As we have seen, the ability of calculation to unhide secrets was one of its greatest perceived merits for eighteenth-century political calculators. Recognizing these continuities between political calculation in that earlier period and later quantitative practices in colonial settings may shed new light on the assumptions and aspirations that lay behind colonial quantification projects, and perhaps on the kinds of damage they wrought.

Political Calculation in the American Republic

Ironically, the calculative politics of eighteenth-century Britain may not have left their greatest legacy to Britain itself. Of all modern polities, including Britain, it is the United States that has come to rely most heavily upon quantitative modes of accounting, evaluation, and decision-making in political processes—a fact which has been extensively documented by historians and STS scholars.[9] Olivier Zunz notes, for example, that in the twentieth-century United States, "new ideas about statistical distribution . . . were to flourish in ways unfathomable in Europe." This quantitative tendency in American civic epistemology has been identified across a remarkable range of different domains in the twentieth and twenty-first centuries: in the rigorous, quantitative procedures of cost-benefit analysis implemented to evaluate government projects and policies; in the prominence of standardized testing and numerical metrics of intelligence; in American financial institutions' heavy reliance on formal calculations, like the "FICO" credit score, in evaluating individual "risks"; in Americans' mass fascination with social-survey data and public opinion polls as guides to the experiences and attitudes of the "average" citizen; in the use of statistical methodologies in the evaluation and licensing of pharmaceuticals; in the "hankering for objectivity based on numerical calculations" evident in American approaches to regulating environmental, public health, and safety risk in the late twentieth century.[10] For all of these problems of judgment—making government investment decisions, assessing intelligence, evaluating financial borrowers, measuring public risk—American political actors and institutions turned to numerical calculations while other polities often found alternative means, like delegating such problems to the discretion of credentialed experts.

This extensive reliance upon quantitative techniques is not only cited as a quintessential characteristic of American civic epistemology; it holds a paradigmatic status within scholarship on civic epistemology in general. It exemplifies the kind of durable habits of public thinking and knowing that the category "civic epistemology" seeks to describe.[11] Yet historical explanations for how and why calculation came to take such a strong hold in American civic epistemology remain incomplete. I contend that, to a large degree, the characteristic trust in numbers evident in the United States can be understood as an inheritance—and intensification—of earlier quantitative habits within British political culture.

To begin, the history told in this book immediately sheds light on the political beginnings of American quantitative culture. The most extensive study of numbers and calculation in early America remains Patricia Cline Cohen's remarkable *A Calculating People* (1982). Cohen explained that study as an attempt to make sense of her observation that, beginning around the 1820s and 1830s, "there suddenly appeared many types of quantitative materials and documents that previously had been quite rare."[12] Cohen sought to explain this observation through a social history of numeracy, by tracing how numerical reck-oning became a sufficiently widespread skill that, by the Jacksonian era, observers could claim Americans were a distinctly "calculating people." Cohen's study richly shows how the development of numeracy was closely connected to other historical developments, including in commerce and politics. Yet it understates the degree to which Americans' distinctive calculating culture was conditioned not only by rising numeracy but by older political habits.

There is ready evidence that American political practice and culture contained an essential quantitative thread extending well back into the eighteenth century. Many of the foremost leaders of America's "founding" generation were defini-tively calculating thinkers, for whom political arithmetic was a natural part of how they thought about their young nation's problems. Benjamin Franklin, for instance, crafted pioneering work in the political arithmetic of population growth and was one of the foremost prophets of the powers of compound interest. Alexander Hamilton read British financial and economic literature extensively and consciously drew upon earlier precedents, including the sinking fund model, in trying to engineer his own "financial revolution" in the 1790s. His famed *First Report on the Public Credit* (1790), in which he presented detailed calculations modeling several different strategies for addressing the nation's Revolution-era debts, is an exemplar of the eighteenth-century calculating style, reminiscent of the multi-scenario models deployed by Hutcheson and Walpole. Thomas Jefferson, who once claimed you could measure the degree of a nation's corruption by calculating the proportion of farmers to non-farmers, was an ardent supporter of arithmetical education.[13]

Such canonical figures were hardly the only early Americans with calculating inclinations. Many eighteenth-century Americans found calculation useful for the very same political purposes their British predecessors had—for leveling cri-tiques, advancing arguments, rebutting opponents, and rendering futures. Con-sider Thomas Cooper (1759–1839) of Northumberland, Pennsylvania. A close associate of Joseph Priestley and ally of Jefferson, Cooper was a vocal spokesman for Republican causes and a critic of the Federalists. During the late 1790s, he wrote multiple articles attacking the John Adams administration and champi-oning a Jeffersonian, agriculture-focused approach to political economy. Two such articles came under the heading "Political Arithmetic" and contained crit-ical calculations showing, for example, that the economic benefits of American overseas trade did not compensate for the burdensome costs of protecting those trades with naval support. Tellingly, Cooper explained that he felt motivated to produce his calculations because he needed to respond to numerical arguments being made on the other side by members of Adams's cabinet—"the numerous and minute calculations submitted to the wisdom of Congress by Messrs. Pinckney, Woolcot, McHenry, and Co."[14]

Or consider, as a second example, Noah Webster (1758–1843) of Connecticut. Though he would achieve renown in the next century as a lexicographer, Webster was most prominent in the late eighteenth century as a newspaper editor and Federalist polemicist. He was also a prominent advocate for the abolition of slavery. In a 1793 pamphlet on the *Effects of Slavery on Morals and Industry,* published in Hartford, Webster produced a creative exercise in "political arithmetic" to show the damaging economic costs of slave-based production. Webster calculated that, by relying entirely upon enslaved labor to cultivate lands, rather than laboring themselves, some "10,000 families of nobility or planters" effectively wasted over $1.3 million annually in productive labor. That amount, if invested productively at 6 percent compound interest, would be worth over $420 million in one hundred years—so much lost to the future wealth of the polity.[15]

Though calculations like Cooper's and Webster's spoke to specifically American problems, they closely resembled the British political calculations discussed in this book in style and spirit: in the use of calculation to critique entrenched institutions and practices; in the creative synthesis of empirics and estimates from a variety of sources (public accounts, market prices, demographic approximations); in the way certain, specific calculative techniques, like compound interest, were used. Cooper and Webster offer just two evocative examples of what, I strongly suspect, was an extensive and vibrant culture of political calculation in early America, spanning a variety of geographic sites, institutional settings, and political questions. More research needs to be done, of course, to recover this calculative culture and examine how much it built upon, as well as diverged from, British precedents.

For the moment, I will offer one hypothesis. This book has stressed the use of numerical calculation in post-1688 Britain as an instrument of political critique, especially by calculators fostering "Country," republican political values like Davenant, Hutcheson, and Price. Bernard Bailyn and others have famously argued that this same Country tradition was massively influential in British North America, and indeed central to the "ideological origins of the American Revolution."[16] It seems likely that, along with many other aspects of Country republicanism—an adversarial conception of politics, a fear of arbitrary government, vigilant attention to the specter of corruption—Americans also inherited a respect for calculation as a defense against the abuse of power.[17] Observe, for instance, the esteem in which many early American political figures held Richard Price, perhaps the most outspoken advocate for both Country thinking *and* calculative thinking in late eighteenth-century Britain. Though he did not take up the invitation to migrate to the United States in 1778, Price corresponded frequently with Franklin, Jefferson, John Adams, Benjamin Rush, and Jonathan Trumbull.[18] What Patricia Cline Cohen calls the "republican arithmetic" of early America may have been even more republican than she realized. In fact, I would suggest that the quantitative habit in American civic epistemology may be better understood as a product of American *republicanism* than of American *democracy,* as it is often framed by scholars.[19]

Beyond these more direct, eighteenth-century connections, the quantitative civic epistemology that took hold in nineteenth- and twentieth-century America would continue to resemble that of the eighteenth century in at least two essential ways. First, the elevation of calculation in American public life would continue to be closely entangled with a deep strain of suspicion about individual discretion, elite authority, and centralized power in American political culture. A central argument of Porter's seminal study *Trust in Numbers* is that the fashion for quantitative forms of accounting, analysis, and communication in American bureaucracy can be attributed to what Porter calls a "political context of systematic distrust."[20] Post-1688 Britain was also a political environment defined by pervasive distrust, in which numerical thinking attained special authority as a consequence of that distrust. There were differences between those calculating cultures, to be sure. For example, I contend that early modern British calculators tended to use calculations in an offensive mode, as an instrument of interrogation, whereas Porter emphasizes how American bureaucrats tended to pick up calculation in a defensive mode, as a means to forestall critique. Yet I think the commonalities between those cultures are, in many ways, more telling than the differences. To understand how and why quantitative methods became such potent instruments of bureaucratic defense in modern America, it is crucial to understand the long-standing values and expectations that were already attached to numbers within Anglo-American civic epistemology.

A second parallel between the early modern British and modern American cases concerns the importance of partisan contestation. There is good reason to believe that, as in Britain, party politics and public dispute were essential drivers of the authority of calculation in American civic epistemology. This certainly seems to be the case in the era of the early republic, for the likes of Hamilton and Jefferson, Thomas Cooper and Noah Webster. Calculations would continue to be an object and instrument of public controversy in the nineteenth century as well. Cohen, for instance, has recounted the heated public controversies that arose surrounding the results of the 1840 census.[21] Arwen Mohun has shown how nineteenth-century labor leaders and political reformers marshaled quantitative data on industrial accidents in order to push for new governmental policies protecting workers' safety. Crucially, Mohun argues that these statistical inquiries were originally driven by the imperatives of public political dispute, developed "to justify a legislative path of action, to convince legislators to pass relevant laws, and as a means for mediating conflicting claims about levels of risk between interested parties."[22]

Alain Desrosières has, in fact, argued that one of the distinguishing features of American statistical practice was its connection to adversarial argumentation. He cites the case of an 1857 analysis by "Helper," an economist from the northern United States, who marshaled a battery of statistics on production, wealth, and social conditions to demonstrate the economic and social superiority of northern states over southern. Helper's calculations, like Webster's, were designed to highlight the dire costs of slavery, an argument which provoked stern computational counterattacks from southerners.[23] Desrosières's description is worth quoting at length:

> The fact that, in the United States, statistics was so closely and so rapidly involved in an arena of contradictory debates stimulated critical thinking and encouraged diversity in the uses and interpretations of this tool. More than in other countries, statistical references were linked to the process of argumentation rather than to some truth presumed to be superior to the diverse camps facing off. This way of thinking seems typical of American democracy.[24]

It is remarkable how well this description applies to the calculative politics of post-1688 Britain. It suggests that this way of thinking may have had as much to do with British—and indeed American—republicanism as "American democracy."

Reckoning with Political Calculations in the Twenty-First Century

Given the dense genealogical connections between the post-1688 culture of calculation reconstructed in this book and later calculating culture in Britain and the United States, it is worth asking how those early modern episodes can help us better understand the politics of calculation in the present. So what instruction, or inspiration, do the messy beginnings of Britain's quantitative age have to offer? It goes without saying that there is much about eighteenth-century numerical politics that is not worthy of emulation—the personal insults, the extreme partisan nastiness, the sometimes irresponsible, cynical, and deceptive ways that numerical data was used. (Not to mention the more profound political vices of that era, notably the exclusion of women and most working people from political participation, and the engagement of British state power and domestic politics in destructive imperial enterprises overseas.) Yet I believe that the history of how calculations were used in eighteenth-century politics can help us to recalibrate contemporary understandings of how calculative thinking operates in political life. The stories told in this book offer an evocative archive against which to compare, and thus rethink, political calculations in our own time.

In the numerical stories that make up this book, calculative thinking was overtly, intimately, unavoidably entangled with the pursuit of political agendas. Political calculators in that era could do little to hide the fact that their political calculations were, frankly, *political*—that they were motivated by partisan agendas, that they were laden with personal stakes, and that they encoded a certain vision of how the polity ought to be. Those political imperatives gave calculative thinking its power. The *frank politics* of calculation evident in the post-1688 era unsettles—and thus helps to reconceive—common ideas about the relationship between quantitative thinking and political action in the twenty-first century.

In the present moment, conversations about quantification and politics, both scholarly and public, tend to be organized around what might be called the *anti-politics of calculation*. For many advocates of quantitative thinking—in academia, government, popular media, business, philanthropy, and beyond—the greatest virtue of quantitative data and calculative analysis is precisely its ability to *avoid* or even transcend politics. The perceived ability of quantification, more than any other way of knowing, to create knowledge that is impartial, impersonal, and *apolitical* is the source of its authority. Immunity from politics is, for many, the essence of objectivity. On the other side, critics of quantification—from critical humanistic scholars to skeptical citizens to anti-"establishment" politicians—often point out that behind supposedly apolitical numbers are, in fact, an array of political assumptions, values, and interests. (To be clear, I do not mean to draw too close an analogy between these different modes of quantitative critique: between thoughtful, rigorous scholarly analyses of the shortcomings of quantification on the one hand, and uncritical, conspiratorial, or demagogic dismissals of "rigged" statistics on the other. Yet there are certain commonalities in their lines of criticism that merit attention.) In many such critiques, the goal seems to be to unveil the embedded politics of calculation as a way to debunk quantification's virtue, revealing a kind of hypocrisy. Quantitative data, metrics, and indicators are inherently political when they claim not to be, the argument goes, and therefore are undeserving of the authority with which they are invested.

Neither of these positions, either for or against quantification, gives much credit to the notion that, as a mode of public knowledge, calculation might be both *political* and *virtuous*. The standard opposition between quantitative thinking and political interest strikes me as a false and even dangerous one. Some of the greatest risks both *from* and *to* the use of calculation in public life stem not from the vices of calculative thinking in itself, but rather from this misperception that calculation can, and should, completely transcend politics. This is true both for the quantitative enthusiasts and skeptics. Becoming more frank about the messiness of political calculation—acknowledging that most politically relevant numbers are as much political arguments as they are political answers—can help present and future political communities to avoid threats in both directions.

First, one of the greatest dangers of overconfident quantification is that it yields numbers that deny and obscure their own political underpinnings and effects. Critical scholarship on quantification has been right to call attention to this pernicious tendency. In a recent collection of studies on the use of quantitative metrics in global health initiatives, anthropologist Vincanne Adams observes: "Quantification strategies and the metrics we rely on to *avoid* politics often do not avoid politics at all; they become a form of politics in their own right, augmenting the political stakes and political underpinnings of health projects in a manner that is frequently invisible to those who believe in these exercises in calculation and counting."[25] By obscuring their own politics, such anti-political calculations actually pursue a covert and illusive kind of politics.

Particularly worrisome is that this kind of calculative sleight-of-hand might, in fact, shut down the possibility for meaningful political discourse and engagement, by transforming disputes about political values into putatively "technical" problems and thus "removing them from the realm of contestable politics."[26] In the twenty-first century, one of the greatest threats to a healthy quantitative culture is that the use of quantitative data and analysis will continue to grow in its intensity, influence, and sophistication within many areas of political life, but that such calculative enterprise will take place within exclusive institutions and produce numerical knowledge that is ever more illegible and inaccessible to the public. There is already distressing evidence of this tendency. Sociologist of technology Zeynep Tufekci, for example, observes the growing use of "big data" analytics by political campaigns to target, persuade, and mobilize potential voters. She notes that these new practices of "computational politics" rely on the exploitation of information asymmetries, as major campaigns have access to data about citizens that citizens themselves do not have. Such techniques threaten to erode open political discourse by allowing campaigns to "profile and interact *individually* with voters outside the public sphere." William Davies has recently cautioned that a decline in public trust in traditional statistics has arisen alongside an alternative calculative culture, in which social phenomena are analyzed anonymously and secretly by massive corporations like Google and Facebook using big data techniques. "Few social findings arising from this kind of data analytics ever end up in the public domain," Davies notes.[27]

The stories told in *Calculated Values* offer valuable conceptual resources for checking these excluding tendencies in quantitative culture. They offer a reminder that quantitative data, analyses, and practices can—and should—be subject to public scrutiny, whatever the pull to remove quantitative questions from the "realm of contestable politics." Crucially, they also show that subjecting quantitative facts and figures to political contestation does not automatically undermine their credibility, destroy their usefulness as instruments of empirical knowledge or political deliberation, or produce a widespread crisis of epistemic relativism. Antagonistic contestation over numbers might produce a certain amount of uncomfortable instability around certain key numerical data-points and hinder the achievement of consensus answers. But it also can stimulate critical reflection on quantitative methods and inspire deeper investigation.

Acknowledging the contestability of political numbers may also help to make numerical discourse more democratic. One of the striking features of the calculative politics of the eighteenth century, at least compared to today's, was its relative openness and diversity. It was not yet a thoroughly disciplined enterprise, with established standards of practice and clear hierarchies of expertise. Calculators took many different paths in forging their political lives. Among the major characters in this story: Charles Davenant began his early career in the theater business; David Gregory was a born academic, the nephew of a mathematician; John Crookshanks started as a bookkeeper in Italy (and maybe a pirate); Archibald Hutcheson began his career as a lawyer and imperial official in the Caribbean; Robert Walpole honed his calculating skills managing the family estate in Norfolk; Richard Price came upon his interest in calculation presumably from his study of moral philosophy. Yet the diversity of eighteenth-century calculating culture extended beyond the different biographies of the era's elite calculators. The practice of political calculation was remarkably varied, flexible, and open-ended. Different individuals and parties identified different epistemic virtues in calculation. Different people had different ideas of what made a good calculation.

I believe there is something to be learned from the visible flexibility of calculative discourse in the eighteenth century. It challenges us to imagine what a more participatory kind of quantitative culture might look like—one that seriously acknowledges and even embraces that numbers might speak differently to different people and that the most methodologically sophisticated calculations are not the only useful ones. Much of the scholarship on the history and sociology of quantification emphasizes the rigid and rule-bound character of quantitative techniques, locating their authority in their perceived ability to limit subjectivity, discretion, and creativity. But the case of the eighteenth century suggests something different: that calculation affords a constrained flexibility that allows participants with different viewpoints, methodologies, and epistemic values to engage in meaningful disagreement. What can we learn by seeing this essential flexibility as a virtue rather than a vice—a feature, not a bug? I do not mean to suggest that pressing quantitative questions about politics and society can never be solved in a meaningful sense. Answers can be found; but in the real-time of politics, in which time, data, and attention are often in short supply, such answers are often extremely hard to come by. I also do not mean to make a vulgarly relativistic point about how all calculations are equally correct, justifiable, or persuasive. They are not. Even in the eighteenth century, contemporaries would often have had plenty of good reasons to believe that some calculations were better than others. My point is rather that there is rarely only one good way to think about a problem quantitatively, and promoting a variety of different computational approaches will almost certainly lead to richer computational inquiry.

Embracing the frank politics of numbers might also lead the way to more productive political conversations. Quantitative argumentation can be seen as a powerful means for facilitating what political theorist Teresa Bejan calls "mere civility": that baseline level of mutual respect that, Bejan argues, is necessary for conducting political conversations in pluralistic polities, even in contexts of severe partisan antagonism.[28] Calculative disputes can be incredibly mean-spirited and full of bad acting, yet the sheer fact that partisans on one side agree to answer the other side's numbers with numbers of their own conveys that modicum of recognition that makes meaningful disagreement possible. To try to out-calculate an opponent is, at least, to acknowledge common participation in a space of quantitative reason.

One powerful implication of acknowledging the frank politics of calculation, and embracing its (constrained) flexibility, is that it opens up new possibilities for using calculation as a tool of critique, reform, even protest. One of the most unexpected, and indeed heartening, findings of this book is that Britain's quantitative revolution was powered by critics who wanted to use calculations to make their polity better. Among the most enterprising and innovative calculations of the period, the preponderance came from relative political outsiders using numbers to interrogate powerful individuals and reform degraded institutions. Harley and the Accounts Commissioners accounted for the corruption of Crown agents, Davenant measured government incompetence, and Hutcheson quantified the mysteries of the South Sea Company. In 1713, when the Tories and Whigs squared off over the balance of trade, calculators on both sides imagined that their numbers were necessary to combat the pernicious influence of entrenched authorities: for the Tories, the danger came from self-serving business interests; for the Whigs, it came from an overreaching government. For the radical Richard Price, the light of numbers promised not only to flush out public corruption but to cultivate more rational and moral citizens.

This book is not meant to be a celebration of quantitative "heroes" of the past. Nearly every character was self-serving, unscrupulous, and prejudiced, and they operated within a political world that was unjust in many ways. Yet there is inspiration to be found in these stories all the same. Much of the scholarship on quantification and its history casts calculative rationality as a product of the drive for either economic profit or state control—as part of an unjust assemblage that, as anthropologist Diane M. Nelson has recently put it, includes "capitalism, colonialism, [and] the military-industrial cybernetic debt complex."[29] Calculation is an instrument for accruing power, it seems, more than questioning it. Yet the stories in this book suggest that this need not necessarily be the case. They offer a vivid reminder of the capacity of calculation to illuminate overlooked problems, to unsettle established prejudices, and to demand public reckoning. In fact, in the political realm, calculation may be far better at performing these critical tasks than at actually settling questions. The beginnings of Britain's quantitative age offer a vital early example of the statistical activism (*statactivism*) that scholars have identified to be a powerful mode of social action in more recent history—from high-profile, social-scientific experts like Thomas Piketty calling attention to global income inequality, to activist-quantifiers seeking to demonstrate economic injustices in Guadeloupe. Indeed, this book suggests that such statactivism was, in fact, essential to how quantification gained the authority it holds in the modern world.[30]

By offering models of how quantification might be deployed as a virtuous yet frankly political tool, attending to the history of Britain's quantitative age may thus help to stave off the creeping tendency toward a more exclusive, secretive, and uncontestable calculative culture evident in the twenty-first century—a tendency that is nourished, I believe, by the belief that quantitative thinking is only virtuous insofar as it is a- or *anti*-political. I believe that acknowledging how calculation may be both political and virtuous is also necessary for combating a second—and perhaps far more pernicious—threat: a wholesale rejection of quantitative facts and analysis on the part of a significant portion of the public. Of course, this book has pointed out that suspicions about quantitative thinking in public life are by no means new. Some, like the notion that calculation was a form of conjuring, predate the widespread authority of numbers; others, like Hume's notions about the malleability of "facts and calculations," developed in tandem with that very authority. Yet, as discussed in the Preface, there is some evidence that such anti-quantitative sentiment is as strong now as it has ever been (at the same time that confidence in quantification is arguably as strong as *it* has ever been). For many, these current anti-quantitative sentiments are also aligned with an anti-establishment, "populist right" ideology profoundly distrusting of government, experts, powerful institutions, and traditional media. As Davies has recently observed, such quantitative skeptics feel that "grounding politics in statistics is elitist, undemocratic, and oblivious to people's emotional investments in their community and nation," and fear that purportedly neutral numerical claims are in fact a rigged form of "establishment" propaganda.[31] While twenty-first-century quantitative optimists continue to seek to promote the value of calculative thinking by touting its anti-political character, doubters observe (rightly) that political calculations are political things—and reject them as a result.

It remains to be seen whether this particular populist, anti-quantitative sentiment is truly distinctive from previous precedents, and I think it is certainly too early to proclaim that we have entered a new age of "post-truth" politics. But it is possible to project a distressing trajectory—in which the trustworthiness of quantitative knowledge itself becomes a thoroughly partisan matter, quantitative believers and quantitative nonbelievers line up on opposite sides of the political aisle, and numbers effectively cease to function as a form of interpartisan discourse. Notably, this trajectory is very much commensurable with the further development of technically intensive and exclusive practices of quantification and computation. The truly distressing prospect is a political world in which some significant fraction of the population continues to double-down on the authority of quantitative expertise and computational technologies for resolving political questions, and another significant fraction refuses to engage that mode of thinking altogether.

But such fracture need not come to pass. Again, the solution for preventing it may lie, at least in part, in recalibrating our perceptions of political calculation— in revising the expectation that quantification produces definitive, uncontestable, apolitical answers; in acknowledging the (often productive) flexibility of certain political numbers; in embracing numbers as something *worth arguing about*. Populistic distrust of statistics and calculations appears to stem from an image of numbers as elitist, emotionally disengaged, ideologically biased, and—perhaps most of all—hypocritical, in that they purport to be apolitical when, in fact, they simply perform a hidden (establishment) politics. The calculative politics of the eighteenth century offer an alternative image of what functions calculation might perform in political life, which challenges both quantitative supporters and skeptics to think about quantitative thinking differently. In that period, calculation was not the exclusive province of aloof, elitist experts, but a more diverse and public tool. It was not emotionally disengaged, but rather a means of expressing and contesting deeply held feelings about the fate of the nation. It was not monopolized by one ideological faction, but was rather a strikingly trans-partisan tool used across the political spectrum. And crucially, it did not hide its politics behind a forbidding veneer of technicality or objectivity.

There is something empowering in the model of quantitative politics offered by the eighteenth century, I think. It urges us to imagine ways that calculation might be more diverse, contentious, emotionally resonant, and politically frank. It also urges us to be more thoughtful, and indeed humble, about the political limitations of quantitative knowledge: to acknowledge that most numerical facts in political life are actually the result of complicated calculations that contain many discretionary choices and debatable assumptions; to accept that, owing to the sufficient "give" in many calculations, it is often possible to make a plausible numerical case defending contrasting political viewpoints, especially when it comes to those political questions we care about the most; to appreciate that calculations are often creative tools for argument rather than rigid mechanisms for getting answers.[32] Greater frankness and humility about the nature of calculative thinking in political life can, I believe, help defend against the corrosive perception that numerical calculation is an exclusive, elitist, and inaccessible form of reasoning. Writing about one of the great political questions of his day, the balance of trade, Hume noted in 1752 that "every man, who has ever reason'd on this subject, has always prov'd his theory, whatever it was, by facts and calculations." Hume meant this as a criticism, but perhaps we can read it as an aspiration. It is far better to live in a polity where every citizen (man or woman) feels empowered to engage in quantitative conversations, than one in which no one does.

Archival Collections Cited

Note: the items listed below include those archives that have been cited in the book. Other archival materials consulted in the course of the project are not specifically listed below. Abbreviations for specific archives and archival collections referenced in notes are identified in *italics*.

The British Library, London—*BL*
Additional Manuscripts—*Add. MS*
Harley Manuscripts
Lansdowne Manuscripts
Sloane Manuscripts
Stowe Manuscripts

Cambridge University Library, Cambridge, UK—*Cambridge U. L.*
Cholmondeley (Houghton) Papers—*Ch(H) Papers*
Cholmondeley (Houghton) Letters—*Ch(H) Letters*

William L. Clements Library, University of Michigan, Ann Arbor, MI
William Petty, 1st Marquis of Lansdowne, 2nd Earl of Shelburne Papers
Charles Townshend Papers

Edinburgh University Library, Centre for Research Collections, Edinburgh—*Edinburgh U. L.*

Gregory Papers
Laing Manuscripts

The Folger Library, Washington, DC—*Folger*
Folger Library Manuscripts

The Huntington Library, San Marino, CA
Huntington Manuscripts—mss*HM*
Stowe Brydges Manuscripts—mss*ST*

The National Archives, Kew—*TNA*
State Papers—*SP*
Treasury Papers—*T*

The National Library of Scotland, Edinburgh—*NLS*
Advocates Manuscripts—*Adv. MS*
Fletcher of Saltoun Papers

National Records of Scotland, Edinburgh—*NRS*
Papers of the Clerk Family of Penicuik, Midlothian—*GD 18*
Papers of the Graham Family, Dukes of Montrose (Montrose
    Muniments)—*GD220*
Papers of the Hamilton-Dalrymple Family of North Berwick—*GD110*
Papers of the Ogilvy Family of Inverquharity—*GD205*

Royal Bank of Scotland Archives, Edinburgh—*RBS*
Equivalent Papers—*EQ*

Notes on Printed Sources

Many English printed texts of the period have lengthy titles, often with multiple parts or subtitles. For reasons of space and clarity, I have often truncated these titles when discussing them in the text or referencing them in the notes, though I have sought to include sufficient title information to avoid any possible ambiguity about sources. While contemporary title pages often displayed a great variety of typefaces and capitalization conventions, I have standardized the form of all titles according to American English conventions.

This study relies heavily on ephemeral print sources, for which publication data (date, publisher, place of publication, etc.) is often incomplete. Furthermore, many of the texts reviewed here were published anonymously or under pseudonyms. Though subsequent research has made it possible to attribute specific authors to many such texts, attribution remains a contested issue. (This is particularly true in the case of well-known literary figures like Daniel Defoe, on which see P. N. Furbank and W. R. Owens, *The Canonisation of Daniel Defoe* [New Haven, CT: Yale University Press, 1988].) Unless otherwise stated, I have chosen to follow the attribution and publication data provided in the English Short Title Catalogue (ESTC), available online at estc.bl.uk. In cases where I believe the ESTC information is questionable, I have indicated as much in the notes. Where texts were originally published anonymously or pseudonymously but authorship has subsequently been attributed, I have indicated the author's name in brackets in the first citation in the Notes (e.g., [William Paterson]). Where no author has been attributed to a given text, I have noted the author as [Anon.].

In general, all materials are dated according to the nominal date provided within the text. In specific situations in which a publication is explicitly dated in one year (e.g., 1714), but for which context indicates a publication date sometime between January 1 and March 25 of the following year (1715), I have indicated the date as such (1714 / 5).

This project has benefited tremendously from the extensive digital archives of seventeenth- and eighteenth-century English print that have been compiled since about 2000. The three most useful for this study have been *Eighteenth Century Collections Online*; *The Making of the Modern World: The Goldsmiths'-Kress Library of Economic Literature 1450–1850*; and *Early English Books Online*. Many useful primary sources on British political history can be found on *British History Online* (www.british-history.ac.uk), including references to the *Calendar of State Papers,* the *Calendar of Treasury Books*, the *Calendar of Treasury Papers*, Grey's *Debates*, the *History and Proceedings of the House of Commons*, the *Journal of the House of Commons* (1547–1699), and the *Journal of the House of Lords*. Many key reference works relevant to the period are also available online. Articles in the *History of Parliament* series can be found at www.historyofparliamentonline.org. Articles in the *Oxford Dictionary of National Biography*, ed. H. C. G. Matthew and Brian Harrison (Oxford: Oxford University Press, 2004), can be found at www.oxforddnb.com. Because nearly all contemporary printed sources used in this study have been accessed through one of the digital archives above, I have not separately noted when sources were reviewed in digital form. For digital sources for which a URL is provided, URL information is correct as of August 31, 2017, unless otherwise noted.

Abbreviations

BL	British Library
Cambridge U. L.	Cambridge University Library, Special Collections
Clements Lib.	Clements Library, University of Michigan
CSP	*Calendar of State Papers*
CTB	*Calendar of Treasury Books*
CTP	*Calendar of Treasury Papers*
DNB	*Dictionary of National Biography*
DSB	*Dictionary of Scientific Biography*
Edinburgh U. L.	University of Edinburgh Library
HMC	Historical Manuscripts Commission
HMSO	Her / His Majesty's Stationery Office
Huntington	Huntington Library
NLS	National Library of Scotland
NRS	National Records of Scotland
RBS	Royal Bank of Scotland
TNA	The National Archives (United Kingdom), Kew

Notes

Preface: Quantification and Its Discontents

1. Steven D. Levitt and Stephen J. Dubner, *Freakonomics: A Rogue Economist Explores the Hidden Side of Everything* (New York: William Morrow, 2005), 14.
2. Carmen M. Reinhart and Kenneth S. Rogoff, "Growth in a Time of Debt," *American Economic Review* 100, no. 2 (May 2010): 573–578; Thomas Herndon, Michael Ash, and Robert Pollin, "Does High Public Debt Consistently Stifle Economic Growth? A Critique of Reinhart and Rogoff," *Cambridge Journal of Economics* 38, no. 2 (Mar. 2014): 257–279; Thomas Piketty, *Capital in the Twenty-First Century,* trans. Arthur Goldhammer (Cambridge, MA: Belknap Press of Harvard University Press, 2014); Chris Giles and Ferdinando Giugliano, "Thomas Piketty's Exhaustive Inequality Data Turn Out to Be Flawed," *Financial Times,* May 23, 2014, http://www.ft.com/cms/s/0/c9ce1a54-e281-11e3-89fd-00144feabdc0.html; Louis Woodhill, "Thomas Piketty Gets the Numbers Wrong," *Forbes,* May 6, 2014, http://www.forbes.com/sites/louiswoodhill/2014/05/06/thomas-piketty-gets-the-numbers-wrong. For Piketty's response, see "My Response to the Financial Times," *Huffpost* (blog), *Huffington Post,* May 29, 2014, http://www.huffingtonpost.com/thomas-piketty/response-to-financial-times_b_5412853.html.
3. Hal R. Varian, "Recalculating the Costs of Global Climate Change," *New York Times,* Dec. 14, 2006; William D. Nordhaus, "A Review of the *Stern Review on the Economics of Global Warming,*" *Journal of Economic Literature* 45 (Sept. 2007): 686–702; Nicholas Stern, "The Economics of Climate Change," *American Economic Review* 98, no. 2 (2008): 1–37.

4. Robert Pear, "Obama-Ryan Battle Intensifies over Medicare Savings," *New York Times,* Aug. 14, 2012; Trudy Lieberman, "Medicare and the $716 Billion Bogeyman," *Columbia Journalism Review,* Aug. 22, 2012, http://www.cjr.org/campaign_desk /medicare_and_the_716_billion_b.php.

5. Jacob Soll, "Greece Owes Less than Europe Says: How Its Creditors Use Their Political Clout to Keep the Debt Number High," *Politico,* Jul. 2, 2015, http://www. politico.com/agenda/story/2015/07/greece-owes-less-than-europe-says-000132.

6. Bureau of Labor Statistics, "How the Government Measures Unemployment," last modified Oct. 8, 2015, https://www.bls.gov/cps/cps_htgm.htm; Adam Davidson, "Trump and the Truth: The Unemployment-Rate Hoax," *New Yorker,* Sep. 10, 2016.

7. "Lies, Damned Lies, and Statistics," Department of Mathematics, University of York, last modified Jul. 19, 2012, http://www.york.ac.uk/depts/maths/histstat/lies .htm; Darrell Huff, *How to Lie with Statistics* (New York: W. W. Norton, 1954); J. Michael Steele, "Darrell Huff and Fifty Years of *How to Lie with Statistics,*" *Statistical Science* 20, no. 3 (2005): 205–209; R. H. Coase, "How Should Economists Choose?," in *Essays on Economics and Economists* (Chicago: University of Chicago Press, 1994), 15–33, on 27; Gary Smith, *Standard Deviations: Flawed Assumptions, Tortured Data, and Other Ways to Lie with Statistics* (New York: Overlook / Duckworth, 2014); Cathy O'Neil, *Weapons of Math Destruction: How Big Data Increases Inequality and Threatens Democracy* (New York: Crown, 2016).

8. Nate Silver, *The Signal and the Noise: Why So Many Predictions Fail—But Some Don't* (New York: Penguin, 2012).

9. William Davies, "How Statistics Lost Their Power—and Why We Should Fear What Comes Next," *Guardian,* Jan. 19, 2017; Scott Clement, "Survey Says: Polls are Biased," *Washington Post,* Sep. 4, 2013; Joel Faulkner Rogers, "Are Conspiracy Theories for (Political) Losers?" *YouGov UK,* Feb. 13, 2015, https://yougov.co.uk /news/2015/02/13/are-conspiracy-theories-political-losers/; Jill Lepore, "Politics and the New Machine," *New Yorker,* Nov. 16, 2015; Kai Ryssdal, "Poll Finds Americans' Economic Anxiety Reaches New High," *Marketplace,* Oct. 13, 2016, http://www.marketplace.org/2016/10/13/economy/americans-economic-anxiety -has-reached-new-high; Catherine Rampell, "When the Facts Don't Matter, How Can Democracy Survive?," *Washington Post,* Oct. 17, 2016.

10. For one discussion of how such "complex valuation" problems transpire in legal settings, see Anthony J. Casey and Julia Simon-Kerr, "A Simple Theory of Complex Valuation," *Michigan Law Review* 113, no. 7 (2015): 1175–1218.

11. In focusing on calculation as a practice, I draw inspiration from scholars in several subfields, including critical accounting studies and the history of the mathematical sciences. For accounting, see Peter Miller, "Governing by Numbers: Why Calculative Practices Matter," *Social Research* 68, no. 2 (Summer 2001): 379–396. For the history of science, see Andrew Warwick, *Masters of Theory: Cambridge and the Rise of Mathematical Physics* (Chicago: University of Chicago Press, 2003); David Kaiser, *Drawing Theories Apart: The Dispersions of Feynman Diagrams in Postwar Physics* (Chicago: University of Chicago Press, 2005).

12. On information and its political history, specifically in Britain, see Daniel Headrick, *When Information Came of Age: Technologies of Knowledge in the Age of Reason and Revolution, 1700–1850* (New York: Oxford University Press, 2000); Edward Higgs, *The Information State in England: The Central Collection of Information on Citizens since 1500* (Basingstoke, UK: Palgrave Macmillan, 2004); Paul Slack, "Government and Information in Seventeenth-Century England," *Past & Present* 184, no. 1 (Aug. 2004): 33–68; Slack, *The Invention of Improvement: Information and Material Progress in Seventeenth-Century England* (Oxford: Oxford University Press, 2015); Joanna Innes, *Inferior Politics: Social Problems and Social Policies in Eighteenth-Century Britain* (Oxford: Oxford University Press, 2009), chap. 4.

13. Lisa Gitelman, ed., *"Raw Data" Is an Oxymoron* (Cambridge, MA: MIT Press, 2013).

14. *Oxford English Dictionary Online*, s.v. "calculate" and "calculating," www.oed.com.

Introduction: Political Calculations

1. *Facts and Figures, a Periodical Record of Statistics Applied to Current Questions*, no. 1, Oct. 1, 1841 (London: H. Hooper), 1.

2. William Pulteney, *A State of the National Debt, As It Stood December the 24th, 1716. With the Payments Made towards the Discharge of It out of the Sinking Fund, &c. Compared with the Debt at Michaelmas, 1725* (London: Printed for R. Francklin, 1727), 15. The Oxford English Dictionary cites a quotation by Abel Boyer as the term's oldest usage. See Abel Boyer, *Political State of Great Britain*, vol. 34 (London: Printed for the Author and Sold by T. Warner, 1727), 130. Boyer's text was a serial news compendium, quoting from Pulteney's original. Searches in Google Books and Eighteenth-Century Collections Online show no earlier uses. Pulteney may have been putting a numerical spin on another adage: "facts are stubborn things." See Garson O'Toole, "Facts Are Stubborn Things," *Quote Investigator*, June 18, 2010, http://quoteinvestigator.com/2010/06/18/facts-stubborn.

3. Pulteney, *State of the National Debt*, 1.

4. A series of important studies has explored the emergence of the "fact" in British culture and epistemology during the early modern period. This book focuses on the related—but distinctive—history of "figures." See Lorraine Daston, "Baconian Facts, Academic Civility, and the Prehistory of Objectivity," *Annals of Scholarship* 8, nos. 3–4 (1991): 337–363; Mary Poovey, *A History of the Modern Fact: Problems of Knowledge in the Sciences of Wealth and Society* (Chicago: University of Chicago Press, 1998); Barbara Shapiro, *A Culture of Fact: England, 1550–1720* (Ithaca, NY: Cornell University Press, 2000).

5. J. R. McCulloch, *A Statistical Account of the British Empire: Exhibiting Its Extent, Physical Capacities, Population, Industry, and Civil and Religious Institutions*, vol. 1 (London: Printed for Charles Knight and Co., 1837), [v]; also quoted in Innes, *Inferior Politics*, 109.

6. Ian Hacking, "Biopolitics and the Avalanche of Printed Numbers," *Humanities in Society* 5, no. 3–4 (Summer / Fall 1982): 279–295; Hacking, *The Taming of Chance* (Cambridge: Cambridge University Press, 1990); Theodore M. Porter, *The Rise of*

Statistical Thinking, 1820–1900 (Princeton, NJ: Princeton University Press, 1986). For a recent example of a study that reaffirms this organizing chronology, see Jean-Guy Prévost and Jean-Pierre Beaud, *Statistics, Public Debate and the State, 1800–1945: A Social, Political and Intellectual History of Numbers* (London: Pickering & Chatto, 2012), 23. Scholars examining social and political uses of quantification in later periods often cite Hacking's "avalanche" as a starting point, for instance: Wendy Nelson Espeland and Michael Sauder, "Rankings and Reactivity: How Public Measures Recreate Social Worlds," *American Journal of Sociology* 113, no. 1 (July 2007): 1–40, on 4; Sally Engle Merry, "Measuring the World: Indicators, Human Rights, and Global Governance," *Current Anthropology* 52, no. S3 (Apr. 2011): S83–S95, on S85. Other studies have pushed this chronology back slightly, observing a distinctive "quantifying spirit" in place by the late eighteenth century, but not before. See Tore Frängsmyr, J. L. Heilbron, and Robin E. Rider, eds., *The Quantifying Spirit in the Eighteenth Century* (Berkeley: University of California Press, 1990); M. Norton Wise, ed., *The Values of Precision* (Princeton, NJ: Princeton University Press, 1995); Lars Behrisch, "Statistics and Politics in the 18th Century," *Historical Social Research / Historische Sozialforschung* 41, no. 2 (2016): 238–257, esp. 238.

7. This book complements a growing collection of studies by British political and social historians that have explored the history of numerical practices during the long eighteenth century, bringing those studies into conversation with STS scholarship on quantification and civic epistemology. See Colin Brooks, "Projecting, Political Arithmetic and the Act of 1695," *English Historical Review* 97, no. 382 (Jan. 1982): 31–53; John Brewer, *The Sinews of Power: War, Money, and the English State, 1688–1783,* paperback ed. (Cambridge, MA: Harvard University Press, 1990 [1989]), chap. 8; Julian Hoppit, "Political Arithmetic in Eighteenth-Century England," *Economic History Review* 49, no. 3 (Aug. 1996): 516–540; Andrea Rusnock, *Vital Accounts: Quantifying Health and Population in Eighteenth-Century England and France* (Cambridge: Cambridge University Press, 2002); Innes, *Inferior Politics,* chap. 4; Philip Loft, "Political Arithmetic and the English Land Tax in the Reign of William III," *Historical Journal* 56, no. 2 (June 2013): 321–343; Slack, *Invention of Improvement,* chap. 6.

8. [Robert Walpole], *Some Considerations Concerning the Publick Funds, the Publick Revenues, and the Annual Supplies, Granted by Parliament. Occasion'd by a Late Pamphlet, Intitled,* An Enquiry into the Conduct of our Domestick Affairs (London: Printed for J. Roberts, 1735), 7.

9. Lorraine Daston and Peter Galison observe that the historical emergence of new attitudes about knowledge does not necessitate the destruction of older ones: "Although they may sometimes collide, epistemic virtues do not annihilate one another like rival armies." See *Objectivity* (New York: Zone Books, 2007), 363.

10. After Pulteney's first usage, "facts and figures" began to show up with regularity in eighteenth-century texts, usually amid political-economic debates. For example, see John Bennet, *Two Letters and Several Calculations on the Sugar Colonies and Trade* (London: Printed for J. Montagu, 1738), 50; *Daily Gazetteer,* no. 2076, Feb. 11, 1742; [John Perceval, Earl of Egmont], *Faction Detected, by the Evidence of*

Facts (London: Printed for J. Roberts, 1743), 136; [Anon.], *Considerations on the Revenues of Ireland: Shewing, the Right, Justice, and Necessity, of Now Applying the Duties Granted There for Guarding of the Seas, to Naval Services* (London: Printed for M. Cooper, 1757), 18; "The New Stamp Duty Pernicious to Charity and Heroism," *London Magazine, or Gentleman's Monthly Intelligencer,* vol. 26 (London: Printed for R. Baldwin, 1757), 184.

11. This book seeks to add a new, historical chapter to a growing literature on the social studies of quantification. For an overview of this literature, see Wendy Nelson Espeland and Mitchell L. Stevens, "A Sociology of Quantification," *European Journal of Sociology* 49, no. 3 (Dec. 2008): 401–436; Rainer Diaz-Bone and Emmanuel Didier, "Introduction: The Sociology of Quantification—Perspectives on an Emerging Field in the Social Sciences," *Historical Social Research/Historische Sozialforschung* 41, no. 2 (2016): 7–26.

12. The seminal study is Theodore M. Porter, *Trust in Numbers: The Pursuit of Objectivity in Science and Public Life* (Princeton, NJ: Princeton University Press, 1995).

13. Clark A. Miller, "Civic Epistemologies: Constituting Knowledge and Order in Political Communities," *Sociology Compass* 2, no. 6 (2008): 1896–1919, on 1896. The foundational discussion is Sheila Jasanoff, *Designs on Nature: Science and Democracy in Europe and the United States* (Princeton, NJ: Princeton University Press, 2005), chap. 10. See also Jasanoff, "Acceptable Evidence in a Pluralistic Society," in *Acceptable Evidence: Science and Values in Risk Management,* ed. Deborah G. Mayo and Rachelle D. Hollander (Oxford: Oxford University Press, 1991), 29–47; Jasanoff, "Cosmopolitan Knowledge: Climate Science and Global Civic Epistemology," in *The Oxford Handbook of Climate Change and Society,* ed. John S. Dryzek, Richard B. Norgaard, and David Schlosberg (Oxford: Oxford University Press, 2011), 129–143; Rebecca Slayton, *Arguments That Count: Physics, Computing, and Missile Defense, 1949–2012* (Cambridge, MA: MIT Press, 2013).

14. On the comparative method, see Jasanoff, *Designs on Nature,* chap. 1. Several important comparative studies of such themes do not directly adopt the "civic epistemology" label: Porter, *Trust in Numbers;* John Carson, *The Measure of Merit: Talents, Intelligence, and Inequality in the French and American Republics, 1750–1940* (Princeton, NJ: Princeton University Press, 2007); Marion Fourcade, *Economists and Societies: Discipline and Profession in the United States, Britain, and France, 1890s to 1990s* (Princeton, NJ: Princeton University Press, 2009).

15. On Germany, see Jasanoff, *Designs on Nature,* esp. 289; Jasanoff, "Cosmopolitan Knowledge," 137–138. On France, see Porter, *Trust in Numbers,* chap. 6; Carson, *Measure of Merit,* 3–6 and chap. 4; Fourcade, *Economists and Societies,* 50–59. On the United States, see Porter, *Trust in Numbers,* chaps. 7–8; Miller, "Civic Epistemologies," 1899. ("In the case of the United States," Miller writes, "the predominance of quantification and statistical forms of knowledge can be seen across broad aspects of administrative, regulatory, and legal knowledge-orders.")

16. Scholars have compellingly examined how calculative practices have come to function as powerful forms of rhetoric. See James A. Aho, "Rhetoric and the Invention of Double Entry Bookkeeping," *Rhetorica: A Journal of the History of Rhetoric* 3,

no. 1 (Winter 1985): 21–43; Deirdre McCloskey, *The Rhetoric of Economics*, 2nd ed. (Madison: University of Wisconsin Press, 1998 [1985]), chaps. 3, 7; Bruce G. Carruthers and Wendy Nelson Espeland, "Accounting for Rationality: Double-Entry Bookkeeping and the Rhetoric of Economic Rationality," *American Journal of Sociology* 97, no. 1 (July 1991): 31–69; Poovey, *History of the Modern Fact*.

17. Compare Alain Desrosières's discussion of the "artifices" used by early modern statistical thinkers: *The Politics of Large Numbers: A History of Statistical Reasoning,* trans. Camille Naish (Cambridge, MA: Harvard University Press, 1998 [1993]), 28–29.

18. These exercises in computational critique can be seen as early examples of the "statactivism" evident in certain modern social movements. See Isabelle Bruno, Emmanuel Didier, and Tommaso Vitale, "Statactivism: Forms of Action between Disclosure and Affirmation," *Partecipazione & Conflitto* 7, no. 2 (2014): 198–220.

19. Espeland and Stevens, "Sociology of Quantification," 403, 407; Lorraine Daston, "The Moral Economy of Science," in *Constructing Knowledge in the History of Science,* ed. Arnold Thackray, *Osiris* 10 (1995): 2–24, esp. 8–12.

20. Slack, *Invention of Improvement,* 175.

21. Desrosières, *Politics of Large Numbers,* 16. On Desrosières's influence, see Emmanuel Didier, "Alain Desrosières and the Parisian Flock. Social Studies of Quantification in France since the 1970s," *Historical Social Research/Historische Sozialforschung* 41, no. 2 (2016): 27–47. Recently, scholars like Tom Crook and Glen O'Hara have pointed out that scholarship on quantification has tended to overemphasize state imperatives: "the dominant concern remains with how numbers 'objectified' society, thereby furnishing the epistemological 'conditions of possibility' for the emergence of the modern state and bureaucratic power." See "The 'Torrent of Numbers': Statistics and the Public Sphere in Britain, c. 1800–2000," in *Statistics and the Public Sphere: Numbers and People in Modern Britain, c. 1800–2000,* ed. Tom Crook and Glen O'Hara (New York: Routledge, 2011), 1–32, on 3.

22. One key inspiration for this line of inquiry was Michel Foucault, who gestured at the central role statistics (especially demographic statistics) play in modern forms of *bio-power.* The most extended discussion is: *Security, Territory, Population: Lectures at the Collège de France, 1977–78,* trans. Graham Burchell, ed. Michel Senellart (Basingstoke, UK: Palgrave Macmillan, 2009), chaps. 1–4, 10, 12. On Foucault and quantification: Diaz-Bone and Didier, "Introduction: The Sociology of Quantification," 13–15.

23. On calculation and colonial power, see, to begin: Bernard S. Cohn, "The Census, Social Structure and Objectification in South Asia," in *An Anthropologist Among the Historians and Other Essays* (Oxford: Oxford University Press, 1987): 224–255; Arjun Appadurai, "Number in the Colonial Imagination," in *Orientalism and the Postcolonial Predicament: Perspectives on South Asia,* ed. Carol A. Breckenridge and Peter van der Veer (Philadelphia: University of Pennsylvania Press, 1993), 314–339; James C. Scott, *Seeing Like a State: How Certain Schemes to Improve the Human Condition Have Failed* (New Haven, CT: Yale University Press, 1998), 76–83; Timothy Mitchell, *Rule of Experts: Egypt, Techno-politics, Modernity* (Berkeley: University of California Press, 2002), chap. 3.

24. Appadurai, "Number in the Colonial Imagination," 317.
25. On innovative uses of numbers and calculation by the state in eighteenth-century Britain, see William J. Ashworth, *Customs and Excise: Trade, Production, and Consumption, 1640–1815* (Oxford: Oxford University Press, 2003), chaps. 14, 15; Matthew Neufeld, "The Biopolitics of Manning the Royal Navy in Late Stuart England," *Journal of British Studies* 56, no. 3 (July 2017): 506–531. Note, however, that both of those studies complement, rather than contradict, the account told in this book in important ways. Many of the distinctive advances in state quantification described by Ashworth took place only in the mid- to late eighteenth century, toward the end of the period covered here. Neufeld's study of the use of political arithmetic in managing naval manpower suggests that such arithmetical thinking was at least as important as a tool of public political argumentation and state propaganda as it was a tool of state administrative control.
26. Porter, *Trust in Numbers,* viii.
27. Ibid., chap. 7.
28. Jacob Soll, *The Reckoning: Financial Accountability and the Rise and Fall of Nations* (New York: Basic Books, 2014), xii. See also Carruthers and Espeland, "Accounting for Rationality"; Poovey, *History of the Modern Fact,* 29–30, 63–65.
29. For example, Bruno Latour suggests that the ability of numerical devices to travel well without losing their meaning—to function as "immutable mobiles"—is a crucial source of their power in modern society. See *Science in Action: How to Follow Scientists and Engineers through Society* (Cambridge, MA: Harvard University Press, 1987), esp. 237–241. See also Wise, *Values of Precision,* 6–7, 92–99. For recent studies of the challenges of transmitting knowledge reliably, see Peter Howlett and Mary S. Morgan, eds., *How Well Do Facts Travel? The Dissemination of Reliable Knowledge* (Cambridge: Cambridge University Press, 2010); Miller, "New Civic Epistemologies," 427.
30. Porter, *Trust in Numbers,* 194.
31. For example, Sarah Igo's *The Averaged American: Surveys, Citizens, and the Making of a Mass Public* (Cambridge, MA: Harvard University Press, 2007) links the emergence of a new fashion for quantitative social-survey data in mid-twentieth-century America to trends in popular culture, citizenship, and identity formation, rather than trust or control.
32. Slack, *Invention of Improvement,* 43–52. For Fish's text, see *A Supplicacyon for the Beggars,* ed. Frederick J. Furnivall, in *Four Supplications. 1529–1533 A.D.* (London: Published for the Early English Text Society by N. Trüber & Co., 1871), 1–18.
33. D'Maris Coffman, "Credibility, Transparency, Accountability and the Public Credit under the Long Parliament and Commonwealth, 1643–1653," in *Questioning Credible Commitment: Perspectives on the Rise of Financial Capitalism,* ed. D'Maris Coffman, Adrian Leonard, and Larry Neal (Cambridge: Cambridge University Press, 2013), 76–103; Coffman, *Excise Taxation and the Origins of Public Debt* (Basingstoke, UK: Palgrave Macmillan, 2013).
34. Keith Wrightson, *English Society 1580–1680,* rev. ed. (Abingdon, UK: Routledge, 2003 [1982]), 142–143.

35. See, for example: William Petty, "Political Arithmetick," in *The Economic Writings of Sir William Petty,* ed. Charles Henry Hull, reprint ed. (Fairfield, NJ: Augustus Kelly, 1986), 244. On Petty's political arithmetic, see, especially, Ted McCormick, *William Petty and the Ambitions of Political Arithmetic* (Oxford: Oxford University Press, 2009). Also see Peter Buck, "Seventeenth-Century Political Arithmetic: Civil Strife and Vital Statistics," *Isis* 68, no. 1 (Mar. 1977): 67–84; Poovey, *History of the Modern Fact,* 120–138; Rusnock, *Vital Accounts,* chap. 1; Slack, *Invention of Improvement,* 116–128.

36. Olive Coleman, "What Figures? Some Thoughts on the Use of Information by Medieval Governments," in *Trade, Government and Economy in Pre-Industrial England: Essays Presented to F. J. Fisher,* ed. D. C. Coleman and A. H. John (London: Weidenfeld and Nicolson, 1976), 96–112, on 110; Slack, "Government and Information," 58.

37. Terence Hutchison, *Before Adam Smith: The Emergence of Political Economy, 1662–1776* (Oxford: Blackwell, 1988), 54.

38. McCormick, *William Petty,* 287.

39. Roger Ascham, "The Scholemaster," in *The Whole Works of Roger Ascham,* ed. Rev. Dr. Giles, vol. 3 (London: John Russell Smith, 1864), 100; A. J. Turner, "Mathematical Instruments and the Education of Gentlemen," *Annals of Science* 30, no. 1 (1973): 51–88, on 54; A. G. Howson, *A History of Mathematics Education in England* (Cambridge: Cambridge University Press, 1982), 24–25, 31, 244 n. 5; Keith Thomas, "Numeracy in Early Modern England," *Transactions of the Royal Historical Society* 37 (1987): 103–132, esp. 109–111.

40. William Shakespeare, *The Tragedy of Othello, Moor of Venice* (1604), I.1.19–20; Patricia Parker, "Cassio, Cash, and the 'Infidel o': Arithmetic, Double-Entry Bookkeeping, and *Othello*'s Unfaithful Accounts," in *A Companion to the Global Renaissance: English Literature and Culture in the Era of Expansion,* ed. Jyotsna G. Singh (Chichester, UK: Wiley-Blackwell, 2009), 223–241.

41. Francis Osborne, *Advice to a Son. The Second Part* (London: Printed for Tho. Robinson, 1658), 40, 42.

42. Steven Shapin, *A Social History of Truth: Civility and Science in Seventeenth-Century England* (Chicago: University of Chicago Press, 1994), chap. 7, quotation on 336. Compare the more contentious mode of intellectual civility, and the greater emphasis on mathematics, that prevailed at Jacques Le Pailleur's informal academy in France around the same time. See Matthew Jones, *The Good Life in the Scientific Revolution: Descartes, Pascal, Leibniz, and the Cultivation of Virtue* (Chicago: University of Chicago Press, 2006), 91–95.

43. Porter has forcefully shown why we should not be content with the "usual answer" that "quantification became a desideratum of social and economic investigation as a result of its successes in the study of nature" (*Trust in Numbers,* viii).

44. Francis Osborne, *Advice to a Son, or, Directions for a Better Conduct. The Sixt[h] Edition* (Oxford: Printed by H. H. for Tho. Robinson, 1658 [1655]), 8.

45. J. Peter Zetterberg, "The Mistaking 'the Mathematicks' for Magic in Tudor and Stuart England," *Sixteenth Century Journal* 11, no. 1 (Spring 1980): 83–97; Stephen

Johnston, "Mathematical Practitioners and Instruments in Elizabethan England," *Annals of Science* 48 (1991): 319–344, esp. 320; Katherine Neal, "The Rhetoric of Utility: Avoiding Occult Associations for Mathematics through Profitability and Pleasure," *History of Science* 37, no. 2 (June 1999): 151–178, esp. 154–158.

46. On the fascinating figure of John Dee and mathematics, see, to begin, Jenny Rampling, "The Elizabethan Mathematics of Everything: John Dee's 'Mathematical Praeface' to Euclid's *Elements,*" *BHSM Bulletin: Journal of the British Society for the History of Mathematics* 26, no. 3 (2011): 135–146.

47. Benjamin Wardhaugh, *Poor Robin's Prophecies: A Curious Almanac, and the Everyday Mathematics of Georgian Britain* (Oxford: Oxford University Press, 2012), 7, 9. The literature on astrology is vast. See, to begin, Harry Rusche, "Merlini Anglici: Astrology and Propaganda from 1644 to 1651," *English History Review* 80, no. 315 (Apr. 1965): 322–333; Nicolas H. Nelson, "Astrology, *Hudibras,* and the Puritans," *Journal of the History of Ideas* 37, no. 3 (July–Sept. 1976): 521–536; Bernard S. Capp, *English Almanacs, 1500–1800: Astrology and the Popular Press* (Ithaca, NY: Cornell University Press, 1979), 101; Patrick Curry, *Prophecy and Power: Astrology in Early Modern England* (Princeton, NJ: Princeton University Press, 1989); Ann Geneva, *Astrology and the Seventeenth-Century Mind: William Lilly and the Language of the Stars* (Manchester: Manchester University Press, 1995), 9.

48. For example, Wardhaugh, *Poor Robin's Prophecies,* 206–210.

49. Evidence on the growing prominence of *calculate* in English usage is based on two separate bibliometric analyses. One is based on the author's analysis of keyword searches for the root *calcul-* in Early English Books Online (1600–1700) and Eighteenth-Century Collections Online (1700–1800). Analysis examined the percentage of total texts containing a *calcul-* term at least once, taken in five-year intervals (1600–1604, 1605–1609, etc.), conducted June 2014. The other is based on a query of the GoogleBooks corpus for a collection of *calcul-* terms using Google "nGrams," conducted August 2016. The two analyses yield similar conclusions.

50. On numeracy in the early modern period, the classic study is Thomas, "Numeracy," on 128. More comprehensive is the excellent recent study by Jessica Marie Otis, "By the Numbers: Understanding the World in Early Modern England" (Ph.D. diss., University of Virginia, 2013).

51. See Ted McCormick, "Political Arithmetic's 18th Century Histories: Quantification in Politics, Religion, and the Public Sphere," *History Compass* 12, no. 3 (Mar. 2014): 239–251.

52. For example, Edward Cocker, *Cocker's Decimal Arithmetick,* 6th ed. (London: Printed for J. Darby et al., 1729); J. S., *The Shepherd's Kalender: or, The Citizen's And Country Man's Daily Companion; Treating of Many Things That Are Useful and Profitable to Man-Kind,* 3rd ed. (London: Printed for Tho. Norris, [1725?]); John Partridge, *Merlinus Liberatus: An Almanack for the Year of Our Redemption 1727* (London: Printed for J. Roberts, 1727); *Comes Commercii: or, the Trader's-Companion,* 5th ed. (London: Printed by T. W. for J. and J. Knapton et al., 1727); Edward Oldenburgh, *A Calculation of Foreign Exchanges, as Transacted on the Royal Exchange of London* (London: Sold by J. Brotherton, W. Hinchcliffe, and J. Clarke, 1729); Edward Hatton, *The Gauger's*

Guide; or, Excise-Officer Instructed (London: Printed for D. Midwinter, 1729); George Clerke, *The Dealers in Stock's Assistant: or, a Calculation of the Value of any Parcel of Stocks* (London: Printed for Edward Symon, 1725); John Collier, *Compendium Artis Nauticæ. Being the Daily Practice of the Whole Art of Navigation* (London: Sold by J. Harbin et al., 1729); William Halfpenny, *The Art of Sound Building; Demonstrated in Geometrical Problems* (London: Printed for Sam. Birt [and] B. Motte, 1725); Charles Leadbetter, *Astronomy of the Satellites of the Earth, Jupiter and Saturn: Grounded upon Sir Isaac Newton's Theory of the Earth's Satellite* (London: Printed for J. Wilcox, 1729); James Hamilton, Earl of Abercorn, *Calculations and Tables Relating to the Attractive Virtue of Loadstones* ([London?]: *s.n.*, 1729); Giovanni Rizzetti, *The Knowledge of Play, Written for Public Benefit, and the Entertainment of All Fair Players* (London: Printed for E. Curll, 1729); Richard Roach, *The Great Crisis: or, the Mystery of the Times and Seasons Unfolded* (London: Printed and Sold by N. Blandford, 1725).

53. Larry Stewart, *The Rise of Public Science: Rhetoric, Technology, and Natural Philosophy in Newtonian Britain, 1660–1750* (Cambridge: Cambridge University Press, 1992); Simon Schaffer, "Machine Philosophy: Demonstration Devices in Georgian Mechanics," *Instruments*, ed. Albert Van Helden and Thomas L. Hankins, *Osiris* 9 (1994): 157–182; Simon Schaffer, "The Show That Never Ends: Perpetual Motion in the Early Eighteenth Century," *British Journal for the History of Science* 28, no. 2 (June 1995): 157–189; Stewart, "Other Centres of Calculation, or, Where the Royal Society Didn't Count: Commerce, Coffee-Houses, and Natural Philosophy in Early Modern London," *British Journal for the History of Science* 32, no. 2 (June 1999): 133–153; Shelly Costa, "Marketing Mathematics in Early Eighteenth-Century England: Henry Beighton, Certainty, and the Public Sphere," *History of Science* 40, no. 2 (June 2002): 211–232; Jeffrey R. Wigelsworth, *Selling Science in the Age of Newton: Advertising and the Commoditization of Knowledge* (Farnham, UK: Ashgate, 2012).

54. Stephen Stigler, "John Craig and the Probability of History: From the Death of Christ to the Birth of Laplace," *Journal of the American Statistical Association* 81, no. 396 (Dec. 1986): 879–887; Jed Z. Buchwald and Mordechai Feingold, *Newton and the Origin of Civilization* (Princeton, NJ: Princeton University Press, 2013), chaps. 2, 5; Ted McCormick, "Political Arithmetic and Sacred History: Population Thought in the English Enlightenment, 1660–1750," *Journal of British Studies* 52, no. 4 (Oct. 2013): 829–857; McCormick, "Statistics in the Hands of an Angry God? John Graunt's *Observations* in Cotton Mather's New England," *William and Mary Quarterly* 72, no. 4 (Oct. 2015): 563–586.

55. Wardhaugh, *Poor Robin's Prophecies*, 6; Shelley Costa, "The *Ladies' Diary:* Gender, Mathematics, and Civil Society in Early-Eighteenth-Century England," *Science and Civil Society*, ed. Lynn K. Nyhart and Thomas Broman, *Osiris* 17 (2002): 49–73.

56. Tim Harris, *Revolution: The Great Crisis of the British Monarchy, 1685–1720* (London: Allen Lane, 2006); Steve Pincus, *1688: The First Modern Revolution* (New Haven, CT: Yale University Press, 2009).

57. Brewer, *Sinews of Power,* 231.

58. Geoffrey Holmes, *British Politics in the Age of Anne* (London / New York: Macmillan / St. Martin's, 1967), quotation on 8; Pincus, *1688,* chap. 10.

59. On the organizing function of Whig-Tory partisanship in Augustan politics, see Aaron Graham, *Corruption, Party, and Government in Britain, 1702–1713* (Oxford: Oxford University Press, 2015).

60. Jason Peacey, "The Print Culture of Parliament 1600–1800," *Parliamentary History* 26, no. 1 (Mar. 2007): 1–16. In general, see C. John Sommerville, *The News Revolution in England: Cultural Dynamics of Daily Information* (New York: Oxford University Press, 1996), which offers a particularly polemical take on the deleterious effects of periodic news; Joad Raymond, *Pamphlets and Pamphleteering in Early Modern Britain* (Cambridge: Cambridge University Press, 2003).

61. Mark Knights, *Representation and Misrepresentation in Later Stuart Britain: Partisanship and Political Culture* (Oxford: Oxford University Press, 2005), 15–17.

62. Holmes, *British Politics in the Age of Anne,* 30–33; W. A. Speck, "Political Propaganda in Augustan England," *Transactions of the Royal Historical Society,* 22 (1972): 17–32; J. A. Downie, *Robert Harley and the Press: Propaganda and Public Opinion in the Age of Swift and Defoe* (Cambridge: Cambridge University Press, 1979); P. B. J. Hyland, "Liberty and Libel: Government and the Press during the Succession Crisis in Britain, 1712–1716," *English Historical Review* 101, no. 401 (Oct. 1986): 863–888.

63. Habermas writes: "a public sphere that functioned in the political realm arose first in Great Britain at the turn of the eighteenth century." See *The Structural Transformation of the Public Sphere: An Inquiry into a Category of Bourgeois Society,* trans. Thomas Burger (Cambridge, MA: MIT Press, 1989), 57–67, on 57.

64. The scholarship on Britain's public sphere is extensive. To begin, see Steven Pincus, "'Coffee Politicians Does Create': Coffeehouses and Restoration Political Culture," *Journal of Modern History,* 67, no. 4 (Dec. 1995), 807–834; Miles Ogborn, *Spaces of Modernity: London's Geographies, 1680–1780* (New York: Guilford Press, 1998), chap. 3; Peter Lake and Steve Pincus, "Rethinking the Public Sphere in Early Modern England," *Journal of British Studies* 45, no. 2 (Apr. 2006): 270–292; Peter Lake and Steven Pincus, eds., *The Politics of the Public Sphere in Early Modern England* (Manchester: Manchester University Press, 2007); Brian Cowan, "Geoffrey Holmes and the Public Sphere: Augustan Historiography from Post-Namierite to the Post-Habermasian," *Parliamentary History* 28, no. 1 (Feb. 2009): 166–178.

65. Brian Cowan, "Mr. Spectator and the Coffeehouse Public Sphere," *Eighteenth-Century Studies* 37, no. 3 (Spring, 2004): 345–366; Cowan, *The Social Life of Coffee: The Emergence of the British Coffeehouse* (New Haven, CT: Yale University Press, 2005), 148–151; Knights, "How Rational was the Later Stuart Public Sphere?," in *The Politics of the Public Sphere,* ed. Lake and Pincus, 252–267.

66. On "politeness," see, to begin, Lawrence E. Klein, "Politeness and the Interpretation of the British Eighteenth Century," *Historical Journal* 45, no. 4 (Dec. 2002): 869–898.

67. Habermas, *Structural Transformation,* 27.

68. On deception in general, see Mark Knights, *The Devil in Disguise: Deception, Delusion, and Fanaticism in the Early English Enlightenment* (Oxford: Oxford University Press, 2011).

69. Gordon S. Wood, "Conspiracy and the Paranoid Style: Causality and Deceit in the Eighteenth Century," *William and Mary Quarterly* 39, no. 3 (July 1982): 401–441.

70. Pat Rogers, "Gulliver and the Engineers," *Modern Language Review* 70, no. 2 (Apr. 1975): 260–270; Stewart, *Rise of Public Science*, chap. 5; Schaffer, "The Show That Never Ends"; Randall McGowen, "Knowing the Hand: Forgery and the Proof of Writing in Eighteenth-Century England," *Historical Reflections / Réflexions Historiques* 24, no. 3 (Fall 1998): 385–414; Koji Yamamoto, "Reformation and the Distrust of the Projector in the Hartlib Circle," *Historical Journal* 55, no. 2 (June 2012): 375–397.

71. On fear of deception and print culture: Adrian Johns, *The Nature of the Book: Print and Knowledge in the Making* (Chicago: University of Chicago Press, 1998); Kate Loveman, *Reading Fictions, 1600–1740: Deception in English Literary and Political Culture* (Aldershot, UK: Ashgate, 2008); Dror Wahrman, *Mr. Collier's Letter Racks: A Tale of Art and Illusion at the Threshold of the Modern Information Age* (New York: Oxford University Press, 2012). Literary scholars have been especially attentive to the parallels between paper finance and fictional writing: Colin Nicholson, *Writing and the Rise of Finance: Capital Satires of the Early Eighteenth Century* (Cambridge: Cambridge University Press, 1994); Sandra Sherman, *Finance and Fictionality: Accounting for Defoe* (Cambridge: Cambridge University Press, 1996); Catherine Ingrassia, *Authorship, Commerce, and Gender in Early Eighteenth-Century England: A Culture of Paper Credit* (Cambridge: Cambridge University Press, 1998).

72. A collection of superb recent studies has examined the entanglements of politics and epistemology in this period: Knights, *Representation and Misrepresentation*, 6; Knights, "The Tory Interpretation of History in the Rage of Parties," *Huntington Library Quarterly* 68, nos. 1–2 (Mar. 2005): 353–373; Carl Wennerlind, *Casualties of Credit: The English Financial Revolution, 1620–1720* (Cambridge, MA: Harvard University Press, 2011), chap. 3; Rachel Weil, *Conspiracy and Political Trust in William III's England* (New Haven, CT: Yale University Press, 2014). These historians have paid limited attention to the period's vibrant quantitative culture, though. See, for example, Knights, *Representation and Misrepresentation*, 61–62, 337–339.

73. Recent scholarship on seventeenth-century England has developed a practice-oriented approach to studying politics: Jason Peacey, *Print and Public Politics in the English Revolution* (Cambridge: Cambridge University Press, 2013), 15–20; Noah Millstone, *Manuscript Circulation and the Invention of Politics in Early Stuart England* (Cambridge: Cambridge University Press, 2016), 15–17.

74. The foundational studies are: P. G. M. Dickson, *The Financial Revolution in England: A Study in the Development of Public Credit* (London / New York: Macmillan / St. Martin's Press, 1967); Brewer, *Sinews of Power*. See also Colin Brooks, "Public Finance and Political Stability: The Administration of the Land Tax,"

Historical Journal 17, no. 2 (June 1974): 281–300; Larry Neal, *The Rise of Financial Capitalism: International Capital Markets in the Age of Reason* (Cambridge: Cambridge University Press, 1990); Henry Roseveare, *The Financial Revolution 1660–1760* (London: Longman, 1991); Anne L. Murphy, *The Origins of English Financial Markets: Investment and Speculation before the South Sea Bubble* (Cambridge: Cambridge University Press, 2009); Wennerlind, *Casualties of Credit;* Coffman, *Excise Taxation.*

75. On the politicization and publicity of economic debates after 1688, see Brodie Waddell, "The Politics of Economic Distress in the Aftermath of the Glorious Revolution," *English Historical Review* 130, no. 543 (Apr. 2015): 318–351.

76. James Bateman to Richard Hill, Apr. 2, 1697, Huntington mss *HM* 78,001.

77. This confusion was not just a feature of finance in the early modern period. Ethnographers have recently emphasized that, *pace* the familiar depiction of finance as a domain of rational actors following a common logic of profit maximization, financial activities are rife with disagreement, misunderstanding, and doubt. See David Stark, *The Sense of Dissonance: Accounts of Worth in Economic Life* (Princeton, NJ: Princeton University Press, 2011), chap. 4; Vincent Antonin Lépinay, *Codes of Finance: Engineering Derivatives in a Global Bank* (Princeton, NJ: Princeton University Press, 2011); Hirokazu Miyazaki, *Arbitraging Japan: Dreams of Capitalism at the End of Finance* (Berkeley: University of California Press, 2013).

78. In seeking to historicize foundational concepts of financial value, this book contributes to a growing body of scholarship on "valuation" as a social process: Marion Fourcade, "Cents and Sensibility: Economic Valuation and the Nature of 'Nature,'" *American Journal of Sociology* 116, no. 6 (May 2011): 1721–1777; Michèle Lamont, "Toward a Comparative Sociology of Valuation and Evaluation," *Annual Review of Sociology* 38 (2012): 201–221; Claes-Fredrik Helgesson and Fabian Muniesa, "For What It's Worth: An Introduction to Valuation Studies," *Valuation Studies* 1, no. 1 (2013): 1–10.

79. On the political, and political-philosophical, problems brought by the financial revolution, the work of J. G. A. Pocock is foundational. See *The Machiavellian Moment: Florentine Political Thought and the Atlantic Republican Tradition* (Princeton, NJ: Princeton University Press, 1975), chaps. 13–14; *Virtue, Commerce and History: Essays on Political Thought and History* (Cambridge: Cambridge University Press, 1985), chaps. 5–6.

80. In a widely cited 1989 article, economists Douglass C. North and Barry R. Weingast argued that the financial revolution was enabled by the Revolution of 1688 and the constitutional reconfiguration that resulted ("Constitutions and Commitment: The Evolution of Institutions Governing Public Choice in Seventeenth-Century England," *Journal of Economic History* 49, no. 4 [Dec. 1989]: 803–832). A growing body of scholarship has shown, contrary to North and Weingast, that Britain's financial revolution and the growth of public credit cannot be explained simply as a mechanistic story about institutional structures and responses to economic incentives. Many other developments—including in partisan politics, administrative practices, private financial activity, and economic thinking—were

essential. For some key critiques, see Bruce Carruthers, *City of Capital: Politics and Markets in the English Financial Revolution* (Princeton, NJ: Princeton University Press, 1996); David Stasavage, "Partisan Politics and Public Debt: The Importance of the 'Whig Supremacy' for Britain's Financial Revolution," *European Review of Economic History* 11, no. 1 (2007): 123–153; Wennerlind, *Casualties of Credit;* Anne L. Murphy, "Demanding 'Credible Commitment': Public Reactions to the Failures of the Early Financial Revolution," *Economic History Review* 66, no. 1 (2013): 178–197; Coffman, Leonard, and Neal, eds., *Questioning Credible Commitments.*

81. Broadly, this book resonates with recent attempts by anthropologists to understand numbers and numerical practices as an "inventive frontier." See Jane Guyer et al., "Introduction: Number as Inventive Frontier," *Anthropological Theory* 10, no. 1–2 (Mar. 2010): 36–61.

82. This literature is rapidly growing. Foundational statements include: Michel Callon, "Introduction: The Embeddedness of Economic Markets in Economics," in *The Laws of the Markets,* ed. Callon (Oxford: Blackwell Publishing / Sociological Review, 1998), 1–57; Donald MacKenzie, *An Engine, Not a Camera: How Financial Models Shape Markets* (Cambridge, MA: MIT Press, 2006); Donald MacKenzie, Fabian Muniesa, and Lucia Siu, eds., *Do Economists Make Markets? On the Performativity of Economics* (Princeton, NJ: Princeton University Press, 2007).

83. On financial crises and the "madness of crowds," see Charles Mackay, *Memoirs of Extraordinary Popular Delusions* (London: R. Bentley, 1841). For an example of a financial crisis told as a morality tale, see John Kenneth Galbraith, *The Great Crash 1929,* later ed. (Boston: Mariner Books, 2009 [1954]), 3. For financial crisis told as political thriller, see John Carswell's definitive history: *The South Sea Bubble,* 2nd ed. (Dover, NH: Sutton, 1993 [1960]).

84. The economic literature on crises is of course enormous. Among the most influential and comprehensive studies are: Charles P. Kindleberger, *Mania, Panics, and Crashes: A History of Financial Crises* (New York: Basic Books, 1978); Carmen Reinhart and Kenneth Rogoff, *This Time Is Different: Eight Centuries of Financial Folly* (Princeton, NJ: Princeton University Press, 2009), on xxv. For an example of using historical bubbles to test market rationality, see Peter Garber, *Famous First Bubbles: The Fundamentals of Early Manias* (Cambridge, MA: MIT Press, 2000). On economists' use of bubbles as test cases, see William Deringer, "For What It's Worth: Historical Financial Bubbles and the Boundaries of Economic Rationality," *Isis* 106, no. 3 (Sep. 2015): 646–656.

85. Plato, *Republic,* VII, 525a–d, trans. G. M. A. Grube, rev. C. D. C. Reeve (Indianapolis: Hackett Publishing, 1992), 197–198.

86. Robert L. Cioffi, "Fuzzy Math: The Place of Numerical Evidence in Cicero *In Verrem* 3.116," *Mnemosyne* 64, no. 4 (Oct. 2011): 645–652.

87. M. Aiken and W. Lu, "Chinese Government Accounting: Historical Perspective and Current Practice," *British Accounting Review* 25, no. 2 (June 1993): 109–129.

88. The relevant research for any of these national contexts is far richer and more extensive than can be addressed here. My hope is that this section will indicate useful points of entry. On finance, politics, and calculation in medieval and renaissance Italy, see Anthony Molho, "The State and Public Finance: A Hypothesis based on the History of Late Medieval Florence," *Journal of Modern History* 67, Supplement: The Origins of the State in Italy, 1300–1600 (Dec. 1995): S97–S135 (on the entanglement of political conflicts and administrative procedures, including accounting); Rebecca Jean Emigh, "Numeracy or Enumeration? The Use of Numbers by States and Societies," *Social Science History* 26, no. 4 (Winter 2002): 653–698 (on the *catasto*); William N. Goetzmann, "Fibonacci and the Financial Revolution," in *The Origins of Value: The Financial Innovations That Created Modern Capital Markets*, ed. William N. Goetzmann and K. Geert Rouwenhorst (Oxford: Oxford University Press, 2005), 123–144; Soll, *The Reckoning*, chaps. 1–3.

89. On the Dutch Republic, see Jacob Soll, "Accounting for Government: Holland and the Rise of Political Economy in Seventeenth-Century Europe," *Journal of Interdisciplinary History* 40, no. 2 (Autumn 2009): 215–238, on 217; Soll, *The Reckoning*, chap. 5; Alexander Bick, "Governing the Free Sea: The Dutch West India Company and Commercial Politics, 1618–1645" (Ph.D. diss., Princeton University, 2012), 223–260.

90. On France, see Soll, *The Information Master: Jean-Baptiste Colbert's Secret State Intelligence System* (Ann Arbor: University of Michigan Press, 2011); Soll, *The Reckoning*, chap. 6; Andrea Rusnock, "Quantification, Precision, and Accuracy: Determinations of Population in the Ancien Régime," in *Values of Precision*, ed. Wise, 17–38; Carol Blum, *Strength in Numbers: Population, Reproduction, and Power in Eighteenth-Century France* (Baltimore, MD: Johns Hopkins University Press, 2002), esp. 7–8.

91. On Germany, see Keith Tribe, *Governing Economy: The Reformation of German Economic Discourse, 1750–1840* (Cambridge: Cambridge University Press, 1988), 33; Hacking, *Taming of Chance*, chap. 3, quotation on 18; Henry E. Lowood, "The Calculating Forester: Quantification, Cameral Science, and the Emergence of Scientific Forestry Management in Germany," in *The Quantifying Spirit in the Eighteenth Century*, ed. Frängsmyr, Heilbron, and Rider, 315–343; Rüdiger Campe, *The Game of Probability: Literature and Calculation from Pascal to Kleist*, trans. Ellwood H. Wiggins, Jr. (Stanford, CA: Stanford University Press, 2012), 206–219; Foucault, *Security, Territory, Population*, 317–318. For a comparison of "uses of statistics in politics" in France and Germany, see Behrisch, "Statistics and Politics."

92. On epistemic virtues, see Daston and Galison, *Objectivity*, 39–42.

93. Desrosières, *Politics of Large Numbers*, 28. See also Foucault, *Security, Territory, Population*, 275.

94. Hacking, *Taming of Chance*, 20.

95. As Jacob Soll writes, "Political economy became highly public in England." ("Accounting for Government," 231.)

CHAPTER ONE
Finding the Money: Public Accounting and
Political Arithmetic after 1688

1. On Jephson, see Stephen B. Baxter, *The Development of the Treasury 1660–1720* (Cambridge, MA: Harvard University Press, 1957), 195–197. Information on Jephson's accounts is reprinted in William A. Shaw, ed., *Calendar of Treasury Books* (*CTB*), vol. 17, *1702* (London: HMSO, 1939), 515–626.

2. On Secret Service, see Shaw, *CTB*, vol. 4, *1672–1675* (London: HMSO, 1909), xlvi–xlvii; C. D. Chandaman, *The English Public Revenue 1660–1688* (Oxford: Clarendon Press, 1975), 244–248, 271–272; Christopher Clay, *Public Finance and Private Wealth: The Career of Sir Stephen Fox, 1627–1716* (Oxford: Clarendon Press, 1978), 25–26.

3. Minutes of the Commissioners of Public Accounts, vol. 1, Mar. 5, 1960 / Sept. 1 to 4, 1691, British Library (BL) Harley MS 1,488, ff. 20^{r-v}, 34^r, 38^r, 52^v–53^r, 64^r.

4. Historical Manuscripts Commission (HMC), *The Manuscripts of the House of Lords, 1690–1691* (London: HMSO, 1892), 363, 399, 420; Anchitell Grey, ed., *The Debates in the House of Commons, From the Year 1667 to the Year 1694,* vol. 10 (London: Printed for D. Henry and R. Cave, and J. Emonson, 1763), 191, 200.

5. There is a considerable literature on the history of double-entry bookkeeping. See Soll, *The Reckoning,* chap. 4 for a recent overview of the technique and its historical development. Recent scholarship has shown the influence of such formalized accounting practices on economic behavior and broader culture in England. See John Richard Edwards, Graeme Dean, and Frank Clarke, "Merchants' Accounts, Performance Assessment and Decision Making in Mercantilist Britain," *Accounting, Organizations & Society* 34, no. 5 (July 2009): 551–570; Adam Smyth, *Autobiography in Early Modern England* (Cambridge: Cambridge University Press, 2010), 61–72; Amy Froide, "Learning to Invest: Women's Education in Arithmetic and Accounting in Early-Modern England," *Early Modern Women: An Interdisciplinary Journal* 10, no. 1 (Fall 2015): 3–26.

6. Michael Power, *The Audit Society: Rituals of Verification* (Oxford: Oxford University Press, 1997), 9–10, 27–31.

7. Historians of early modern accounting have shown this was often true of private account books as well. There was great variety in both the methods and quality of private account books, and few individuals lived up to textbook ideals. See Basil Yamey, "Some Topics in the History of Financial Accounting in England, 1500–1900," in *Studies in Accounting,* ed. W. T. Baxter and Sidney Davidson, 3rd ed. (London: Institute of Chartered Accountants in England and Wales, 1977), 11–34, esp. 17, 21; Froide, "Learning to Invest," 16–17.

8. My account of the transformation of political arithmetic builds upon the excellent biography by Ted McCormick. See *William Petty,* esp. 299.

9. Writing specifically of quantitative practices in medicine, Andrea Rusnock also observes how "seventeenth- and eighteenth-century political and medical arithmeti-

cians had to convince governments, savants, and the literate public of the value of collecting numbers"; see *Vital Accounts,* 3.

10. Knights, *Representation and Misrepresentation,* 6. See also Brewer, *Sinews of Power,* chap. 8.

11. On the period of the Civil Wars and the Interregnum, see Roseveare, *Financial Revolution 1660–1760,* esp. 6–7; D'Maris Coffman, "The Earl of Southampton and the Lessons of Interregnum Public Finance," in *Royalists and Royalism during the Interregnum,* ed. David Smith and Jason McElligott (Manchester: Manchester University Press, 2010), 235–256; Coffman, "Credibility, Transparency, Accountability." On the Restoration era, see Clayton Roberts, *The Growth of Responsible Government in Stuart England* (Cambridge: Cambridge University Press, 1966), 175–176, 182–183, 261; E. A. Reitan, "From Revenue to Civil List, 1689–1702: The Revolution Settlement and the 'Mixed and Balanced' Constitution," *Historical Journal* 13, no. 4 (Dec. 1970): 571–588; Henry Horwitz, *Parliament, Policy and Politics in the Reign of William III* (Newark: University of Delaware Press, 1977), 59; Henry Roseveare, *The Treasury 1660–1870: The Foundations of Control* (London / New York: Allen & Unwin / Barnes & Noble, 1973), 46–74.

12. Reitan, "Revenue to Civil List," 571–572.

13. Clay, *Public Finance and Private Wealth,* 40–111; Roberts, *Responsible Government,* 169; Dickson, *Financial Revolution,* 254.

14. Brewer, *Sinews of Power,* 101–114; Michael J. Braddick, *The Nerves of State: Taxation and the Financing of the English State, 1558–1714* (Manchester: Manchester University Press, 1996), chaps. 3, 5; Miles Ogborn, "The Capacities of the State: Charles Davenant and the Management of the Excise, 1683–1698," *Journal of Historical Geography* 24, no. 3 (July 1998): 289–312; Ashworth, *Customs and Excise.*

15. Chandaman, *English Public Revenue,* 6–7, 287–295.

16. On Howard, see H. J. Oliver, *Sir Robert Howard (1626–1698): A Critical Biography* (Durham, NC: Duke University Press, 1963); Florence R. Scott, "Sir Robert Howard as a Financier," *PMLA* 52, no. 4 (Dec. 1937): 1094–1100.

17. For a contemporary depiction of how Exchequer procedures ideally worked, see William Lowndes, "The Course of th^e Exchequer on th^e Receipt side," 1685, Hyde Papers, BL Add. MS 15,898, ff. 100–109, esp. 106^v–107^r. For an example of Wardour's General Declarations, see BL Lansdowne MS 1,215, ff. 29–39. See also Chandaman, *English Public Revenue,* 281–302; Roseveare, *Treasury: Foundations,* 46–51; Baxter, *Development of the Treasury,* 120–121.

18. Jennifer Carter, "The Revolution and the Constitution," in *Britain after the Glorious Revolution 1689–1714,* ed. Geoffrey Holmes (London / New York: Macmillan / St. Martin's Press, 1969), 39–58; Clayton Roberts, "The Constitutional Significance of the Financial Settlement of 1690," *Historical Journal* 20, no. 1 (Mar. 1977): 59–76; Reitan, "Revenue to Civil List."

19. Grey, *Debates,* 9:123–125.

20. William Cobbett, ed., *The Parliamentary History of England, from the Earliest Period to the Year 1803,* vol. 5, *1688–1702* (London: Printed for T. Hansard, 1809), cols. 150–151, 187–191; *Journal of the House of Commons,* vol. 10, *1688–1693* (London: HMSO,

1802) 37–38, 55; Grey, *Debates,* 9:157. See also Shaw, *CTB,* vol. 9, *1689–92* (London: HMSO, 1931), xxiv, xxxviii; Reitan, "Revenue to Civil List," 575; Roberts, "Financial Settlement," 70.

21. *Commons Journal,* 10:56; Grey, *Debates,* 9:176–180. On the challenges of classification in contemporary accounting, see Donald MacKenzie, "Measuring Profit," in *Material Markets: How Economic Agents Are Constructed* (Oxford: Oxford University Press, 2009), 109–136, esp. 112–116.

22. Shaw, *CTB,* 9:xxxix, xliv–xlvi.

23. Grey, *Debates,* 10:13.

24. On the tensions between the Lords and Commons over the Commission, see Robert Harley for Sir Edward Harley, Dec. 30, 1690 and Jan. 1, 1690 / 1, HMC, *The Manuscripts of His Grace the Duke of Portland, Preserved at Welbeck Abbey,* vol. 3 (London: HMSO, 1891), 456; *Journal of the House of Lords,* vol. 14, *1685–1691* (London: HMSO, 1767–1830), 52. See also Colin Brooks, "Taxation, Finance and Public Opinion, 1688–1714" (Ph.D. diss., University of Cambridge, 1970), 118.

25. The most thorough treatment of the early history of the Commission is J. A. Downie, "The Commission of Public Accounts and the Formation of the Country Party," *The English Historical Review* 91, no. 358 (Jan. 1976): 33–51, esp. 34–37. See also Reitan, "Revenue to Civil List," 577–579, 582; Horwitz, *Parliament, Policy, and Politics,* 38, 59–62, 64; B. W. Hill, *Robert Harley: Speaker, Secretary of State and Premier Minister* (New Haven, CT: Yale University Press, 1988), 20.

26. Robert Harley to Sir Edward Harley, Mar. 7, 1960 / Mar. 1 and 12, 1690 / 1, HMC *Portland,* 3:459; Hill, *Robert Harley,* 24–27; Roseveare, *Treasury: Foundations,* 58; Downie, "Commission of Public Accounts," 34–37.

27. The Commission's minute-books are preserved as BL Harley MS 1,488–1,495. The personal diary of Commissioner Peter Colleton is BL Harley MS 6,837, ff. 164–206. See in particular BL Harley MS 1,488, ff. 3^v–4^v, 72^v, 176^v–177^r; BL Harley MS 6,837, ff. 172^r–173^r, 174$^{r\text{-}v}$.

28. BL Harley MS 1,488, f. 68^v; BL Harley MS 1,489, ff. 3^v–4^r, 13$^{r\text{-}v}$, 27^v–28^r, 38^r; BL Harley MS 6,837, ff. 184^v, 190^v–191^r.

29. BL Harley MS 1,488, ff. 109^r–110^r; BL Harley MS 6,837, ff. 195$^{r\text{-}v}$; Paula Watson and Sonya Wynne, "Jones, Richard, 1st Earl of Ranelagh (1641–1712)," in *The History of Parliament: The House of Commons 1690–1715,* ed. David Hayton, Eveline Cruickshanks, and Stuart Handley (Cambridge: Cambridge University Press for the History of Parliament Trust, 2002).

30. BL Harley MS 1,488, f. 64^r.

31. Ibid., ff. 124^r, 127^v, 129^v, 130^r, 133^v, 135^v, 143^v; BL Harley MS 1,489, ff. 14^v, 56^r; BL Harley MS 6,837, ff. 168^r, 169^v, 184^v. On Van der Esch, see Baxter, *Development of the Treasury,* 55–56. See also Braddick, *Nerves of State,* 23–25.

32. BL Harley MS 1,488, f. 52^v.

33. BL Harley MS 6,837, f. 165^r; BL Harley MS 1,488, f. 103^r. Baxter suggests Squib knew more than he let on (*Development of the Treasury,* 235–236). See also Brewer, *Sinews of Power,* 223.

34. BL Harley MS 1,489, f. 11ʳ, 21ᵛ; Shaw, *CTB*, 9:1313–1314; HMC *Lords, 1690–1691*, 405. Based on the published records of Jephson's Secret Service accounts, it appears that the Chancellor of the Exchequer sporadically reviewed Jephson's Secret Service Accounts before passing them on for royal approval. (See Shaw, *CTB*, 17:528, 534, 547, 560, and 626.)

35. BL Harley MS 1,488, ff. 22ʳ, 28ʳ, 32ʳ, 113ʳ, 153ᵛ, 170ʳ⁻ᵛ; BL Harley MS 1,489, f. 14ᵛ.

36. BL Harley MS 1,489, f. 58ʳ. The report is reprinted in HMC *Lords, 1690–1691*, 356–399; also 392, 400, 433.

37. HMC *Lords, 1690–1691*, 400–401, 410–411, 433.

38. Shaw, *CTB*, 9:clxiii.

39. Narcissus Luttrell, *The Parliamentary Diary of Narcissus Luttrell, 1691–1693*, ed. Henry Horwitz (Oxford: Clarendon Press, 1972), 420–421. For various appraisals of the Commission's impact, see Shaw, *CTB*, 9:clxxii; Roseveare, *Treasury: Foundations*, 56–59; Roseveare, *The Treasury: The Evolution of a British Institution* (London: Allen Lane, 1969), chap. 3; Brooks, "Taxation," 117–126.

40. Treasury Papers, The National Archives (United Kingdom), Kew (TNA) T 30; Shaw, *CTB*, 9:clxxiv; Dickson, *Financial Revolution*, 46; Porter, *Trust in Numbers*, 4, 89, 194.

41. [Anon.], *An Impartial State of the Case of the Earl of Danby, in a Letter to a Member of the House of Commons* (London: *s.n.*, 1679); [Anon.], *An Examination of* The Impartial State of the Case of the Earl of Danby. *In a Letter to a Member of the House of Commons* (London: *s.n.*, 1680); Earl of Danby, *The Answer of the Right Honourable the Earl of Danby to a Late Pamphlet, Entituled,* An Examination of the Impartial State of the Case of the Earl of Danby (London: *s.n.*, 1680); Robert Howard, *An Account of the State of His Majesties Revenue, As It Was Left by the Earl of Danby, at Lady-day, 1679* (London: Printed for Thomas Fox, 1681).

42. On the connections between republican thought and public accounting in a different national context, see Jacob Soll, "From Virtue to Surplus: Jacques Necker's *Compte rendu* (1781) and the Origins of Modern Political Rhetoric," *Representations* 134, no. 1 (Spring 2016): 29–63.

43. Downie, "Commission of Public Accounts." On Court-Country politics more broadly, see Dennis Rubini, *Court and Country, 1688–1702* (London: Hart-Davis, 1968).

44. The literature on republicanism and the "Commonwealth" tradition in Anglophone political culture is vast. Pocock's classic statement is *The Machiavellian Moment*. Other foundational studies include: Caroline Robbins, *The Eighteenth-Century Commonwealthman: Studies in the Transmission, Development and Circumstance of English Liberal Thought from the Restoration of Charles II until the War with the Thirteen Colonies* (Cambridge, MA: Harvard University Press, 1959); Bernard Bailyn, *The Ideological Origins of the American Revolution,* enlarged ed. (Cambridge, MA: Harvard University Press, 1992 [1968]), which traces the influence of the English republican tradition in North America; and Quentin Skinner, *Liberty Before Liberalism* (Cambridge: Cambridge University Press, 1998), which focuses on republican notions of liberty and emphasizes Roman sources. Recently,

scholars have emphasized the eclecticism of the "Atlantic republican tradition" and its sources, pointing to the influence of Protestant religiosity and Greek thought. See, for example, Jonathan Scott, "What Were Commonwealth Principles?," *Historical Journal* 47, no. 3 (Sept. 2004): 591–613; Michael P. Winship, "Algernon Sidney's Calvinist Republicanism," *Journal of British Studies* 49, no. 4 (Oct. 2010): 753–773; Eric Nelson, *The Greek Tradition in Republican Thought* (Cambridge: Cambridge University Press, 2004), esp. chap. 3. For overviews of this historiography, see David Wootton, "The Republican Tradition: From Commonwealth to Common Sense," in *Republicanism, Liberty, and Commercial Society, 1649–1776*, ed. David Wootton (Stanford, CA: Stanford University Press, 1994), 1–41; Rachel Hammersley, "Introduction: The Historiography of Republicanism and Republican Exchanges," *History of European Ideas* 38, no. 3 (Sept. 2012): 323–337.

45. On the moral dimension in particular, see David Hayton, "Moral Reform and Country Politics in the Late Seventeenth-Century House of Commons," *Past & Present* 128 (Aug. 1990): 48–91.

46. J. G. A. Pocock, "Machiavelli, Harrington, and English Political Ideologies in the Eighteenth Century," in *Politics, Language, and Time: Essays on Political Thought and History* (Chicago: University of Chicago Press, 1989 [1971]), 104–147, on 122.

47. Grey, *Debates*, 10:191. Thompson would serve on the Commission of Public Accounts from 1695 to 1696. See J. S. Crossette, "Thompson, Sir John, 1st Bt. (1648–1710)," in *The History of Parliament: the House of Commons 1660–1690*, ed. B. D. Henning (London: Published for the History of Parliament Trust by Secker & Warburg, 1983).

48. Robert Harley to Sir Edward Harley, Feb. 9, 1692/3, Portland Papers, BL Add. MS 70,017, f. 22r. See also Brooks, "Taxation," 124–125; Hill, *Robert Harley*, 32.

49. "A Computation of the Charges for Land and Sea Services from the 5 of Novr 1688 to the Last of Decr 1690," Lowndes Papers, TNA T 48/87, ff. 210–213, on f. 210; Grey, *Debates*, 10:168–169; Shaw, *CTB*, 9:cxiii–cxxiv, cxxix–cxxxvii; Brooks, "Taxation," 115; Downie, "Commission of Public Accounts," 35; Horwitz, *Parliament, Policy, and Politics*, 71–73; Hill, *Robert Harley*, 27–28.

50. HMC, *The Manuscripts of the House of Lords, 1693–1695*, new series, vol. 1 (London: HMSO, 1900), 12–29. On naval finance procedures in this period, and the role of Parliament therein, see Daniel Baugh, "Parliament, Naval Spending and the Public: Contrasting Legacies of Two Exhausting Wars, 1689–1713," *Histoire & Mesure* 30, no. 2 (2015): 23–50.

51. HMC *Lords, 1693–5*, 27. For the Treasury Lords' calculation, see p. 17.

52. Compare Espeland and Sauder's observation that "commensuration invites reflection on what numbers represent." See "Rankings and Reactivity," 21–24.

53. Robert Howard, "Paper of Exceptions to the Commissioners of Accompts delivered to the House of Commons by Sr Robert Howard on Thursday the 16th of February 1692[3]," 1693, BL Harley MS 7,019, ff. 5–8, on f. 5^v, 8^r. Note that around the same time, another creative English calculator—Isaac Newton—was also experimenting with using numerical averages as a way to improve the quality of empirical observations and avoid problems of observational error. See Buchwald and Feingold, *Newton and the Origin of Civilization*, 5–6, 90–106.

54. *A Collection of the Debates and Proceedings in Parliament, in 1694, and 1695. Upon the Inquiry into the Late Briberies and Corrupt Practices* (London: *s.n.*, 1695), 12–13, 28, 36. See also *A Supplement to the Collection of the Debates and Proceedings in Parliament, in 1694 and 1695. Upon the Inquiry into the Late Briberies and Corrupt Practices* (London: *s.n.*, 1695).

55. James Brydges, "A Journal of My Daily Actions beginning Sat. Jan. 16, 1696 / 7," Huntington Stowe Brydges mss*ST* 26, vol. 2. See entries for Jan. 13, 1699 / 1700; Jan. 26, 1699 / 1700; Feb. 22, 1699 / 1700. On Brydges's early career, see also D. W. Hayton, "Brydges, Hon. James (1674–1744)," in *History of Parliament: the House of Commons 1690–1715*. Brydges would go on to a colorful career in his own right: Godfrey Davies, "The Seamy Side of Marlborough's War," *Huntington Library Quarterly* 15, no. 1 (Nov. 1951): 21–44; Stewart, *Rise of Public Science*, chap. 10.

56. Broadly speaking, the 1690s witnessed a flurry of new publications dedicated to financial and commercial matters. See Slack, *Invention of Improvement*, 170–175; Waddell, "Politics of Economic Distress."

57. Charles Davenant, *Discourses on the Publick Revenues, and on the Trade of England*, vol. 1 (London: Printed for J. Knapton, 1698), 2.

58. On Davenant's life, see D. Waddell, "Charles Davenant (1656–1714)—a Biographical Sketch," *Economic History Review* 11, no. 2 (1958): 279–288; Julian Hoppit, "Davenant, Charles (1656–1714)," *Oxford Dictionary of National Biography* (*DNB*); D. W. Hayton, "Davenant, Charles (1656–1714), of Red Lion Square, Mdx.," in *History of Parliament: the House of Commons 1690–1715*. On Davenant and the excise, see Ogborn, "Capacities of the State."

59. *Lords Journal*, vol. 14, *1691–1696*, 52; Hayton, "Davenant."

60. Charles Davenant, "A Memorial Concerning the East India Trade," 1696, BL Harley MS 1,223, ff. 158^r–168^r; Davenant, *Essay on the East India Trade* (London: *s.n.*, 1696); D. Waddell, "Charles Davenant and the East India Company," *Economica* 23, no. 91 (Aug. 1956): 261–264.

61. Brydges, "Journal," Huntington Stowe Brydges mss*ST* 26, vol. 1, entries for: Feb. 3, 1696 / 7; Mar. 18, 1696 / 7; Mar. 23, 1696 / 7; Apr. 12, 1697; Apr. 17, 1697.

62. On Davenant as republican thinker, see Pocock, *Machiavellian Moment*, 429–454; Kustaa Multamäki, *Towards Great Britain: Commerce & Conquest in the Thought of Algernon Sidney and Charles Davenant* (Tuusula: Finnish Academy of Science and Letters, 1999), 149–208; Andrea Finkelstein, *Harmony and the Balance: An Intellectual History of Seventeenth-Century English Economic Thought* (Ann Arbor: University of Michigan Press, 2000), chap. 14; Seiichiro Ito, "Charles Davenant's Politics and Political Arithmetic," *History of Economic Ideas* 13, no. 1 (Oct. 2005): 9–36. On Davenant and Harley, see Downie, *Robert Harley and the Press*, 37.

63. Charles Davenant, "Essay on Publick Virtue," unpublished MS, 1696, BL Harley MS 1,223, ff. 7–79, esp. ff. 8^v, 21^v, 14^v, 38^r–39^r.

64. [Charles Davenant], *An Essay upon Ways and Means of Supplying the War* (London: Printed for Jacob Tonson, 1695), 8–9, 18–19.

65. Charles Davenant, "Memorial Concerning a Council of Trade," unpublished MS, 1696, BL Harley MS 1,223, ff. 184ʳ–189ᵛ; Ogborn, "Capacities of State," 292–295.

66. Davenant, *Discourses on the Publick Revenues*, 1:266.

67. McCormick, *William Petty*, 285–302, quotations on 299, 300, 302.

68. Davenant, *Essay upon Ways and Means*, 46–55. On public acceptance of taxation in the period, see Brooks, "Public Finance and Political Stability."

69. Davenant, *Discourses on the Publick Revenues*, 1:119–121.

70. See insert in ibid., after p. 74.

71. Ibid., 95–96.

72. Joseph A. Schumpeter explicitly described Petty and Davenant as "econometricians." See *History of Economic Analysis*, ed. Elizabeth Booty Schumpeter (New York: Oxford University Press, 1961), 209; Richard Stone, "Some Seventeenth Century Econometrics: Consumer's Behavior," *Revue Européenne des Sciences Sociales* 26, no. 81 (1988): 19–41.

73. Davenant, *Discourses on the Publick Revenues*, 1:1–5.

74. Davenant, *"Memorial Concerning the East India Trade"*; Davenant, *Essay on the East India Trade*. On this conflict, see W. Darrell Stump, "An Economic Consequence of 1688," *Albion* 6, no. 1 (Spring 1974): 26–35; Henry Horwitz, "The East India Trade, the Politicians, and the Constitution: 1689–1702," *Journal of British Studies* 17, no. 2 (Spring 1978): 1–15; Gary Stuart De Krey, *A Fractured Society: The Politics of London in the First Age of Party 1688–1715* (Oxford: Clarendon Press, 1985), chap. 4; Philip J. Stern, "'A Politie of Civill and Military Power': Political Thought and the Late Seventeenth-Century Foundations of the East-India Company State," *Journal of British Studies* 47, no. 2 (Apr. 2008): 253–283; Pincus, *1688*, 372–381.

75. Davenant, *Essay on the East India Trade*, 15–17.

76. T.[homas] S.[mith], *Reasons Humbly Offered for the Passing a Bill for the Hindering the Home Consumption of East-India Silks, Bengals &c.* (London: Printed for J. Bradford, 1697), 10; [John Pollexfen], *England and East-India Inconsistent in Their Manufactures* (London: s.n., 1697), 5.

77. [Charles Davenant?], *Some Reflections on a Pamphlet, Intituled,* England and East-India Inconsistent in Their Manufactures (London: s.n., 1696), 7–8.

78. Davenant, *Discourses on the Publick Revenues*, 1:11.

79. Gregory King, "Natural and Political Observations and Conclusions upon the State and Condition of England," in King, *Two Tracts*, ed. George E. Barnett (Baltimore, MD: John Hopkins University Press 1936 [1696]), 13. On King and his motivations, see G. S. Holmes, "Gregory King and the Social Structure of Pre-Industrial England," *Transactions of the Royal Historical Society* 27 (1977): 41–68; Brooks, "Projecting, Political Arithmetic and the Act of 1695."

80. Quoted in Rampling, "Elizabethan Mathematics of Everything," 139.

81. On probability in this period, foundational are: Ian Hacking, *The Emergence of Probability: A Philosophical Study of Early Ideas about Probability, Induction, and Statistical Inference*, 2nd ed. (Cambridge: Cambridge University Press, 2006 [1975]); Barbara J. Shapiro, *Probability and Certainty in Seventeenth-Century*

England: A Study of the Relationships between Natural Science, Religion, History, Law, and Literature (Princeton, NJ: Princeton University Press, 1983); Lorraine Daston, *Classical Probability in the Enlightenment* (Princeton, NJ: Princeton University Press, 1988). In addition, see Carl Wennerlind's discussion of how "qualitative probabilistic reasoning" empowered new thinking about credit during Britain's financial revolution: *Casualties of Credit*, 3–4, 83–92.

82. McCormick notes that Petty did not intend political arithmetic to be a rigorous "statistical form of social analysis." Statistician Anders Hald writes that "Petty did not contribute new methods of statistical analysis." See McCormick, *William Petty*, 206, 220–221, also see 177–178, 302; Hald, *A History of Probability and Statistics and Their Applications before 1750* (New York: John Wiley & Sons, 1990), 105.

83. Davenant, *Discourses on the Publick Revenues*, 1:27.

84. [Anon.], *Remarks upon Some Wrong Computations and Conclusions, Contained in a Late Tract, Entitled,* Discourses on the Publick Revenues, and on the Trade of England. *In a Letter to Mr. D. S.* (London: Printed for W. Keblewhite, 1698).

85. Ibid., 3–4.

86. Ibid., 4, 28; also 5–8, 13.

87. Ibid., 23, 31, 41, 47.

88. Ibid., 34.

CHAPTER TWO

The Great Project of the Equivalent: A Story of the Number 398,085½

1. Historical scholarship on the Anglo-Scottish union is voluminous and contentious. For overviews, see Bob Harris, "The Anglo-Scottish Treaty of Union, 1707 in 2007: Defending the Revolution, Defeating the Jacobites," *Journal of British Studies* 49, no. 1 (Jan. 2010): 28–46; Clare Jackson, "Union Historiographies," in *The Oxford Handbook of Modern Scottish History*, ed. T. M. Devine and Jenny Wormald (Oxford: Oxford University Press, 2012), 338–354.

2. [John Clerk], *An Essay upon the XV. Article of the Treaty of Union, Wherein the Difficulties That Arise upon the Equivalents, Are Fully Cleared and Explained* ([Edinburgh?]: *s.n.*, 1706), 3.

3. George Lockhart, *Memoirs Concerning the Affairs of Scotland, from Queen Anne's Accession to the Throne, to the Commencement of the Union of the Two Kingdoms of Scotland and England, in May, 1707* (London: Printed and Sold by J. Baker, 1714), 212–213; Robert Burns, "Such a Parcel of Rogues in a Nation" [1791], in *The Poetical Works of Robert Burns* (Edinburgh / London: William and Robert Chambers / W. S. Orr and Company, 1838), 111.

4. For different interpretations of the Equivalent, see P. W. J. Riley, *The Union of England and Scotland: A Study in Anglo-Scottish Politics of the Eighteenth Century* (Manchester: Manchester University Press, 1978), 239–240 (confidence trick); William Ferguson, "The Making of the Treaty of Union of 1707," *Scottish Historical Review* 43, no. 136 (Oct. 1964): 89–110, on 104, 110 (consolation prize); Alan I. Macinnes, *Union and Empire: The Making of the United Kingdom in 1707* (Cambridge:

Cambridge University Press, 2007), 320, also 94–95; Macinnes, "The Treaty of Union: Made in England," in *Scotland and the Union 1707–2007*, ed. T. M. Devine (Edinburgh: Edinburgh University Press, 2008), 54–70 (inept negotiation). For a less skeptical view, see Christopher A. Whatley, "The Making of the Union of 1707: History with a History," in Devine, *Scotland and the Union 1707–2007*, 23–38, esp. 34–35.

5. Douglas Watt, *The Price of Scotland: Darien, Union and the Wealth of Nations* (Edinburgh: Luath Press, 2007), 220, 222, 229–230, 238; Crawford Spence, "Accounting for the Dissolution of a Nation State: Scotland and the Treaty of Union," *Accounting, Organizations & Society* 35, no. 3 (Apr. 2010): 377–392.

6. John Clerk, "Journal of Proceedings of the Commissioners for the Treaty of Union," vol. 1, Papers of Clerk Family of Penicuik, National Records of Scotland (NRS) GD18 / 3132, f. 21.

7. For another study of how complex political (and legal) negotiations about compensation have been rendered as problems of calculation, compare Marion Fourcade, "Price and Prejudice: On Economics and the Enchantment (and Disenchantment) of Nature," in *The Worth of Goods: Valuation and Pricing in the Economy*, ed. Jens Beckert and Patrik Aspers (Oxford: Oxford University Press, 2011), 41–62; Fourcade, "Cents and Sensibility."

8. On "reasonableness" in the eighteenth century, compare Daston, *Classical Probability*, 68–106.

9. Macinnes, *Union and Empire*, 11.

10. On the contested role of economic factors in the Scottish interest in union, see T. C. Smout, "The Anglo-Scottish Union of 1707: I. The Economic Background," *Economic History Review* 16, no. 3 (Apr. 1964): 455–467; Christopher A. Whatley, "Economic Causes and Consequences of the Union of 1707: A Survey," *Scottish Historical Review* 68, no. 186, part 2 (Oct. 1989): 150–181; Christopher A. Whatley, "The Issues Facing Scotland in 1707," *Scottish Historical Review* 87, Supplement, *Union of 1707: New Dimensions*, ed. Stewart J. Brown and Christopher A. Whatley (2008): 1–30; David Armitage, *The Ideological Origins of the British Empire* (Cambridge: Cambridge University Press, 2000), chap. 5; Macinnes, *Union and Empire*, part 3.

11. [Abel Boyer], *The History of the Reign of Queen Anne, Digested into Annals. Year the First* (London: Printed for A. Roper [and] F. Coggan, 1703), 158–159. On Equivalent discussions during the 1702–1703 negotiations, see Watt, *Price of Scotland*, 222–225; Macinnes, *Union and Empire*, 94–95.

12. For example, in 1697, the banker James Bateman reported that English observers were concerned that the Holy Roman Emperor would "not condesend to take an Equivalent" for the contested city of Strasbourg: James Bateman to Richard Hill, 3 Sept. 1697, Huntington mss*HM* 78,012.

13. Christopher Storrs, *War, Diplomacy and the Rise of Savoy, 1690–1720* (Cambridge: Cambridge University Press, 1999), chap. 2; Peter H. Wilson, "Prussia as a Fiscal-Military State, 1640–1806," in *The Fiscal Military State in Eighteenth-Century Europe*, ed. Christopher Storrs (Abingdon, UK: Ashgate, 2009), 95–125, on 114.

14. *Commons Journal*, vol. 12, *1697–1699*, 330–331.

15. [George Savile, Marquess of Halifax], *The Anatomy of an Equivalent* ([London?]: s.n., 1688); [Anon.], *The Anatomy of an Equivalent by the Marquess of Halifax: Adapted to the Equivalent in the Present Articles, 1706* ([Edinburgh]: s.n., 1706); H. C. Coxcroft, "The Works of George Savile, First Marquis of Halifax," *English Historical Review* 11, no. 44 (Oct. 1896): 703–730, on 719; Mark N. Brown, "The Works of George Savile Marquis of Halifax: Dates and Circumstances of Composition," *Huntington Library Quarterly* 35, no. 2 (Feb. 1972): 143–157, esp. 148–149; Watt, *Price of Scotland,* 222.

16. Boyer, *History of the Reign of Queen Anne . . . Year the First,* 159; Charles Davenant, "A Memorial Relating to an Union between England & Scotland," MS Copy (dated Jan. 31, 1704/5), Portland Papers, BL Add. MS 70,038, ff. 19–22, on f. 19ᵛ.

17. "Heads Proposed for an Union between the Kingdom of England and Scotland," 1705, BL Stowe MS 222, ff. 343–344. Allan Macinnes recently called attention to this "proposition paper": *Union and Empire,* 279–280. Due to Old Style dating, the paper could date as late as March 25, 1706.

18. "Heads Proposed for an Union," BL Stowe MS 222, f. 343ᵛ.

19. Two pieces of evidence suggest Paterson's authorship. First, the "fund" for Scottish economic improvement described under Head 12 of the proposal very closely echoed similar plans for a "National Fund" that Paterson had been promoting for several years. Second, and more compelling, Paterson's pamphlet *An Inquiry into the Reasonableness and Consequences of an Union with Scotland,* dated Apr. 9, 1706, repeated the twelve-point Stowe MS plan almost verbatim. (See "Lewis Medway" [William Paterson], *An Inquiry into the Reasonableness and Consequences of an Union With Scotland* [London: Printed and Sold by Ben. Bragg, 1706], 91–94.) Paterson conceivably could have been copying someone else's work in his 1706 pamphlet, though it seems vastly more likely he authored the original manuscript himself.

20. For Paterson's biography, see Saxe Bannister, ed., *The Writings of William Paterson, Founder of the Bank of England,* 1st ed., vol. 1 (London: Effingham Wilson, 1858), ix–cxliv, esp. xxvii–xxviii; David Armitage, "Paterson, William (1658–1719)," *Oxford DNB.* On Paterson's ideas about credit and banking, see Wennerlind, *Casualties of Credit,* 110–112. On Paterson and Darien, see David Armitage, "The Scottish Vision of Empire: Intellectual Origins of the Darien Venture," in *A Union for Empire: Political Thought and the British Union of 1707,* ed. John Robertson (Cambridge: Cambridge University Press, 1995), 97–118; Macinnes, *Union and Empire,* 174–175.

21. Armitage, "Scottish Vision of Empire," 101, 118.

22. William Paterson, "Letter to King William III," unpublished MS, Nov.–Dec. 1701, BL Add. MS 88,618 B. See also Armitage, "Scottish Vision of Empire," 112–114.

23. Paterson, "Letter to King William III," BL Add. MS 88,618 B, f. 3ʳ.

24. [William Paterson], *Proposals & Reasons for Constituting a Council of Trade* (Edinburgh: s.n., 1701), "Introduction," (unpaginated), [ix].

25. Paterson, "Letter to King William III," BL Add. MS 88,618 B, f. 11ʳ⁻ᵛ.

26. Paterson, *Proposals & Reasons for Constituting a Council of Trade,* [v], [vii], 43, 58. On the intellectual context of Paterson's pamphlet, see John Robertson, *The Case for Enlightenment: Scotland and Naples 1680–1750* (Cambridge: Cambridge

University Press, 2005), chap. 4, esp. p. 171. This pamphlet has been occasionally misattributed to John Law, though that seems irreconcilable with Law's own biography. On the misattribution to Law, see Saxe Bannister, ed., *Writings of William Paterson,* vol. 1, "Preface," [unpaginated]. On Law's life during the period, see Antoin E. Murphy, *John Law: Economic Theorist and Policy-Maker* (Oxford: Clarendon Press, 1997), chap. 5.

27. Paterson, *Proposals & Reasons for Constituting a Council of Trade,* [iii], 51–54, 62–65.

28. Saxe Bannister, ed., *The Writings of William Paterson,* 2nd ed., vol. 3 (London: Judd & Glass, 1859), 47–74.

29. "T. W. Philopatris" [William Paterson], *An Essay Concerning Inland and Foreign, Publick and Private Trade; Together with Some Overtures, Shewing How a Company or National Trade, May be Constituted in Scotland, with the Advantages Which Will Result Therefrom* ([Edinburgh]: *s.n.,* [1704 or 1705?]), 1. See also Macinnes, *Union and Empire,* 229, n. 72.

30. Editor Saxe Bannister suggests that Paterson had also published a 1695 pamphlet set at the "Wednesday Club at Friday Street," entitled *Conferences on the Public Debts* (London, 1695). See Bannister, ed., *Writings of William Paterson,* 2nd ed., vol. 3, "Preface" (unpaginated), [v]. His source is John Ramsey McCulloch, *The Literature of Political Economy: A Classified Catalogue* (London: Printed for Longman, Brown, Green, and Longmans, 1845), 159. Neither the English Short Title Catalogue nor WorldCat contains a record of that text.

31. Paterson, *Inquiry,* "Preface" (unpaginated), [i].

32. Ibid., 90–94.

33. Ibid., 102–109.

34. Ibid., 108–109.

35. I discuss the history of present value calculations in the early modern period in greater detail elsewhere: "Pricing the Future in the Seventeenth Century: Calculating Technologies in Competition," *Technology & Culture* 58, no. 2 (Apr. 2017): 506–528.

36. When landed property changed hands for cash, the price was usually determined by multiplying the property's annual rents by a multiplier, the *years purchase.* For example, if land in Yorkshire was customarily selling for twenty years purchase, a property in Yorkshire yielding £100 in annual rent would cost £2,000 outright. The years-purchase multiplier was determined by local custom and negotiation. For a very early example, see J. D. Alsop, "A Late Medieval Guide to Land Purchase," *Agricultural History* 57, no. 2 (Apr. 1983): 161–164.

37. Paterson, *Inquiry,* 82.

38. Ibid., 109–110.

39. Annuity valuation was key to the Equivalent calculation for two reasons. First, the English national debt was largely composed of long-term annuities. Second, the Equivalent was meant to offer upfront compensation for various streams of future Scottish tax revenues, which effectively reduced to a problem about valuing annuities.

40. On the union and Britishness, see Linda Colley, *Britons: Forging a Nation, 1707–1837* (New Haven, CT: Yale University Press, 1992), 11–18. On Gregory's sense of na-

tional identity, see P. D. Lawrence and A. G. Molland, "David Gregory's Inaugural Lecture at Oxford," *Notes and Records of the Royal Society of London* 25, no. 2 (Dec. 1970): 143–178, esp. 146–147.

41. On James Gregory, see D. T. Whiteside, "Gregory (More Correctly Gregorie), James," in *The Complete Dictionary of Scientific Biography* (*DSB*), ed. Charles Coulston Gillipsie et al., updated ed., vol. 5 (Detroit: Charles Scribner's Sons, 2008 [1972]), 524–530.

42. For an overview of David Gregory's life, and especially his move to Oxford, see Whiteside, "Gregory, David," in *DSB*, 5:520–522, quotation on 520; Lawrence and Molland, "David Gregory's Inaugural Lecture," 144–146.

43. Anita Guerrini, "The Tory Newtonians: Gregory, Pitcairne, and their Circle," *Journal of British Studies* 25, no. 3 (July 1986): 288–311, esp. 295–298, 302, 309–310; Christina Eagles, "David Gregory and Newtonian Science," *British Journal for the History of Science* 10, no. 3 (Nov. 1977): 216–225; Guerrini, "James Keill, George Cheyne, and Newtonian Physiology, 1690–1740," *Journal of the History of Biology* 18, no. 2 (Summer 1985): 247–266; Simon Schaffer, "The Glorious Revolution and Medicine in Britain and the Netherlands," *Notes and Records of the Royal Society of London* 43, no. 2 (July 1989): 167–190, esp. 172–179; Stigler, "John Craig and the Probability of History."

44. Schaffer, "The Glorious Revolution and Medicine," 172–180; [Edward Eizat], *Apollo Mathematicus: or the Art of Curing Diseases by the Mathematicks, According to the Principles of Dr Pitcairn* ([Edinburgh]: *s.n.*, 1695), 18, 105; Stephen M. Stigler, "Apollo Mathematicus: A Story of Resistance to Quantification in the Seventeenth Century," *Proceedings of the American Philosophical Society* 136, no. 1 (Mar. 1992): 93–126.

45. [John Arbuthnot], *An Essay on the Usefulness of Mathematical Learning, in a Letter from a Gentleman in the City to his Friend in Oxford* (Oxford: Printed for Anth. Peisley, 1701), 1, 28. On Gregory-Arbuthnot connections, see Michael Fry, *The Union: England, Scotland and the Treaty of 1707* (Edinburgh: Birlinn, 2006), 256–257. On Gregory's pedagogy, see Eagles, "David Gregory and Newtonian Science," 222–224.

46. David Gregory, "To the Committee of Parliament for Visiting Schools and Coledges" (1687), Gregory Papers, Folio C, Edinburgh U. L. Dc. 1. 61, item no. 215, ff. 747–50, on f. 749.

47. On partisan tensions and the union of 1707, see P. W. J. Riley, "The Union of 1707 as an Episode in English Politics," *English Historical Review* 84, no. 332 (July 1969): 498–527.

48. Clerk, "Journal," vol. 1, NRS GD18 / 3132, ff. 22–29; *The Minutes of the Proceedings of the Lords Commissioners for the Union of the Kingdoms of England and Scotland, The Treaty for Which Began on the Sixteenth Day of April, 1706 and was Concluded the Twenty Second Day of July Following* (London: Printed for Charles Bill, 1706), 26–27.

49. John Clerk (Younger) to Sir John Clerk (Elder), 11 May 1706, Clerk Papers, NRS GD 18 / 3131 / 13, f. 1^r.

50. Clerk, "Journal," vol. 2, NRS GD18/3132, f. 107.

51. *Minutes of the Proceedings of the Lords Commissioners*, 61–62.

52. "Report from the Six Persons to Whom the Equivalent was Referr'd," [May–June 1706], Copy, Lowndes Papers, TNA T 48/22, ff. 1–2.

53. [David Gregory], "Querys to Sr Humphrey Mackworth June 1706," Gregory Papers, Quarto A, Edinburgh U. L. Dk. 1.2.1, item no. 49.2, f. 50. On Mackworth, see Koji Yamamoto, "Piety, Profit and Public Service in the Financial Revolution," *English Historical Review* 126, no. 521 (Aug. 2011): 806–834.

54. Untitled calculations, Gregory Papers, Quarto A, Edinburgh U. L. Dk. 1.2.1, item no. 49.3, f. 51^{r-v}. It is not clear when the calculations were produced, though it was certainly not before late May or June, after the committee had taken a first cut at the computation. It may have been carried out somewhat later, for example, during the intense scrutiny given to the Equivalent computations during fall 1706. See also Spence, "Accounting for the Dissolution," 384.

55. *Minutes of the Proceedings of the Lords Commissioners*, 66, 68.

56. Compare: [Anon.], *An Essay upon the Equivalent. In a Letter to a Friend* ([Edinburgh]: *s.n.*, 1706), 5.

57. Two notational peculiarities to note: the second denominator, $r^x \times r-1$, would be $r^x \times (r-1)$ in modern notation; and r in Gregory's formula does not represent just the interest rate (as in modern usage) but rather one plus the interest rate.

58. This computation was printed as part of the official versions of the Union Treaty and the Minutes of the Treaty Commissioners, for example: *The Minutes of the Proceedings of the Lords Commissioners*, 61–68.

59. The choice of "discount rate" was based on the going rate for long-term annuities, judged to be fifteen years and three months on a "years purchase" basis.

60. For alternative, hypothetical calculations of the Equivalent, see Deringer, "Calculated Values: The Politics and Epistemology of Economic Numbers in Britain, 1688–1738" (Ph.D. diss., Princeton University, 2012), 410–414.

61. Watt, *Price of Scotland*, 225, 230; Macinnes, *Union and Empire*, 320.

62. I see no evidence to support Watt's contention that the Equivalent calculators "manipulated" the Equivalent calculation "by flexing the discount rate" (*Price of Scotland*, 229). In fact, the discount rate that the Equivalent calculators ultimately used, around 6½ percent, was effectively the same figure Paterson had cited as the appropriate discount rate (based on prevailing Scottish interest rates) in his *Inquiry into the Reasonableness of an Union*, before the Equivalent calculation was carried out.

63. Karin Bowie, "Popular Opinion, Popular Politics and the Union of 1707," *Scottish Historical Review* 82, no. 214, part 2 (Oct. 2003): 226–260; Bowie, *Scottish Public Opinion and the Anglo-Scottish Union, 1699–1707* (Woodbridge, UK and Rochester, NY: Royal Historical Society/Boydell Press, 2008).

64. On public reactions to the Equivalent: Spence, "Accounting for the Dissolution," 384–387; Watt, *Price of Scotland*, 228.

65. David Nairne to the Earl of Mar, Nov. 1, 1706, HMC, *Report on the Manuscripts of the Earl of Mar and Kellie, Preserved at Alloa House, N.B.* (London: HMSO, 1904), 307.

66. [Anon.], *An Essay upon the Equivalent. In a Letter to a Friend* ([Edinburgh]: *s.n.*, 1706), 3.

67. [John Clerk], *A Letter to a Friend, Giving an Account How the Treaty of Union Has Been Received Here* ([Edinburgh]: *s.n.*, 1706), 32; [Clerk], *An Essay upon the XV. Article of the Treaty of Union, Wherein the Difficulties That Arise upon the Equivalents, Are Fully Cleared and Explained* ([Edinburgh?]: *s.n.*, 1706), 19, 23; [Anon.], *A Letter to a Member of Parliament, Anent the Application of the 309885 Lib: 10 Shil: Sterl: Equivalent; With Consideration of Reducing the Coin to the Value and Standard of England* ([Edinburgh]: *s.n.*, 1706). For other mis-printings, see [William Black], *A Short View of the Trade and Taxes, Compar'd with What These Taxes May Amount To After the Union, Even Tho' Our Trade Should Not Augment One Sixpence* ([Edinburgh?]: *s.n.*, [1706]), 3; [James Donaldson?], *Considerations in Relation to Trade Considered, and a* Short View of Our Trade and Taxes, *Compared with What These Taxes May Amount to After the Union, &c. Reviewed* ([Edinburgh?]: *s.n.*, 1706), 18.

68. Compare Latour, *Science in Action*, 237–241.

69. The inability of calculators to control the meaning of their figures once they have escaped their grasp is a key theme in recent studies of modern quantitative cultures, including in twentieth-century America: Igo, *Averaged American*, 241, 283; Dan Bouk, *How Our Days Became Numbered: Risk and the Rise of the Statistical Individual* (Chicago: University of Chicago Press, 2015), 164. Compare also Kaiser, *Drawing Theories Apart*.

70. [Anon.], *The Equivalent Explain'd* ([Edinburgh?]: *s.n.*, [1706]), 1–2.

71. [Andrew Fletcher], *State of the Controversy Betwixt United and Separate Parliaments* ([Edinburgh]: *s.n.*, 1706), 17–18; [Roderick Mackenzie], *A Full and Exact Account of the Proceedings of the Court of Directors and Council-General of the Company of Scotland Trading to Africa and the Indies, With Relation to the Treaty of Union, Now Under the Parliament's Consideration* ([Edinburgh]: *s.n.*, 1706); [Anon.], *An Essay upon the Equivalent*, 5–6; [William Black?], *Some Considerations in Relation to Trade, Humbly Offered to His Grace Her Majesty's High Commissioner and the Estates of Parliament* ([Edinburgh?]: *s.n.*, 1706), 11; [Robert Wylie], *A Letter Concerning the Union, with Sir George Mackenzie's Observations and Sir John Nisbet's Opinion upon the Same Subject* ([Edinburgh?]: *s.n.*, 1706), [6].

72. Daniel Defoe, *A History of the Union between England and Scotland*, later ed. (London: Printed for John Stockdale, 1786 [1709]), 296–297; *Minuts of the Proceedings in Parliament* (Edinburgh: Printed by the Heirs and Successors of Andrew Anderson, 1707), nos. 6–8 (Oct. 19–23, 1706) and no. 34 (Dec. 7, 1706).

73. [Anon.], *An Essay upon the Equivalent*, 7–8. For the source of the *Essay's* figures, see *Minutes of the Proceedings of the Lords Commissioners*, 66–67. See also Spence, "Accounting for the Dissolution," 385–386.

74. Black, *Some Considerations in Relation to Trade*, 15. There is strong manuscript evidence that Fletcher was closely involved with the calculation in *Some Considerations*. A manuscript in the Fletcher of Saltoun papers reproduces the calculation exactly, under the title "A Calcull How the Equivalent of 398085.10.0 is Exhausted." The document appears to be a working draft, not a reproduction. See Fletcher of Saltoun Papers, National Library of Scotland (NLS) MS 17,499, f. 144.

75. For example, Cris Shore and Susan Wright, "Audit Culture Revisited: Rankings, Ratings, and the Reassembling of Society," *Current Anthropology* 56, no. 3 (June 2015): 421–444, esp. 421.

76. Black, *Some Considerations in Relation to Trade,* 16; Black, *Short View of our Trade and Taxes;* Donaldson, *Considerations in Relation to Trade Considered,* 3, 13–14; [Black], *Remarks upon a Pamphlet, Intitled* The Considerations in Relation to Trade Considered, and a Short View of Our Present Trade and Taxes Reviewed ([Edinburgh?]: *s.n., 1706*), [1], 3. On the attribution of Donaldson's pamphlet, see P. N. Furbank and W. R. Owens, "Defoe and the Tippony Ale," *Scottish Historical Review* 72, no. 193, part 1 (Apr. 1993): 86–89.

77. Daniel Defoe [to Robert Harley], Nov. 26, 1706, Letter no. 69, in *The Letters of Daniel Defoe,* ed. George Harris Healey (Oxford: Clarendon Press, 1955), 160–162.

78. Nairne to the Earl of Mar, Nov. 1, 1706, HMC *Mar & Kellie,* 307; Earl of Mar to Nairne, Nov. 3, 1706, HMC *Mar & Kellie,* 310.

79. Nairne to the Earl of Mar, Nov. 1, 1706, HMC *Mar & Kellie,* 307.

80. Jasanoff, *Designs on Nature,* 250–255, 270–271.

81. John R. Young, "The Scottish Parliament and the Politics of Empire: Parliament and the Darien Project, 1695–1707," *Parliaments, Estates & Representation* 27, no. 1 (2007): 175–190.

82. English excise protocols specified two different tiers of excise taxes on beer and ale. Negotiations led to the creation of a third, intermediate tier to accommodate the medium-strength "two-penny" or "tippony" ale common in Scotland. See Daniel Defoe to Robert Harley, Nov. [13?], 1706, Letter no. 62, *Letters of Daniel Defoe,* ed. Healey, 145–147; Defoe to Harley, Nov. 22, 1706, Letter no. 67, ibid., 155–158; Furbank and Owens, "Defoe and the Tippony Ale"; Bowie, *Scottish Public Opinion,* 136–137, 157.

83. [James Smollett?] to [?], late 1707, Papers of the Ogilvy Family of Inverquharity, NRS GD205/36/13; Spence, "Accounting for the Dissolution," 381, 387–389; Watt, *Price of Scotland,* 1, 241.

84. "A List of the Committee upon the Equivalent," 1715, Royal Bank of Scotland (RBS) Archives EQ/23/5. The complications that arose concerning the Equivalent in 1715 are also discussed in [Sir David Dalrymple] to [Hugh Dalyrmple] (unsigned), Aug. 23, 1715, Papers of the Hamilton-Dalrymple Family of North Berwick, NRS GD110/1254/8.

85. [William Paterson], *A State of Mr. Paterson's Claim upon the Equivalent* ([London?]: *s.n.,* [1708?]).

CHAPTER THREE

The Balance-of-Trade Battle and the Party Politics
of Calculation in 1713–1714

1. Note that although it took place after the 1707 union, the technical debate about the balance of trade was discussed almost entirely in terms of the balance between England and France, not Britain and France. This stemmed from many factors,

including the relative dominance of English interests in British politics and commerce and the greater availability of historical information about English trade.

2. While Defoe has long been cited as the editor of the *Mercator* and is still cited as such in the English Short Title Catalogue, recent scholarship has identified Davenant as the journal's leading voice, at least through October 1713. See Doohwan Ahn, "The Anglo-French Treaty of Commerce of 1713: Tory Trade Politics and the Question of Dutch Decline," *History of European Ideas* 36 (2010): 167–180, esp. 176. The clearest evidence of Davenant's influence is simply the very high level of quantitative acuity displayed in the journal, particularly over the first six months.

3. *Mercator: or Commerce Retrieved*, no. 1 (May 26, 1713), recto.

4. D. C. Coleman, "Politics and Economics in the Age of Anne: the Case of the Anglo-French Trade Treaty of 1713," in *Trade, Government and Economy in Pre-Industrial England: Essays Presented to F. J. Fisher,* ed. D. C. Coleman and A. H. John (London: Weidenfeld and Nicolson, 1976), 187–211, on 196.

5. Slack, *Invention of Improvement*, 185. On the 1713 Commerce Treaty debate generally, the best overview is Perry Gauci, *The Politics of Trade: The Overseas Merchant in State and Society, 1660–1720* (Oxford: Oxford University Press, 2001), chap. 6. See also D. A. E. Harkness, "The Opposition to the 8th and 9th Articles of the Commercial Treaty of Utrecht," *Scottish Historical Review* 21, no. 83 (Apr. 1924): 219–226; Joyce Oldham Appleby, *Economic Thought and Ideology in Seventeenth-Century England* (Princeton, NJ: Princeton University Press, 1978), chap. 9; Geoffrey Holmes and Clyve Jones, "Trade, the Scots, and the Parliamentary Crisis of 1713," *Parliamentary History* 1 (1982): 44–77; Christopher Dudley, "Party Politics, Political Economy, and Economic Development in Eighteenth-Century Britain," *Economic History Review* 66, no. 4 (Dec. 2013): 1084–1100; Ahn, "Anglo-French Treaty of Commerce."

6. Charles King, ed., *The British Merchant, or; Commerce Preserv'd*, 3 vols. (London: Printed for J. Darby, 1721); G. N. Clark, *Guide to English Commercial Statistics, 1696–1782* (London: Office of the Royal Historical Society, 1938), 18.

7. Coleman, "Politics and Economics in the Age of Anne," 196.

8. Historian Joyce Appleby makes this case especially strongly, though she overstates the matter in arguing that the balance of trade was "moribund" by 1713. See Appleby, *Economic Thought and Ideology,* 248, 269.

9. Jean Lave, "The Values of Quantification," *Sociological Review* 32, no. S1, special issue, *Power, Action, and Belief: A New Sociology of Knowledge,* ed. John Law (May 1984): 88–111; Daston, "Moral Economy," 8–12.

10. On "undeceiving," see *Mercator,* no. 1 (May 26, 1713), recto; *Mercator,* no. 47 (Sept. 8–10, 1713), verso. The trope was also evident in Defoe's pro-union writings in 1706. See Karin Bowie, "Popular Opinion," 256.

11. Adam Smith, *An Inquiry into the Nature and Causes of the Wealth of Nations,* vol. 2 (London: Printed for W. Strahan and T. Cadell, 1776), 76 (section IV.3.31).

12. On the diversity and disorder of "mercantilist" thinking, see Steve Pincus, "Rethinking Mercantilism: Political Economy, the British Empire, and the Atlantic World in the Seventeenth and Eighteenth Centuries," *William and Mary Quarterly*

69, no. 1 (Jan. 2012): 3–34; Philip J. Stern and Carl Wennerlind, *Mercantilism Reimagined: Political Economy in Early Modern Britain and Its Empire* (Oxford: Oxford University Press, 2014). For arguments favoring the coherence of "mercantilism," see Lars Magnusson, *Mercantilism: the Shaping of an Economic Language* (London: Routledge, 1994); Jonathan Barth, "Reconstructing Mercantilism: Consensus and Conflict in British Imperial Economy in the Seventeenth and Eighteenth Centuries," *William and Mary Quarterly* 73, no. 2 (Apr. 2016): 257–290.

13. Later analysts like William Petty and Abraham Hill cite an early calculation from 1354, likely passed on by Edward Misselden around the 1620s. See "Calculations on the Wool Trade," Abraham Hill Papers, BL Sloane MS 2,902, f. 112; "3 Ballances of the Trade of England," William Petty Papers, BL Add. MS 72,890, ff. 100–105. See also Clark, *Guide to English Commercial Statistics,* xi–xvi; Slack, "Government and Information," 40–41, 51–52; Appleby, *Economic Thought and Ideology,* 38.

14. Appleby, *Economic Thought and Ideology,* 35–51; Poovey, *History of the Modern Fact,* 66–91; Finkelstein, *Harmony and the Balance,* chaps. 2–5.

15. E[dward] M[isselden], *The Circle of Commerce. Or The Balance of Trade, in Defence of Free Trade* (London: Printed by Iohn Dawson for Nicholas Bourne, 1623), 130; Thomas Mun, *England's Treasure by Forraign Trade* (London: Printed by J. G. for Thomas Clark, 1664 [written 1623?]), 208–209; Poovey, *History of the Modern Fact,* 76–78; Finkelstein, *Harmony and the Balance,* 89.

16. Sam[uel]. Fortrey, *Englands Interest and Improvement. Consisting in the Increase of the Store, and Trade of this Kingdom* (Cambridge: Printed by John Field, Printer to the University, 1663), [ii], 1, 25. On balance calculations in the seventeenth century, see Slack, "Government and Information," 52–53.

17. Margaret Priestley, "Anglo-French Trade and the 'Unfavourable Balance' Controversy, 1660–1685," *Economic History Review* 4, no. 1 (1951): 37–52.

18. "An Accompt of the Revenue of His Maties. Customes in the severall Ports of England for One year from Michaelmas 1676 to Michaelmas 1677," BL Add. MS 36,785.

19. The strongest statement about the rise of liberal trade ideas is Appleby, *Economic Thought and Ideology,* chaps. 7–8. See also: William J. Ashley, "The Tory Origin of Free Trade Policy," *Quarterly Journal of Economics* 11, no. 4 (July 1897): 335–371; G. L. Cherry, "The Development of the English Free-Trade Movement in Parliament, 1689–1702," *Journal of Modern History* 25, no. 2 (June 1953): 103–119; Chandaman, *English Public Revenue,* 16–20; Slack, "Government and Information," 53; Slack, "The Politics of Consumption and England's Happiness in the Later Seventeenth Century," *English Historical Review* 122, no. 497 (June 2007): 609–631.

20. Abraham Hill, "Reflections on the Current Coin," 1695, Abraham Hill Papers, BL Sloane MS 2,902, ff. 21–26; "Estimate of the Present State of the Trade of England," 1697, BL Sloane MS 2,902, ff. 115–116; [John Pollexfen], *A Discourse of Trade, Coyn, and Paper Credit: And of Ways and Means to Gain, and Retain Riches* (London: Printed for Brabazon Aylmer, 1697), 2–3, 55–56; Davenant, "Memorial

Concerning a Council of Trade," 1696, BL Harley MS 1,223, f. 188ᵛ; Clark, *Guide to English Commercial Statistics,* 3–17.

21. See, for example, a small portable book containing import-export data, preserved at the faculty of law at the University of Edinburgh: "An Abstract of yᵉ Inspecᵗʳ. Genˡˢ. Accounᵗˢ of Impᵗˢ. & Expᵗˢ. from Xmas 1702 to Xmas 1703," Advocates Library Manuscripts, NLS Adv. MS 34.7.5.

22. Charles Davenant, *A Report to the Honourable the Commissioners for Putting in Execution the Act, Intitled,* An Act, for Taking, Examining, and Stating the Publick Accounts of the Kingdom, Part I (London: *s.n.,* 1712), 45–46; Clark, *Guide to English Commercial Statistics,* 15; Waddell, "Charles Davenant," 287.

23. The literature on partisanship is extensive. To begin, foundational studies are: Holmes, *British Politics in the Age of Anne;* Knights, *Representation and Misrepresentation;* Pincus, *1688.*

24. On the culture of conspiracy, see esp., Rachel Weil, *A Plague of Informers,* 1–15, 217–276. Also see Wood, "Conspiracy and the Paranoid Style"; J. A. W. Gunn, *Beyond Liberty and Property: The Process of Self-Recognition in Eighteenth-Century Political Thought* (Kingston, CA: McGill-Queen's University Press, 1983), 10–15; David Hayton, "Moral Reform."

25. Geoffrey Holmes, "The Sacheverell Riots: The Crowd and Church in Early Eighteenth-Century London," *Past & Present* 72 (Aug. 1976): 55–85; Mark Knights, "Introduction: The View from 1710," *Parliamentary History* 31, no. 1, special issue, *Faction Displayed: Reconsidering the Impeachment of Dr. Henry Sacheverell,* ed. Knights (Feb. 2012): 1–15, quotation on 7.

26. Downie, *Robert Harley and the Press,* chap. 8; Hyland, "Liberty and Libel"; Heinz-Joachim Müllenbrock, *The Culture of Contention: A Rhetorical Analysis of the Public Controversy about the Ending of the War of Spanish Succession, 1710–1713* (Munich: Wilhelm Fink Verlag, 1997).

27. [Daniel Defoe], *The Re-Representation: or, A Modest Search After the Great Plunderers of the Nation* (London: *s.n.,* 1711).

28. B. W. Hill, "Oxford, Bolingbroke, and the Peace of Utrecht," *Historical Journal* 16, no. 2 (June 1973): 241–263; Pincus, "Addison's Empire: Whig Conceptions of Empire in the Early 18th Century," *Parliamentary History* 31, no. 1 (Feb. 2012): 99–117, esp. 102–105; Ahn, "Anglo-French Treaty of Commerce," 175–180; Dudley, "Party Politics, Political Economy, and Economic Development," 1090–1096.

29. For a detailed account of the political contest over the Commerce Treaty, see Gauci, *Politics of Trade,* 236–253.

30. De Krey, *Fractured Society,* chap. 3; Pincus, *1688,* chap. 12; Pincus, "Addison's Empire"; Dudley, "Party Politics," 1088–1090.

31. Perry Gauci, "The Clash of Interests: Commerce and the Politics of Trade in the Age of Anne," *Parliamentary History* 28, no. 1 (Feb. 2009): 115–125.

32. Brewer, *Sinews of Power,* 231–249; Rosemary Sweet, "Local Identities and a National Parliament, c. 1688–1835," in *Parliaments, Nations and Identities in Britain and Ireland, 1660–1860,* ed. Julian Hoppit (Manchester: Manchester University Press, 2003),

48–64; John Beckett, "The Glorious Revolution, Parliament, and the Making of the First Industrial Nation," *Parliamentary History* 33, no. 1 (Feb. 2014): 36–53.

33. Gauci, *Politics of Trade*, 239; Harkness, "Opposition to the 8th and 9th Articles."

34. Gauci, "Clash of Interest," 121–123.

35. Gauci, *Politics of Trade*, 165; Pincus, "Addison's Empire," 101.

36. Graham, *Corruption, Party, and Government*, 3.

37. "Letter from Mr Secry S[t]. John of the 8th Ins[t]. relating to a Tarif and Treaty of Commerce with France," to the Lords Commissioners for Trade & Plantations, May 9, 1712, Copy, TNA CO 389/23, ff. 102–108, esp. f. 108.

38. *Lords Journal*, vol. 19, *1709–14*, 542–545; Gauci, *Politics of Trade*, 244–245.

39. HMC *Lords, 1712–14*, New Series, 10:95–96; Rob[er]t Paul to W[illia]m Lowndes, May 30, 1713, TNA T 64/274/116.

40. Holmes and Jones, "Trade, the Scots, and Parliamentary Crisis"; Gauci, *Politics of Trade*, 248–253.

41. Daston and Galison, *Objectivity*, 372–374; Harry M. Marks, "Trust and Mistrust in the Marketplace: Statistics and Clinical Research, 1945–1960," *History of Science* 38 (2000): 343–355, on 349.

42. *Mercator*, no. 47 (Sept. 8–10, 1713), verso; *Mercator*, no. 1 (May 26, 1713), recto. I discuss the alternative Tory and Whig epistemologies of commerce at greater length elsewhere: "'It Was Their Business to Know': British Merchants and Mercantile Epistemology in the Eighteenth Century," *History of Political Economy* 49, no. 2 (June 2017): 177–206. On the fear of misrepresentation and its centrality to the era's politics, see Knights, *Representation and Misrepresentation*.

43. *Mercator*, no. 1 (May 26, 1713), recto; Davenant, *Report to the Honourable the Commissioners*, 47.

44. [Daniel Defoe], *A General History of Trade, and Especially Consider'd As It Respects The British Commerce*, no. 1 (June 1713), 13, 18.

45. King, ed., *The British Merchant*, 1:xiv–xix; Gauci, *Politics of Trade*, 248, 256.

46. [Anon.], *A Letter to the Honourable A—r M—re, Com—r of Trade and Plantation* (London: Printed for J. Roberts, 1714), 18; [John Oldmixon], *Torism and Trade Can Never Agree. To Which is Added, an Account and Character of the Mercator* (London: Printed and Sold by A. Baldwin, 1713), 32; *British Merchant*, no. 1 (Aug. 7, 1713), recto; *British Merchant*, no. 33 (Nov. 24–27, 1713), verso; *British Merchant*, no. 21 (Oct. 13–16, 1713), recto; King, ed., *British Merchant*, 1:x, xii.

47. [Anon.], *Remarks on a Scandalous Libel, Entitil'd* A Letter from a Member of Parliament, &c., Relating to the Bill of Commerce (London: Printed for A. Baldwin, 1713), 4, 12.

48. Anthony Ashley Cooper, 3rd Earl of Shaftesbury, "Sensus Communis, an Essay on the Freedom of Wit and Humour in a Letter to a Friend," *Characteristics of Men, Manners, Opinions, Times*, ed. Lawrence Klein (Cambridge: Cambridge University Press, 1999), 29–69, on 61; Lawrence Klein, *Shaftesbury and the Culture of Politeness: Moral Discourse and Cultural Politics in Early Eighteenth-Century England* (Cambridge: Cambridge University Press, 1994), chap. 7; Sophia Rosen-

feld, *Common Sense: A Political History* (Cambridge, MA: Harvard University Press, 2011), chap. 1; Shapiro, *Culture of Fact,* 25–26.

49. [Anon.], *Remarks on a Scandalous Libel,* 4.

50. Jonathan Swift, *Proposal for Correcting, Improving, and Ascertaining the English Tongue* (London: Printed for Benj. Tooke, 1712), 18, 22, 29; [John Oldmixon], *Reflections on Dr. Swift's Letter to the Earl of Oxford, about the English Tongue* (London: Sold by A. Baldwin, 1712), 2; [Oldmixon], *The British Academy: Being a New-Erected Society for the Advancement of Wit and Learning* (London: *s.n.,* 1712).

51. *Mercator,* no. 1 (May 26, 1713), recto.

52. Steven Shapin and Simon Schaffer, *Leviathan and the Air-Pump: Hobbes, Boyle, and the Experimental Life,* paperback ed. (Princeton, NJ: Princeton University Press, 1989 [1985]), chaps. 2, 4 (on experimental matters of fact); Shapiro, *Culture of Fact,* 9–12 (on facts and the law); Poovey, *History of the Modern Fact,* chap. 2 (on facts and accounting).

53. [Anon.], *A Letter to a West-Country Clothier and Freeholder, Concerning the Parliament's Rejecting The French Treaty of Commerce* (London: Sold by J. Baker, 1713), 8; [Joseph Addison], *The Late Tryal and Conviction of Count Tariff* (London: Printed for A. Baldwin, 1713), 1–2, 10. On the plain-spokenness of facts, see Poovey, *History of the Modern Fact,* 66–91.

54. Davenant, *Discourses on the Publick Revenues,* 1:29–31.

55. [Anon.], *Remarks on a Scandalous Libel,* 4; Oldmixon, *Torism and Trade,* 25–26.

56. Richard Steele, *Guardian,* no. 6 (Mar. 18, 1713); Steele, *The Spectator,* ed. Gregory Smith, vol. 2 (New York: Dutton, 1970), 16–17; Poovey, *History of the Modern Fact,* 144–146.

57. Compare Thomas Hobbes's emphasis on method in his dispute with Robert Boyle over Boyle's air-pump (Shapin and Schaffer, *Leviathan and the Air-Pump,* 145).

58. *Mercator,* no. 2 (May 26–28, 1713).

59. *Mercator,* no. 2 (May 26–28, 1713); no. 11 (June 16–18, 1713).

60. *Mercator,* no. 27 (July 23–25, 1713), recto.

61. *Mercator,* no. 9 (June 11–13, 1713); no. 10 (June 13–16, 1713).

62. *Commons Journal,* vol. 17, *1711–1714,* 422–424; Ahn, "Anglo-French Treaty of Commerce," 170.

63. *Mercator,* no. 11 (June 16–18, 1713), verso.

64. *Mercator,* no. 32 (Aug. 4–6, 1713), recto; no. 35 (Aug. 11–13, 1713), recto; no. 42 (Aug. 27–29), recto; no. 66 (Oct. 22–24), recto; no. 83 (Dec. 1–3), verso. On forgery, see McGowen, "Knowing the Hand"; Wennerlind, *Casualties of Credit,* 84–85, 95–96, 141–142. On the interpretation of quantitative errors as "lies," see also Alain Desrosières, "How Real are Statistics? Four Possible Attitudes," *Social Research* 68, no. 2 (Summer 2001): 339–355, esp. 344–345.

65. See *Mercator,* nos. 32–47, esp. no. 36 (Aug. 13–16), recto.

66. *Mercator,* no. 11 (June 16–18, 1713), verso.

67. *Mercator,* no. 49 (Sept. 12–15, 1713), verso; no. 55 (Sept. 26–29, 1713), verso.

68. *British Merchant,* no. 1 (Aug. 7, 1713), recto; no. 3 (Aug. 11–14, 1713), recto; no. 4 (Aug. 14–18, 1713), verso. Müllenbrock notes that it was a common Whig tactic to attack political journalists; see *Culture of Contention,* 167.

69. For example, *British Merchant*, no. 13 (Sept. 15–18, 1713), recto.

70. *British Merchant*, no. 20 (Oct. 9–13, 1713), recto.

71. *British Merchant*, no. 12 (Sept. 11–15, 1713), verso.

72. *British Merchant*, no. 14 (Sept. 18–22, 1713); no. 15 (Sept. 22–25, 1713).

73. On smuggling in the period, see Ashworth, *Customs and Excise,* chap. 10; Linda Colley, *The Ordeal of Elizabeth Marsh: A Woman in World History* (New York: Pantheon, 2007), 94–113; William Farrell, "Smuggling Silks into Eighteenth-Century Britain: Geography, Perpetrators, and Consumers," *Journal of British Studies* 55, no. 2 (Apr. 2016): 268–294.

74. *British Merchant*, no. 16 (Sept. 25–29, 1713), verso.

75. *Mercator*, no. 36 (Aug. 13–15, 1713), verso.

76. *Guardian*, no. 170 (Sept. 25, 1713), as compiled in the *Guardian,* vol. 2 (London: Printed for J. Tonson, 1714), 323–329; *Mercator*, no. 58 (Oct. 3–6, 1713), recto; no. 60 (Oct. 8–10, 1713), recto.

77. *Mercator*, nos. 60–66 (Oct. 8–24, 1713).

78. See *British Merchant*, no. 15 (Sept. 22–25, 1713), recto; no. 21 (Oct. 13–16, 1713), recto; no. 22 (Oct. 16–20, 1713), verso.

79. *British Merchant*, no. 19 (Oct. 6–9, 1713), verso.

80. *Mercator*, no. 64 (Oct. 17–20, 1713), verso.

81. See *Mercator*, no. 82 (Nov. 28–Dec. 1, 1713); no. 84 (Dec. 3–5, 1713).

82. *Mercator*, no. 77 (Nov. 17–19, 1713), verso.

83. *Mercator*, no. 82 (Nov. 28–Dec. 1, 1713), recto.

84. King, ed., *British Merchant*, 2:51. For a similar criticism: [Anon.], *A Letter to the Honourable A—r M—re,* 19.

85. *Mercator*, no. 76 (Nov. 14–17, 1713), verso.

86. Ahn, "Anglo-French Treaty of Commerce," 176.

87. Ibid. See also Gauci, *Politics of Trade,* 259–270.

88. Abel Boyer, *The History of the Life & Reign of Queen Anne* (London: Printed by J. Roberts, 1722), 633–634.

89. Quoted in Gauci, *Politics of Trade,* 270. On the influence of the *British Merchant* and its translations, see Sophus Reinert, *Translating Empire: Emulation and the Origins of Political Economy* (Cambridge, MA: Harvard University Press, 2011), 71; Koen Stapelbroek, "Between Utrecht and the War of Austrian Succession: The Dutch Translation of the *British Merchant* of 1728," *History of European Ideas* 40, no. 8 (2014): 1026–1043; Antonella Alimento, "Beyond the Treaty of Utrecht: Véron de Forbonnais's French Translation of the *British Merchant* (1753)," *History of European Ideas* 40, no. 8 (2014): 1044–1066.

90. Gauci, *Politics of Trade,* 244.

91. Priestley, "Anglo-French Trade," 48, 51; Reinert, *Translating Empire,* 172; Elizabeth Boody Schumpeter, *English Overseas Trade Statistics, 1697–1808* (Oxford: Clarendon Press, 1960), 17–18; Brian R. Mitchell, *British Historical Statistics* (Cambridge: Cambridge University Press, 1988), 442–447.

92. The question of closure has been foundational to scholarship in science studies. For example, H. M. Collins, *Changing Order: Replication and Induction in Scientific Prac-*

tice, new ed. (Chicago: University of Chicago Press, 1992 [1985]); Shapin and Schaffer, *Leviathan and the Air-Pump,* chap 6. For a discussion of "non-closure" in early-modern scientific controversy, see Mario Biagioli, *Galileo, Courtier: the Practice of Science in the Culture of Absolutism* (Chicago: University of Chicago Press, 1993), 207–209.

93. Knights, "How Rational Was the Later Stuart Public Sphere," 262.

94. Henry Martin, "Observations upon the Account of Exports and Imports for 17 Years Ending at Christmas 1714 Delivered in to the Board of Trade 1717/8," in Clark, *Guide to English Commercial Statistics,* 62–69, on 68.

95. [Joshua Gee], *The Trade and Navigation of Great-Britain Considered* (London: Printed by Sam. Buckley, 1729). See also Reinert, *Translating Empire,* 144–146.

96. David Hume, *Political Discourses* (Edinburgh: Printed by R. Fleming for A. Kincaid and A. Donaldson, 1752), 81.

97. Josiah Tucker, *A Brief Essay on the Advantages and Disadvantages which Respectively Attend France and Great Britain, with Respect to Trade,* 2nd ed. (London: Printed for T. Trye, 1750), ii.

CHAPTER FOUR

The Preeminent Bookkeepers in Christendom: Personalities of Calculation

1. [John Crookshanks], *Some Seasonable Remarks on a Book Publish'd in the Month of July, 1718* (London: *s.n.,* 1718), 7.

2. Patricia Cline Cohen, *A Calculating People: the Spread of Numeracy in Early America* (Chicago: University of Chicago Press, 1981), 6–12; Emigh, "Numeracy or Enumeration," 654.

3. Lester M. Beattie, *John Arbuthnot, Mathematician and Satirist* (Cambridge, MA: Harvard University Press, 1935), 29; John A. Dussinger, "Samuel Richardson's 'Elegant Disquisitions': Anonymous Writing in the 'True Briton' and Other Journals," *Studies in Bibliography* 53 (2000): 195–226, esp. 201–202.

4. The role of "calculator" assumed by Hutcheson and Crookshanks was not yet solid enough to be considered a "persona," in the sense described by Lorraine Daston and H. Otto Sibum. See "Introduction: Scientific Personae and Their Histories," *Science in Context* 16, nos. 1–2 (Mar. 2003): 1–8.

5. See, for example, [Daniel Defoe], *An Essay at Removing National Prejudices against a Union with Scotland. To Be Continued during the Treaty Here. Part II* (London: *s.n.,* 1706), 11; Defoe, *An Essay upon Loans* (London: Printed and Sold by the Booksellers, 1710), 6; [Defoe], *The Secret History of the White Staff, Being an Account of Affairs Under the Conduct of Some Late Ministers,* 3rd ed. (London: Printed for J. Baker, 1714/1715), 16.

6. In April 1720, for example, critics of the South Sea Company would assail a pro-Company "calculator." See Archibald Hutcheson, *Some Seasonable Considerations for Those Who Are Desirous, by Subscription, or Purchase, to Become Proprietors of South-Sea Stock* (London: Printed and Sold by J. Morphew, 1720), 6; [Anon.], *Remarks on the Celebrated Calculations of the Value of South-Sea Stock, in the Flying-Post of the 9th of April, 1720* (London: Printed for J. Roberts, 1720), 18.

7. Shapin and Schaffer, *Leviathan and the Air-Pump*, 23; Shapin, *A Social History of Truth*.

8. Porter, *Trust in Numbers*, 8. See also Espeland and Sauder, "Rankings and Reactivity," 4; Merry, "Measuring the World," S85; Robert Salais, "Quantification and Objectivity: From Statistical Conventions to Social Conventions," *Historical Social Research/Historische Sozialforschung* 41, no. 2 (2016): 118–134, esp. 127, 130.

9. Daston, "Moral Economy," 10.

10. Porter, *Trust in Numbers*, chaps. 4–5; Poovey, *History of the Modern Fact*, chap. 2, esp. p. 90; Carruthers and Espeland, "Accounting for Rationality."

11. On mathematics and the cultivation of virtue, see Jones, *Good Life in the Scientific Revolution*.

12. Soll, *The Reckoning*, chap. 2; Smyth, *Autobiography*, chap. 2.

13. Alain Desrosières misstates the case slightly when he asserts that English political arithmeticians fashioned a "new social role . . . : *the expert* with a precise field of competence who suggests techniques to those in power while trying to convince them that, in order to realize their intentions, they must go through him" (*Politics of Large Numbers*, 24, emphasis mine). On the history of expertise in early modern Europe, see Eric H. Ash, ed., *Expertise: Practical Knowledge and the Early Modern State, Osiris* 25 (2010).

14. *The Answer of the Merchants-Petitioners, and Trustees for the Factory at Leghorn, To The Account of Damages, Laid to the Charge of the Great Duke of Toscany, by Sir Alexander Rigby* (London: *s.n.*, 1704), 48.

15. Jasanoff, *Designs on Nature*, 267.

16. Thomas, "Numeracy in Early Modern England"; Otis, "By the Numbers"; Peter Wardley and Pauline White, "The Arithmeticke Project: A Collaborative Research Study of the Diffusion of Hindu-Arabic Numerals," *Family & Community History* 6 (May 2003): 5–17.

17. On mathematical practitioners, see E. G. R. Taylor, *The Mathematical Practitioners of Tudor & Stuart England* (Cambridge: For the Institute of Navigation at the University Press, 1954); Johnston, "Mathematical Practitioners"; Neal, "Rhetoric of Utility."

18. Otis, "By the Numbers," 113–114, 290; Howson, *History of Mathematics Education*, chap. 1; Deborah Harkness, *The Jewel House: Elizabethan London and the Scientific Revolution* (New Haven, CT: Yale University Press, 2007), 103–110; John Denniss, "Learning Arithmetic: Textbooks and their Users in England 1500–1900," in *The Oxford Handbook of the History of Mathematics*, ed. Eleanor Robson and Jacqueline Stedall (Oxford: Oxford University Press, 2009), 448–467; Wardhaugh, *Poor Robin's Prophecies*, chap. 3. On the Rule of Three in another, earlier context, see Michael Baxandall, *Painting and Experience in Fifteenth Century Italy: A Primer in the Social History of Pictorial Style*, 2nd ed. (Oxford: Oxford University Press, 1988 [1972]), 94–110.

19. Otis, "By the Numbers," 106–107. For manuscript arithmetic texts, see, for example, "Treatise on Arithmetic," c. 1660, Huntington mss*HM* 70,980, esp. f. 20; Arithmetic exercise book of Sarah Cole, 1685, Folger Library Manuscripts V.b.292.

20. On gauging texts, see Wardhaugh, *Poor Robin's Prophecies,* 101.

21. *Such as Are Desirous, Eyther Themselues to Learne, or to Haue Theyr Children or Seru-ants Instructed in Any of These Artes and Faculties Heer Vnder Named, It May Please Them to Repayre Vnto The House Of Humfry Baker* ([London?]: [T. Purfoot?], c. 1590).

22. Otis, "By the Numbers," 102–117.

23. H. Spencer Jones, "Forward by the Astronomer Royal," in Taylor, *Mathematical Practitioners of Tudor & Stuart England,* ix; Howson, *History of Mathematics Education,* chap. 2; Thomas, "Numeracy," 111–112; Dennis, "Learning Arithmetic," 455; Otis, "By the Numbers," 127–128.

24. Otis, "By the Numbers," 120.

25. Ibid., 118–136. On mathematical culture in the universities, see Mordechai Fein-gold, *The Mathematicians' Apprenticeship: Science, Universities and Society in England, 1560–1640* (Cambridge: Cambridge University Press, 1984).

26. John Aubrey, *Aubrey on Education: A Hitherto Unpublished Manuscript by the Author of* Brief Lives, ed. J. E. Stephens (London: Routledge, 2012 [1972]), 59.

27. Quoted in Kate Bennett, "John Aubrey and the 'Lives of Our English Mathematical Writers,'" in *Oxford Handbook of the History of Mathematics,* ed. Robson and Stedall, 329–352, on 349.

28. Thomas Foley to Thomas Foley (his father), Feb. 23, 1688 / 89, BL Add. MS 70,227. For a similar sentiment, see John Wallis, "A Letter from a Friend of the Universi-ties, in Reference to the New Project of an Academy for Riding the Great Horse &c.," in C. R. L. Fletcher, ed., *Collectanea,* First Series (Oxford: Printed for the Oxford Historical Society at the Clarendon Press, 1885), 269–337, on 320. See also Turner, "Mathematical Instruments"; Thomas, "Numeracy," 112.

29. Arbuthnot, *Essay on the Usefulness of Mathematical Learning,* 2–3.

30. On Crookshanks, see P. W. J. Riley, *The English Ministers and Scotland 1707–1727* (London: University of London / Athlone Press, 1964), 43, 223, 276; Jacob M. Price, "The Excise Affair Revisited: The Administrative and Colonial Dimensions of a Parliamentary Crisis," in *England's Rise to Greatness,* ed. Stephen B. Baxter (Berkeley: University of California Press, 1983), 257–322, on 267–268; J. D. Alsop, "The Politics of Whig Economics: the National Debt on the Eve of the South Sea Bubble," *Durham University Journal* 77, no. 2 (1985): 211–218.

31. *The Answer of the Merchants-Petitioners,* 48; Captain Thomas James to Sir Edward Nicholas, Jan. 29, 1634 / 5, *Calendar of State Papers* (*CSP*), *Domestic Series, of the Reign of Charles I,* vol. 7, *May 1634–Mar. 1635,* ed. John Bruce (London: HMSO, 1864), vol. 282, no. 104, p. 480 (thanks to Mordechai Levy-Eichel for this reference). On the importance of Scots in the extension of British power overseas, see Colley, *Britons,* 117–32.

32. Alexander Rigby, *The Case of Sir Alexander Rigby, and William Shephard, Complain-ants, as to Their Particular Damages by the Grand Duke of Tuscany's Proceedings, in the Seizure of William Plowman, and of their Effects* (London: s.n., 1704), 17–22; *Answer of the Merchants-Petitioners,* 24; Eveline Cruickshanks and Richard Har-rison, "Rigby, Sir Alexander (c.1663–1717), of Layton, nr. Liverpool, Lancs.," in *History of Parliament: The House of Commons 1690–1715.*

33. John Crookshanks to the Duke of Marlborough, Mar. 27, 1706, Bleinheim Papers, BL Add. MS 61,285, ff. 159–160; [Crookshanks], *Instructions for the Collectors and Other Officers Employ'd in Her Majesties Customs, &c. in the North-Part of Great Britain* (Edinburgh: Printed by the Heirs and Successors of Andrew Anderson, 1707), 4–6; Commissioners of the Customs of North Britain to John Crookshanks, Dec. 30, 1708, TNA T 1/110, ff. 226–227. On military finance in the period, see Davies, "Seamy Side"; Clay, *Public Finance and Private Wealth;* Graham, *Corruption, Party, and Government.*

34. "Presentment of the Commissioners of the Customs," Nov. 27, 1708, TNA T 1/110, f. 126; "The Case and Petition of John Crookshanks," in John Crookshanks to Robert Walpole, Mar. 28, 1727, Cambridge U.L. Ch(H) Letter 1,415; Price, "Excise Affair Revisited," 267, 309–310, n. 29; Jacob M. Price, "Glasgow, the Tobacco Trade, and the Scottish Customs, 1707–1730: Some Commercial, Administrative and Political Implications of the Union," *Scottish Historical Review* 63, no. 175, part 1 (Apr. 1984): 1–36, esp. 2, 8–12.

35. HMC *Portland,* vol. 10. See the following letters: John Crookshanke[s] to the Lord High Treasurer, May 26, 1713, p. 169; Crookshanks to the Lord High Treasurer, June 8, 1713, pp. 175–176; John Crookshanks to the Lord High Treasurer, Apr. 1714, p. 178. See also pp. 163–165, 168–178, 492–495; Riley, *English Ministers,* 223.

36. "Case and Petition of John Crookshanks," Cambridge U.L. Ch(H) Letter 1,415.

37. On the "arising" Equivalent, see "Memorial About the Equivalent," 1711, RBS Archives EQ/23/1; "The Memorial of Sir John Areskin, Sir Patrick Johnston, John Pringle & Alex^r. Abercromby Esq^{rs}. for themselves and in Name & behalf of Other Comm^{rs}. of Equivalent," July 26, 1714, RBS Archives EQ/23/1; "The Memorial and Representation of the Commissrs. of the Equivalent," [1715?], RBS Archives EQ/23/5; "Memorial to the Treasury With an Acco^{tt}. of the Equivalents & Encreases due out of the Excyse of Scotland from 1st May 1707 to 1 May 1714, presented 30 June 1715," RBS Archives EQ/23/5; Crookshanks to Harley, May 29, 1714, and Crookshanks to Harley, June 7, 1714, in HMC *Portland,* 10:178, 493–495; Richard Saville, *Bank of Scotland: A History, 1695–1995* (Edinburgh: Edinburgh University Press, 1996), 57–58, 84–85; Spence, "Accounting for the Dissolution," 387–389.

38. "Case and Petition of John Crookshanks," Cambridge U.L. Ch(H) Letter 1,415; Price, "Excise Affair Revisited," 267.

39. The most detailed biographical information on Hutcheson is by Andrew A. Hanham: "Hutcheson, Archibald (c.1660–1740), of the Middle Temple, and Golden Sq., Westminster," in *History of Parliament: The House of Commons 1690–1715;* Hanham, "Hutcheson, Archibald (c. 1660–1740)," *Oxford DNB.* Eveline Cruickshanks's entry in *The History of Parliament: The House of Commons, 1715–1754,* ed. R. Sedgwick (New York: Published for the History of Parliament Trust by Oxford University Press), emphasizes Hutcheson's later career and his supposed Jacobite connections.

40. Wilfrid Prest, "Legal Education of the Gentry at the Inns of Court, 1560–1640," *Past & Present* 38 (Dec. 1967): 20–39, esp. 38–39.

41. Kelsey Jackson Williams, "Training the Virtuoso: John Aubrey's Education and Early Life," *Seventeenth Century* 27, no. 2 (June 2012): 157–182, on 160; Lotte Mulligan and Glenn Mulligan, "Reconstructing Restoration Science: Styles of Leadership and Social Composition of the Early Royal Society," *Social Studies of Science* 11, no. 3 (Aug. 1981): 327–364, esp. 347–348.

42. On the Leewards' history, see Vere Langford Oliver, *The History of the Island of Antigua, One of the Leeward Caribees in the West Indies, from the First Settlement in 1635 to the Present Time* (London: Mitchell and Hughes, 1894), 1:lxvii; Richard S. Dunn, *Sugar and Slaves: The Rise of the Planter Class in the English West Indies, 1624–1713,* new ed. (Chapel Hill: University of North Carolina Press for the Omohundro Institute, 2000 [1972]), chap. 4; Richard B. Sheridan, *Sugar and Slavery: An Economic History of the British West Indies, 1623–1775* (Baltimore, MD: Johns Hopkins University Press, 1974), chap. 8; Susan Amussen, *Caribbean Exchanges: Slavery and the Transformation of English Society, 1640–1700* (Chapel Hill: University of North Carolina Press, 2007), 6–7, 30.

43. Christian J. Koot, "A 'Dangerous Principle': Free Trade Discourses in Barbados and the English Leeward Islands, 1650–1689," *Early American Studies* 5, no. 1 (Spring 2007): 132–163; Pincus, "Rethinking Mercantilism," 14–20.

44. Governor Sir Nathaniel Johnson to Lords of Trade and Plantations, Mar. 3, 1688, Nevis, in *CSP, Colonial Series, America and West Indies,* ed. J. W. Fortescue, vol. 12, *1685–1688 and Addenda 1653–1687* (London: HMSO, 1899), 512–513; Amussen, *Caribbean Exchanges,* 7.

45. Archibald Hutcheson, "Report of the Attorney General Touching Estates in Montserrat and the Leeward Islands," Apr. 19, 1688, in *CSP Colonial, America and West Indies,* 12:530–532; Governor Sir Nathaniel Johnson to Lords of Trade and Plantations, June 2, 1688, Nevis, in *CSP Colonial, America and West Indies,* 12:552–554.

46. Archibald Hutcheson to William Blathwayt, Apr. 3, 1691, Antigua; Hutcheson to Blathwayt, June 3, 1691, Barbados; Governor Codrington to [Earl of Nottingham?], Aug. 11, 1692, Nevis; Governor Codrington to the Agents for the Leeward Islands, Nevis, Aug. 20, 1692; all in *CSP Colonial, America and West Indies,* vol. 13, *1689–1692,* 402–411, 460–479, 679–692; Andrew J. O'Shaughnessy, "Codrington, Christopher (1639 / 40–1698)," *Oxford DNB.*

47. J. Johns Sonn to Admiral Nevill, Copy, May 4, 1697, *CSP Colonial, America and West Indies,* vol. 15, *1697–1698,* 477. See also, *CSP Colonial, America and West Indies,* vol. 16, *1697–1698,* xxviii–xxx, 125, 335–339, 342, 443.

48. Joseph Redington, ed., *Calendar of Treasury Papers (CTP),* vol. 2, *1697–1702* (London: HMSO, 1871), 333.

49. For scholars' descriptions, see Linda Colley, *In Defiance of Oligarchy: The Tory Party 1714–1760* (Cambridge: Cambridge University Press, 1982), 99; Holmes, *British Politics in the Age of Anne,* 283; Hanham, "Hutcheson, Archibald," in *History of Parliament: The House of Commons 1690–1715.*

50. Also see Hutcheson to Mr. Kraenberg, Frankfort, Aug. 13/2, 1713, Huntington mss*HM* 44,710, ff. 147–148.

51. [Hutcheson], *A Speech Made in the House of Commons, on Tuesday the 24th of April 1716. At the Second Reading of the Bill for Enlarging the Time for Continuance of Parliaments, &c.* (London: Printed for T. Baker and T. Warner, 1716).

52. Archibald Hutcheson, *A Collection of Advertisements, Letters and Papers, and Some Other Facts, Relating to the Last Elections at Westminster and Hasting* (London: Printed for T. Payne, 1722), 5–6.

53. On the Davenant-Hutcheson link, see John M'Arthur, *Financial and Political Facts of the Eighteenth Century; with Comparative Estimates of the Revenue, Expenditure, Debts, Manufactures, and Commerce of Great Britain*, 3rd ed. (London: Printed for J. Wright, 1801), 72.

54. For data on the historical growth of the national debt, see North and Weingast, "Constitutions and Commitment," 822; Nathan Sussman and Yishay Yafeh, "Institutional Reforms, Financial Development and Sovereign Debt: Britain 1690–1790," *Journal of Economic History* 66, no. 5 (Dec. 2006): 1–30, esp. 17.

55. Pocock, *Machiavellian Moment*, 439.

56. [Archibald Hutcheson], *A Proposal for the Payment of the Publick Debts, and an Account of Some Things Mentioned in Parliament on that Occasion* ([London]: *s.n.*, 1714 / 15), 15–18, 20.

57. Hutcheson's sinking fund model had odd implications. Notably, it suggested that the nation would pay off its debt *faster* if the interest rate on the debt were higher. For example, paying £1 million on the debt each year, it would take twenty-nine years to pay off the full debt at 3 percent interest, versus only twenty-three years at 6 percent. This was mathematically true—but it only worked if the nation *already* had the revenue to pay the higher interest rates.

58. Hutcheson, *Proposal for the Payment of the Publick Debts*, 19–20.

59. [Archibald Hutcheson], *Computations Relating to the Publick Debts, Taken from the Abstract Deliver'd into Parliament the 14th of March, 1716* (London: Printed for H. Clements, 1717), 4–5.

60. Ibid., 3–5.

61. [Archibald Hutcheson], *Abstracts of the Number and Yearly Pay of the Landforces, of Horse, Foot and Dragoons in Great Britain, for the Year 1718* (London: *s.n.*, 1718), 14; [Hutcheson], *Some Further Remarks Relating to Half-Pay* ([London]: *s.n.*, [1718]), 3.

62. Archibald Hutcheson, *Some Calculations and Remarks Relating to the Present State of the Publick Debts and Funds* (London: *s.n.*, 1718), 10–11, 14.

63. Abel Boyer, *Animadversions and Observations upon a Treatise Lately Publish'd, Entituled* Some Calculations and Remarks . . . *by Archibald Hutcheson* (London: Printed for J. Roberts, 1718); John Crookshanks to the Earl of Sunderland, Aug. 13, 1718, BL Add. MS 61,605, ff. 32–34, on f. 32ʳ.

64. [John Crookshanks], *Some Seasonable Remarks on a Book Published in the Month of July, 1718, by Archibald Hutcheson, Esq; Relating to the Publick Debts and Fonds* (London: *s.n.*, 1718). On the symbolism of the November 5 date, see James Richard Redmond McConnel, "The 1688 Landing of William of Orange at Torbay: Numerical Dates and Temporal Understanding in Early Modern England," *Journal of Modern History* 84, no. 3 (Sept. 2012): 539–571.

65. Crookshanks, *Some Seasonable Remarks,* 1–2.
66. Ibid., 2.
67. Ibid., 3.
68. Ibid. On contemporary thinking and policies regarding interest rates, see H. J. Habakkuk, "The Long-Term Interest Rate and the Price of Land in the Seventeenth Century," *Economic History Review* 5, no. 1 (1952): 26–45; Peter Temin and Hans-Joachim Voth, "Interest Rate Restrictions in a Natural Experiment: Loan Allocation and the Change in the Usury Laws in 1714," *Economic Journal* 118, no. 528 (Apr. 2008): 743–758.
69. Crookshanks, *Some Seasonable Remarks,* 3–4.
70. Ibid.
71. Ibid., 4.
72. Ibid., 5–7.
73. Ibid., 7.
74. Ibid.
75. Jasanoff, "Acceptable Evidence"; Porter, *Trust in Numbers,* 90.
76. On the continuation of the debate, see Archibald Hutcheson, *Mr. Hutcheson's Answer to Mr. Crookshanks's* Seasonable Remarks (London: *s.n.,* 1719), esp. 20–24; John Crookshanks, *Some Matters of Fact Relating to the Revolution, the Annuities, the Civil List Branches, and National Debts . . . in Reply to Archibald Hutcheson* (London: *s.n.,* 1720). On Hutcheson's purported role in the Jacobite "Atterbury Plot," a polemical take is: Eveline Cruickshanks and Howard Erskine-Hill, *The Atterbury Plot* (Houndmills, UK: Palgrave Macmillan, 2004), 89–90, 126, 136.
77. "The Case and Petition of John Crookshanks," Cambridge U.L. Ch(H) Letter 1,415; Price, "Glasgow," 12; Riley, *English Ministers,* 276.
78. "The Case and Petition of John Crookshanks," Cambridge U.L. Ch(H) Letter 1,415; "Memorial for the Lord President concerning a Debt due by John Cruikshanks Esqʳ.," Hamilton-Dalrymple Papers, NAS GD110/57. At least twenty-one letters from Crookshanks to Walpole remain in Walpole's papers at Cambridge U. L. For example: Oct. 21, 1721, Ch(H) Letters 923; July 24, 1724, Ch(H) Letters 1,150; Feb. 20, 1726/7, Ch(H) Letters 1,405; Crookshanks to Walpole, Apr. 3, 1727, Ch(H) Letters 1,417; Ch(H) Papers 73/10–10a.
79. John Crookshanks to Robert Walpole, Jan. 1, 1732/3, Seville, Ch(H) Letters 1,944.

CHAPTER FIVE
Intrinsic Values: Figuring Out the South Sea Bubble

Epigraph: John Trenchard and Thomas Gordon ("Cato"), "No. 6. How Easily the People Are Bubbled by Deceivers," in *Cato's Letters: Or, Essays on Liberty, Civil and Religious, and Other Important Subjects,* ed. Ronald Hamowy, vol. 1 (Indianapolis: Liberty Fund, 1995), 55–56. Emphasis mine.

1. A rough estimate: By the end of 1720, the South Sea Company had issued roughly 378,000 shares of Stock in total (£37.8 million par value), according to Company

reports. At £1,000 per share, roughly the top market price the Stock reached, this implied a paper value of £378 million. King's calculation of England's capital value was certainly an underestimate, but the general point stands. On King's and other estimates, see Paul Slack, "Measuring the National Wealth in Seventeenth-Century England," *Economic History Review* 57, no. 4 (Nov. 2004): 607–635, on 614.

2. Throughout this and subsequent chapters, the capitalized terms "Company," "Stock," "Scheme," "Bubble," and "Crisis" will refer to the South Sea Company, South Sea Stock, South Sea Scheme, etc.

3. [Anon.], *Considerations on the Present State of the Nation, as to Publick Credit, Stocks, the Landed and Trading Interests* (London: Printed for A. Dodd and J. Roberts, 1720), 1–2. For another example, see [John Trenchard and Thomas Gordon], *Advice and Considerations for the Electors of Great Britain* (London: Printed for J. Peele, 1722), 16–17.

4. On Hutcheson as financial pioneer, see Richard Dale, *The First Crash: Lessons from the South Sea Bubble* (Princeton, NJ: Princeton University Press, 2004), 2; Paul Harrison, "Rational Equity Valuation at the Time of the South Sea Bubble," *History of Political Economy* 33, no. 2 (Summer 2001): 269–281. For a critical response to this celebratory view of Hutcheson, see Helen Paul, "Archibald Hutcheson's Reputation as an Economic Thinker: His Pamphlets, the National Debt and the South Sea Bubble," *Essays in Economic and Business History* 30 (Jan. 2012): 93–104. Much of the recent attention to Hutcheson's calculations has been inspired by economists' interest in using historical bubbles as test cases for modeling financial bubbles in general, and specifically for testing whether such bubbles can be understood as "rational" economic events. I discuss this further in "For What It's Worth: Historical Financial Bubbles and the Boundaries of Economic Rationality," *Isis* 106, no. 3 (Sept. 2015): 646–656.

5. For such accounts in contemporary media, see Felix Salmon, "Recipe for Disaster: The Formula That Killed Wall Street," *Wired* 17, no. 3 (Feb. 23, 2009); Joe Nocera, "Risk Mismanagement," *New York Times Magazine*, Jan. 2, 2009. Recently, social scientists have taken a more nuanced stance on the relationship between financial calculation and financial crisis. See, for example, Donald MacKenzie and Taylor Spears, "'The Formula that Killed Wall Street': The Gaussian Copula and Modelling Practices in Investment Banking," *Social Studies of Science* 44, no. 3 (June 2014): 393–417; MacKenzie and Spears, "'A Device for Being Able to Book P&L': The Organizational Embedding of the Gaussian Copula," *Social Studies of Science* 44, no. 3 (June 2014): 418–440.

6. The definitive modern chronicler of the South Sea Bubble, John Carswell, considered the many computational pamphlets produced during 1720 an "unhealthy" distraction from the fraud and corruption that were the crisis's true cause. Other historians like Benjamin Wardhaugh and Larry Stewart have argued that the South Sea Bubble damaged the reputation of mathematical thinking in eighteenth-century Britain. Wardhaugh writes specifically that "contemporaries tried, and failed, to calculate the 'real value' of South Sea stock" in 1720, visibly demonstrating

mathematical experts' inability to solve pressing public problems." See Carswell, *South Sea Bubble,* 148; Wardhaugh, *Poor Robin's Prophecies,* 41–52, on 45; Larry Stewart, "Others Centre of Calculation," 153.

7. For the public-finance context, see Dickson, *The Financial Revolution in England,* 90–93.

8. [Humphrey Mackworth], *A Proposal for Paying Off the Publick Debts by the Appropriated Funds* (London: *s.n.,* 1720); Richard Steele, *A Nation a Family: Being the Sequel of the Crisis of Property: Or, a Plan for the Improvement of the South-Sea Proposal* (London: Printed for W. Chetwood et al., 1720); [Anon.], *An Essay for Discharging the Debts of the Nation, by Equivalents . . . And the South-Sea Scheme Consider'd* (London: Printed for J. Noon, 1720).

9. On the Company's origins and financial schemes, see Carswell, *South Sea Bubble,* 19–106. On the Company and the slave trade, see Helen Paul, *The South Sea Bubble: An Economic History of Its Origins and Consequences* (London: Routledge, 2011), chap. 6; Wennerlind, *Casualties of Credit,* chap. 6; Abigail L. Swingen, *Competing Visions of Empire: Labor, Slavery, and the Origins of the British Atlantic Empire* (New Haven, CT: Yale University Press, 2015), chap. 7.

10. Scholars continue to debate the Company's primary motivations. For example, Richard Kleer argues that the goal of the Company's Scheme was to establish a dominant position in the banking industry: " 'The Folly of Particulars': The Political Economy of the South Sea Bubble," *Financial History Review* 19, no. 2 (Aug. 2012): 175–197.

11. Technically, investors at the time rarely paid cash up front for the full price of shares when investing in joint-stock companies. Rather, they bought shares on "subscription," paying a fraction of the share price at the time of purchase and then paying the remainder over time. On the mechanics of subscription finance and the potential impact on financial market behavior during 1720, see Gary Shea, "Understanding Financial Derivatives during the South Sea Bubble: The Case of South Sea Subscription Shares," *Oxford Economic Papers* 59, suppl. 1 (Oct. 2007): i73–i104.

12. On contemporary literature, see John G. Sperling, *The South Sea Company: A Historiographical Essay and Bibliographical Finding List* (Boston: Baker Library, Harvard Graduate School of Business Administration, 1962).

13. [Anon.], *Argument to Shew the Disadvantage That Would Accrue to the Publick, from Obliging the South-Sea Company to Fix What Capital Stock They Will Give for the Annuities* (London: Printed for J. Roberts, 1720), 9–10, 14–16; [Anon.], *Remarks upon Several Pamphlets Writ in Opposition to the South-Sea Scheme: Particularly,* An Examination and Explanation of the South-Sea Scheme, *&c.* (London: Printed for J. Roberts, 1720).

14. James Brydges, 1st Duke of Chandos, to Mr. Drum[m]ond, Feb. 8, 1719 / 20, Huntington Stowe Brydges mss*ST* 57, vol. 17, fol. 1. On Chandos's investing activities, see Koji Yamamoto, "Beyond Rational vs Irrational Bubbles: James Brydges the First Duke of Chandos during the South Sea Bubble," in *Le crisi finanziarie: Gestione, implicazioni sociali e conseguenze nell'età preindustriale / The Financial Crises:*

Their Management, Their Social Implications and Their Consequences in Pre-Industrial Times, Atti delle "Settimane di Studi" e altri Convegni, vol. 47 (Florence: Firenze University Press, 2016), 325–358.

15. *An Examination and Explanation of the South-Sea Company's Scheme, for Taking in the Publick Debts* (London: Printed for J. Roberts, 1720), 5–7; [Anon.], *A Farther Examination and Explanation of the South-Sea Company's Scheme* (London: Printed and Sold by J. Roberts, 1720); [Anon.], *Considerations Occasioned by the Bill for Enabling the South-Sea Company to Increase Their Capital Stock, &c.: with Observations of Mr. Law* (London: Printed and Sold by J. Roberts, 1720), 25–26; John Trenchard, *Comparison Between the Proposals of the Bank and South-Sea Company* (London: Printed and Sold by J. Roberts, 1720), 14–17.

16. For details of the South Sea Company's finance operations, see Neal, *Rise of Financial Capitalism,* 235; Dickson, *Financial Revolution,* 124–142.

17. Note that this dividend promise was based on the par value of Stock purchased. So an investor who bought one share of stock, with par value £100, would receive a dividend of £30–£50, regardless of the amount paid for the stock. An investor who paid £1,000 for that one share would in effect be receiving a dividend of 3–5 percent on their invested capital. It was a highly optimistic promise, nonetheless.

18. James Milner, "A Second Letter, Occasion'd by What Has Past Since, in Relation to the South-Sea Company's Bargain," in *Visit to the South-Sea Company and the Bank: in a Letter to a Friend, Concerning the Late Proposals for the Payment of the Nation's Debts,* 2nd ed. (London: Printed for J. Roberts and A. Dodd, 1720), 26. For more on Milner, see Paul, *South Sea Bubble,* 49–50.

19. Schaffer, "The Show That Never Ends."

20. For an overview, see Hong and Stein, "Disagreement and the Stock Market," *Journal of Economic Perspectives* 21, no. 2 (Spring 2007): 109–128.

21. Helen Paul only slightly overstates the point in saying that, in 1720, "there was nothing that amounted to a theory of finance" (*South Sea Bubble,* 7, 55). Paul Harrison offers a more positive judgment of the state of financial knowledge in "Rational Equity Valuation"; "What Can We Learn for Today from 300-Year-Old Writings about Stock Markets?" *History of Political Economy* 36, no. 4 (Winter 2004): 667–688. On didactic financial publications, see John Houghton, *Collection for the Improvement of Husbandry and Trade,* nos. 98–103 (June 15–July 20, 1694), reprint ed., ed. Richard Bradley, vol. 1 (London, 1727), 261–276; Natasha Glaisyer, *The Culture of Commerce in England, 1660–1720* (Woodbridge, UK: Boydell Press, 2006), 100–142; Alex Preda, "The Rise of the Popular Investor: Financial Knowledge and Investing in England and France, 1840–1880," *Sociological Quarterly* 42, no. 2 (Spring 2001): 205–232, esp. 211. On vernacular financial knowledge, see Ann M. Carlos, Jennifer Key, and Jill L. Dupree, "Learning and the Creation of Stock-Market Institutions: Evidence from the Royal African and Hudson's Bay Companies, 1670–1700," *Journal of Economic History* 58, no. 2 (June 1998): 318–344; Murphy, *Origins of English Financial Markets.*

22. Schaffer, "Show that Never Ends"; Stewart, *Rise of Public Science,* 388–393.

23. R. M. Chisholm, "Intrinsic Value," in *Recent Work on Intrinsic Value,* ed. Toni Rønnow-Rasmussen and Michael J. Zimmerman (Dordrecht: Springer, 2005), 2. See also Zimmerman, "Intrinsic vs. Extrinsic Value," *Stanford Encyclopedia of Philosophy* (Winter 2010 ed.), ed. Edward N. Zalta, http://plato.stanford.edu/archives /win2010/entries/value-intrinsic-extrinsic/.

24. On "just price" in the medieval period, see Raymond de Roover, "The Concept of Just Price: Theory and Economic Policy," *Journal of Economic History* 18, no. 4 (Dec. 1958): 418–434, and subsequent discussions by J. A. Raftis and David Herlihy (435–438); John W. Baldwin, "The Medieval Theories of the Just Price: Romanists, Canonists, and Theologians in the Twelfth and Thirteenth Centuries," *Transactions of the American Philosophical Society* 49, no. 4 (1959): 1–92; Joel Kaye, *Economy and Nature in the Fourteenth Century: Money, Market Exchange, and the Emergence of Scientific Thought* (Cambridge: Cambridge University Press, 2000), 87–101.

25. Bert De Munck, "Guilds, Product Quality, and Intrinsic Value. Towards a History of Conventions?" *Historical Social Research / Historische Sozialforschung* 36, no. 4 (2011): 103–124.

26. On land, see, for example, John Graunt, "Natural and Political Observations . . . Made upon the Bills of Mortality" (1662), in Hull, ed., *The Economic Writings of Sir William Petty,* 396.

27. De Munck, "Guilds," 112.

28. Michel Foucault, *The Order of Things: An Archaeology of the Human Sciences* (New York: Pantheon Book, 1970), 174–180.

29. Misselden, *Circle of Commerce,* 97–98; Finkelstein, *Harmony and the Balance,* 1–3, 66, 85–88; Poovey, *History of the Modern Fact,* 69–74. On the deeper history of monetary thinking in England, from the medieval period through the eighteenth century, see Christine Desan, *Making Money: Coin, Currency, and the Coming of Capitalism* (Oxford: Oxford University Press, 2014).

30. Deborah Valenze, *The Social Life of Money in the English Past* (Cambridge: Cambridge University Press, 2006), 39–44.

31. Joyce Appleby, "Locke, Liberalism and the Natural Law of Money," *Past & Present* 71 (May 1976): 43–69, esp. 49–52. More recently, Christine Desan has argued that Locke's contention that the value of silver money stemmed entirely from the intrinsic value of its silver content, and not the official imprimatur given by the state, was not a traditionalist view but rather a substantial rethinking of the nature of money itself, which shifted responsibility for money's value from the polity to the market. See *Making Money,* chap. 9.

32. James Hodges, *The Present State of England as to Coin and Publick Charges* (London: Printed for Andr. Bell, 1697), 142, quoted in Appleby, "Locke, Liberalism, and the Natural Law of Money," 50; [Nicholas Barbon], *A Discourse of Trade* (London: Printed for Tho. Milbourn for the Author, 1690), 20. On Barbon, see Appleby, "Ideology and Theory: The Tension between Economic and Political Liberalism in Seventeenth-Century England," *American Historical Review* 81, no. 3 (June 1976): 499–515, esp. 508–509; Finkelstein, *Harmony and the Balance,* chap. 13; Slack, "Politics of Consumption," 614–618.

33. Valenze, *Social Life of Money,* 42.
34. Seiichiro Ito, "The Making of Institutional Credit in England, 1600 to 1688," *European Journal of History of the Economic Thought* 18, no. 4 (Oct. 2011): 487–519, esp. 513–514.
35. "Philopatris" [Josiah Child], *A Treatise Wherein Is Demonstrated I. That the East-India Trade Is the Most National of All Foreign Trades* [etc.] (London: Printed by J. R. for the Honourable the East-India Company, 1681), 11.
36. Wennerlind, *Casualties of Credit,* 172–189, on 175.
37. "A Merchant" [Simon Clement], *A Discourse of the General Notions of Money, Trade, & Exchanges, As They Stand in Relation to Each Other. Attempted by Way of Aphorism* (London: *s.n.,* 1695), 10; [Clement], *A Vindication of the Faults on Both Sides . . . With a Dissertation on the Nature and Use of Money and Paper-Credit in Trade, and the True Value of Joint-Stocks* (London: Printed and Sold by the Booksellers of London and Westminster, 1710), 15–16.
38. John Trenchard, *Letter of Thanks from the Author of the* Comparison between the Proposals of the Bank and South-Sea *&c., to the Author of the* Argument (London: Printed and Sold by J. Roberts, 1720), 10–11; [Anon.], *Letter to a Conscientious Man: Concerning the Use and the Abuse of Riches, and the Right and the Wrong Ways of Acquiring Them* (London: Printed for W. Boreham, 1720), 12.
39. [Anon.], *Remarks upon Several Pamphlets Writ in Opposition to the South-Sea Scheme,* 7, 11; [Anon.], *Letter to a Conscientious Man,* 20; T. Johnson to Charles Mack[a]y, Aug. 23, 1720, from the Hague, Laing Manuscripts, Edinburgh U. L. La. II. 91, f. 30^v. Compare: Peter Temin and Hans-Joachim Voth, "Riding the South Sea Bubble," *American Economic Review* 94, no. 5 (Dec. 2004): 1654–1668.
40. James, 1st Duke of Montrose, to Mungo Graham of Gorthrie, Mar. 25, 1720, Papers of the Graham Family, Dukes of Montrose (Montrose Muniments), NRS GD220/5/832/20a; [Anon.], *South-Sea Scheme Detected, and the Management Thereof Enquir'd Into,* 2nd ed. (London: Printed for J. Roberts, 1720), 14.
41. "Some Memorandums Relating to Exchange, by an Eminent Merchant," contained in Hutcheson to [Stair?], Nov. 16 (*N.S.*)/5 (*O.S.*), 1718, TNA SP 78/162, ff. 322–325, on f. 324^r. See also: Hutcheson to [Craggs?], Nov. 21/Nov. 10, 1718, TNA SP 78/162, f. 334; Hutcheson to "Labbe' Dubois," Copy, Mar. 21, 1719/[Mar. 10, 1718/9], TNA SP 78/163, ff. 228–229.
42. Archibald Hutcheson, "Some Hints Relating to Trade and Coin," TNA SP 78/162, enclosed in Stair to Craggs, Nov. 22/[Nov. 11], 1718, NA SP 78/162, ff. 335–336, 344–350, quotations on ff. 346^{r-v}. This appears to be a scribal copy of a manuscript Hutcheson had sent to Stair previously (see TNA SP 78/162, ff. 326–330). For Stair's attribution of the manuscript to Hutcheson, see TNA SP 78/162, f. 335.
43. [Archibald Hutcheson], *Some Calculations Relating to the Proposals Made by the South-Sea Company, and the Bank of England, to the House of Commons; Shewing the Loss to the New Subscribers* (London: Printed and Sold by J. Morphew, 1720). The later edition, with the Preface, can be found in *A Collection of Calculations and Remarks Relating to the South Sea Scheme & Stock, Which Have Been Already Published* (London: *s.n.,* 1720), 6–21, on 7–8.

44. Hutcheson, *Collection of Calculations and Remarks,* 7–8.

45. On the history of equity valuation, especially discounted cash flow, see R. H. Parker, "Discounted Cash Flow in Historical Perspective," *Journal of Accounting Research* 6, no. 1 (Spring 1968): 58–71; Janette Rutterford, "From Dividend Yield to Discounted Cash Flow: A History of UK and US Equity Valuation Techniques," *Accounting, Business & Financial History* 14, no. 2 (July 2004): 115–149; Liliana Doganova, "Décompter le Futur: La Formule des Flux Actualisés et le Manager-Investisseur," *Sociétés Contemporaines* 93 (2014): 67–87; Eve Chiapello and Christian Walter, "The Three Ages of Financial Quantification: A Conventionalist Approach to the Financiers' Metrology," *Historical Social Research / Historische Sozialforschung* 41, no. 2 (2016): 155–177.

46. Hutcheson, *Collection of Calculations and Remarks,* 8, 16.

47. *Flying-Post,* or, *Post-master,* no. 4428 (Apr. 7–9, 1720); [Hutcheson], *Some Seasonable Considerations,* [2]. Dated Apr. 21, 1720.

48. Hutcheson, *Some Seasonable Considerations,* 3, 5.

49. [Archibald Hutcheson], *An Estimate of the Value of South-Sea Stock. With Some Remarks Relating Thereto* (London: s.n., 1720), [5], 8–9. Dated Sep. 10, 1720.

50. Hutcheson, *Estimate of the Value,* 9–10. For the original statement of Hudibras's rule, see [Samuel Butler], *Hudibras: The Second Part* (London: Printed by T. R. for John Martyn and James Allestry, 1664 [1663]), 34.

51. Hutcheson, *Estimate of the Value,* 8–10.

52. For example, Dale, *First Crash,* esp. 160–161; Edward Chancellor, *Devil Take the Hindmost: A History of Financial Speculation* (New York: Plume Books / Penguin, 1999), 93–94.

53. House of Commons, *The Reports of the Committee of Secrecy to the Honourable House of Commons, Relating to the Late South-Sea Directors, &c. Carefully Corrected* (London: Printed for Cato, 1721). The pagination is split through the document. See 1st series: 11, 15; and 2nd series: 50, 72–73. See also Carswell, *South Sea Bubble,* 93–94, 97, 219–225.

54. Sir James Lowther to [Henry Howard, 4th Earl of Carlisle], Dec. 29, 1720, in HMC, *The Manuscripts of the Earl of Carlisle, Preserved at Castle Howard* (London: HMSO, 1897), 26. See also C[avers] Douglas to [John Clerk], n.d. (c. Dec. 29, 1720), Clerk Papers, NRS GD18 / 5319 / 11, f. 1^v; Minutes of the Court of Directors of the SSC, vol. 6, Dec. 30, 1720, BL Add. MSS 25,499, ff. 83–103, quotation on 103^r.

55. Archibald Hutcheson, *A Computation of the Value of South-Sea Stock, on the Foot of the Scheme as it Now Subsists* (London: s.n., 1720 / 21), 3. Dated Feb. 6, 1720 / 1721.

56. Ibid., 16.

57. Ibid., 13.

58. *Reports of the Committee of Secrecy,* 16–18; Hutcheson, *A Computation of the Value of South-Sea Stock,* 4, 13. See also Shapiro, *Culture of Fact,* 31.

59. *Reports of the Committee of Secrecy,* 15–21 (first pagination series).

60. [Anon.], *The Nation Preserved; or, The Plot Discovered. Containing an Impartial Account of the Secret Policy of Some of the South-Sea Directors* (London: Printed for the Author, 1720 [1721?]), 4–5.

61. The original edition is *Time Bargains Tryed by the Rules of Equity and Principles of the Civil Law* (London: Printed for Eliz. Morphew, 1720 [1721?]), 13–14. The date given is 1720 (Old Style), though context suggests it was published early 1721.

62. "Timothy Telltruth," *Matter of Fact; or, The Arraignment and Tryal of the DI—RS of the S—S—Company, with The Pleadings of the Counsel on Both Sides* (London: Printed for John Appelbee et al., 1720 [1721?]), 14, 37.

63. *The Director* (London: Printed for W. Boreham and Sold by A. Dodd, 1720–1721), no. 6 (Oct. 21, 1720), no. 11 (Nov. 7, 1720), no. 13 (Nov. 14, 1720).

64. For example, "A Hearty Well-Wisher to Publick Credit," *The South-Sea Scheme Examin'd: And the Reasonableness Thereof Demonstrated* (London: Printed for J. Roberts, 1720).

65. *Director,* no. 30 (Jan. 16, 1720 / 21); *Moderator* (Apr. 26, 1721), as discussed in John Asgill, *The Computation of Advantages Saved to the Publick by the South-Sea Scheme. As Published in the* Moderator *of Wednesday the 26h of April, 1721. Detected to be Fallacious* (London: Printed for J. Roberts, 1721). Note, the *Director* and *Moderator* calculations simply summed up the annual projected interest savings across all future years, rather than doing what Hutcheson and other sophisticated calculators might have done, namely calculate a discounted "present value" for those future expected savings.

66. *Minutes of the Court of Directors of the SSC,* vol. 6, entries for Jan. 3 and Jan. 11, 1720 / 1, BL Add. MS 25,499, f. 108, 119ᵛ.

67. "The Scheme or Calculation upon which the Court of Directors of the South Sea Company Grounded the Resolutions of Making the Dividend of Thirty p[er]Cent at Christmas & not less than 50 p[er]Cent p[er] Ann for 12 Years," Treasury Papers, TNA T 1/231, ff. 163–164; Redington, ed., *CTP,* vol. 6, *1720–1728,* 30.

68. The calculations are mentioned in "The Secret History of the South Sea Scheme," in *A Collection of Several Pieces of Mr. John Toland, Now First Publish'd from His Original Manuscripts,* vol. 1 (London: Printed for J. Peele, 1726), 404–474, on 437–438. Though published under the name John Toland, the "Secret History" was likely the work of a former company director.

69. "The Scheme or Calculation upon which the Court of Directors of the South Sea Company Grounded the Resolutions," TNA T 1/231, f. 164ʳ.

70. "Secret History of the South Sea Scheme," 437.

71. "The Scheme or Calculation upon which the Court of Directors of the South Sea Company Grounded the Resolutions," TNA T 1/231, f. 164ʳ.

72. *Moderator,* no. 18 (June 2, 1721).

73. Scholars have recently highlighted how some modern corporate interests have actively cultivated public attitudes of doubt and ignorance concerning the risks associated with their controversial business practices, including in the realm of finance, as a strategy for navigating public scrutiny. See Naomi Oreskes and Erik M. Conway, *Merchants of Doubt: How a Handful of Scientists Obscured the Truth on Issues from Tobacco Smoke to Climate Change* (New York: Bloomsbury Press, 2010); William Davies and Linsey McGoey, "Rationalities of Ignorance: On Financial

Crisis and the Limits of Neo-Liberal Epistemology," *Economy and Society* 41, no. 1 (Feb. 2012): 64–83.

74. Sheila Jasanoff, "The Idiom of Co-Production," in *States of Knowledge: The Co-Production of Science and Social Order,* ed. Jasanoff (Abingdon, UK: Routledge, 2004), 1–12.

75. Hutcheson, *Collection of Advertisements . . . Relating to the Last Elections,* 5; Eustace Budgell, *Case of the Annuitants and Proprietors of the Redeemable Debts: In a Letter to the Author of the Several Calculations on South-Sea Stock* (London: Printed for J. Roberts, 1720), 3; [Anon.], *An Epistle to Archibald Hutcheson, Esq.* (London: Printed for T. Payne, 1722), 4.

76. Adam Anderson, *An Historical and Chronological Deduction of the Origin of Commerce,* vol. 2 (London: Printed for A. Miller et al., 1764), 288, 302; David Macpherson, *Annals of Commerce, Manufactures, Fisheries, and Navigation of Great Britain and Ireland,* vol. 3 (London: Nichols and Son, 1805), 82, 113 (repeats Anderson's account); Richard Preston, *Further Observations on the State of the Nation, the Means of Employment of Labor, the Sinking Fund and its Application* (London: Printed for A. J. Valpy, 1816), 33–34; J. D. Parry, *An Historical and Descriptive Account of the Coast of Sussex* (Brighton: Published for the Author by Wright & Son, 1833), 230.

77. It was a highly contentious election, and Hutcheson and running mate John Cotton were accused of tampering with the results by riling up the London crowds. See Lowndes Papers, TNA T 48 / 20, no foliation; [Archibald Hutcheson], *The Case of Archibald Hutcheson and John Cotton, Esquires, Members Return'd for the City of Westminster. In Answer to the Petition Against Them by William Lowndes, Esquire* ([London]: *s.n.,* [1722]), 3.

78. [Anon.], *A Letter of Advice to Archibald Hutcheson, Esq; Occasion'd by His Pretended Letters to the Earl of Sunderland* (London: Printed for A. Moore, 1722), 17.

79. Dale, *First Crash,* viii. See also Julian Hoppit, "The Myths of the South Sea Bubble," *Transactions of the Royal Historical Society* 12 (2002): 141–165, esp. 146–47.

80. Mackay relied heavily on Walpole's eighteenth-century biographer William Coxe. Walpole only spoke out against the Company after Hutcheson and others. Many of the prophetic warnings Mackay attributed to Walpole read like a précis of Hutcheson's arguments. See Charles Mackay, *Memoirs of Extraordinary Popular Delusions,* 1:80–81; J. H. Plumb, *Sir Robert Walpole,* vol. 1, *The Making of a Statesman,* reprint ed. (London: Penguin, 1972 [1956]), 302–304.

81. See, for example, Dickson, *Financial Revolution,* 136; Carswell, *South Sea Bubble,* 105, 179. For Paul's critique, see Paul, *South Sea Bubble,* 76; Paul, "Archibald Hutcheson's Reputation."

82. For example, sociologist Alex Preda contends that it was not until the late nineteenth century and the rise of "Chartist" analysis (the graphical analysis of stock price movements) that the stock market was seen to constitute a legitimate object of knowledge. By contrast, he argues that eighteenth-century observers generally believed financial values were "radically unknowable" and outside the realm of

calculation. See Alex Preda, *Framing Finance: The Boundaries of Markets and Modern Capitalism* (Chicago: University of Chicago Press, 2009), 86.

83. It can be useful to think of the failure of the South Sea Scheme as a large-scale, economic version of what sociologist John Downer calls an *epistemic accident*. See John Downer, " '737-Cabriolet': The Limits of Knowledge and the Sociology of Inevitable Failure," *American Journal of Sociology* 117, no. 3 (Nov. 2011): 725–762. Note that the South Sea Bubble (and other financial crises) does not fit Downer's model of epistemic accident perfectly, because human error and malfeasance were certainly key factors.

CHAPTER SIX

Futures Projected: Robert Walpole's Political Calculations

1. Davenant, *Essay upon Ways and Means,* 54–55; Davenant, *Discourses on the Publick Revenues,* 1:101–102.

2. Untitled manuscript calculations, [c. 1733?], Cambridge U. L. Ch(H) Papers 43 / 11 / 2.

3. On Walpole's excise scheme, see Paul Langford, *The Excise Crisis: Society and Politics in the Age of Walpole* (Oxford: Clarendon Press, 1975), esp. 26–31; Price, "Excise Affair Revisited"; Ashworth, *Customs and Excise,* chap. 4. See also Cambridge U. L. Ch(H) Papers 41, esp. 41 / 17; 43 / 11 / 6.

4. Davenant, *Discourses on the Publick Revenues,* 1:6.

5. Michel Foucault, "Governmentality," in *The Foucault Effect: Studies in Governmentality,* ed. Graham Burchell, Colin Gordon, and Peter Miller (Chicago: University of Chicago Press, 1991), 87–104.

6. William Coxe, *Memoirs of the Life and Administration of Sir Robert Walpole, Earl of Orford,* vol. 1 (London: Printed for T. Cadell, Jun. and W. Davies, 1798), 2; Thomas Pownell, "Character of Walpole," originally Oct. 1783, Horace Walpole Papers, BL Add. MS 35,335, ff. 69–81, quotation on f. 76^r.

7. W. R. Scott, *The Constitution and Finance of English, Scottish and Irish Joint-Stock Companies to 1720. Volume I: The General Development of the Joint-Stock System to 1720,* reprint ed. (Gloucester, MA: Peter Smith, 1968 [1912]), 437; Dickson, *Financial Revolution,* 176, 210.

8. *North American Review* 55, no. 116 (July 1842): 6.

9. Plumb, *Sir Robert Walpole,* vol. 1, esp. 293–294, 304, 309–316; J. H. Plumb, *Sir Robert Walpole,* vol. 2, *The King's Minister* (London: Cresset Press, 1960). As evidence, Plumb points to Walpole's failures to recognize the dangers of the South Sea Scheme early on and his own poor personal investment decisions during 1720. See also Plumb, *The Growth of Political Stability in England 1675–1725* (London: Macmillan, 1967), chap. 6. Administrative historians have also criticized the negative impact of Walpole's ministry on the development of a professional, modern state bureaucracy, for example, Henry Roseveare, *Treasury: Foundations of Control,* 58–60.

10. For examples of letters with Gould, see Nathaniel Gould to Robert Walpole, Apr. 18, 1720, Cambridge U. L. Ch(H) Letters 783; Gould to Walpole, Apr. 12, 1733, Cambridge U. L. Ch(H) Letters 1,967.

11. On political temporality in the eighteenth century, see Isaac Kramnick, *Bolingbroke and His Circle: The Politics of Nostalgia in the Age of Walpole* (Cambridge, MA: Harvard University Press, 1968); Pocock, *Virtue, Commerce, and History*, chap. 5.

12. Edward M. Jennings, "The Consequences of Prediction," *Studies on Voltaire and the Eighteenth Century* 153 (1976): 1131–1150, on 1131; William Max Nelson, "The Weapon of Time: Constructing the Future in France, from 1750 to Year 1," (Ph.D. diss., University of California, Los Angeles, 2006); Jan Golinski, *British Weather and the Climate of Enlightenment* (Chicago: University of Chicago Press, 2007), chap. 3; Will Slauter, "Forward-Looking Statements: News and Speculation in the Age of the American Revolution," *Journal of Modern History* 81, no. 4 (Dec. 2009): 759–792. For an alternative take, emphasizing the much older historical roots of predictive and conjectural thinking, see James Franklin, *The Science of Conjecture: Evidence and Probability before Pascal* (Baltimore, MD: Johns Hopkins University Press, 2001). On recent historiographical interest in the future, see Daniel Rosenberg and Susan Harding, eds., *Histories of the Future* (Durham, NC and London: Duke University Press, 2005); David C. Engerman, "Introduction: Histories of the Future and the Futures of History," *American Historical Review* 117, no. 5 (Dec. 2012): 1402–1410.

13. [John Asgill], *Dr. D—nant's Prophecys* (London: Printed for A. Baldwin, 1713), 3; Daniel Defoe, *Review of the Affairs in France,* Dec. 9, 1704, quoted in Slauter, "Forward-Looking Statements," 779. Note the title of Asgill's pamphlet.

14. Slauter and Nelson both locate the rise of a practical orientation toward the future after 1750.

15. Hutcheson, *Collection of Advertisements . . . Relating to the Last Elections,* xii, 5; [Anon.], *Letter of Advice to Archibald Hutcheson,* 17–19.

16. House of Commons, *The Report from the Committee Appointed to Enquire How Far the Several Imprest Accountants Have Passed Their Respective Accounts* (London: Printed for Samuel Keble and Henry Clements, 1711); Defoe, *The Re-Representation.* See also Plumb, *Sir Robert Walpole,* 1:169–182.

17. [Robert Walpole and Arthur Maynwaring], *A State of the Five and Thirty Millions Mention'd in the Report of a Committee of the House of Commons* ([London?]: *s.n.,* [1711?]), 1. See also [Arthur Maynwaring and Robert Walpole], *A Letter to a Friend Concerning the Public Debts, Particularly That of the Navy* ([London?]: *s.n.,* 1711), 1; [Robert Walpole? with Arthur Maynwaring], *An Estimate of the Debt of Her Majesty's Navy on the Heads Hereafter Mention'd, as It Stood on the 30th of September Last* ([London]: *s.n.,* [1712]). On authorship, see Henry L. Snyder, "Daniel Defoe, Arthur Maynwaring, Robert Walpole, and Abel Boyer: Some Considerations of Authorship," *Huntington Library Quarterly* 33, no. 2 (Feb. 1970): 133–153.

18. Maynwaring and Walpole, *Letter to a Friend,* 1; Walpole and Maynwaring, *A State of the Five and Thirty Millions,* 2.

19. Plumb, *Sir Robert Walpole*, 1:172–173.
20. Ibid., 182.
21. Carswell, *South Sea Bubble*, chaps. 13–14; Helen J. Paul, "Limiting the Witch-Hunt: Recovering from the South Sea Bubble," *Past, Present & Policy*, 4th International Conference (Geneva, Feb. 3–4, 2011), online at http://www.norges-bank.no/contentassets/452261373fb743999b50daaa39524994/en/paul-paper.pdf.
22. [Anon.], *Considerations on the Present State of the Nation*, 3. On the economic effects of the Bubble, see Hoppit, "Myths of the South Sea Bubble," 150–158. Compare also Neal, *Rise of Financial Capitalism*, 101–111.
23. On the Bank's suspension of discounting, see Lord Godolphin to Robert Walpole, Oct. [4?], 1720, Cambridge U. L. Ch(H) Letters 847; Robert Jacombe to Robert Walpole, Oct. 13, 1720, Cambridge U. L. Ch(H) Letters 852. See also Dickson, *Financial Revolution*, 160–161.
24. [Sir Robert Walpole?], "Some Thoughts & Considerations Concerning the Present Posture of the South Sea Stock Humbly Laid Before His Majesty," [1720], MS [Copy?], Walpole Papers, BL Add. MS 74,066, item no. 7, ff. 1^{r-v}. See also Henry Vander Esch (Eich?), "Remarks on the South Sea Scheme," [1720], Walpole Papers, BL Add. MS 74,066, item no. 15, f. 5.
25. On the Bank Contract, see Cambridge U. L. Ch(H) Papers 88/20–24 and 8/25/1–8; Dickson, *Financial Revolution*, 162–167.
26. The "Ingraftment Scheme" is first mentioned in Robert Jacombe to Robert Walpole, Oct. 11, 1720, Cambridge U. L. Ch(H) Letters 851. See also: Cambridge U. L. Ch(H) Papers 88/25/10, 88/35/1–2, 88/104/1, 88/107; Dickson, *Financial Revolution*, 170–171.
27. Dickson, *Financial Revolution*, 175–176.
28. The correspondence for the period September–December 1720 is contained in Cambridge U. L. Ch(H) Letters 830–868. For an illustrative example, see Jacombe to Walpole, Oct. 1, 1720, Ch(H) Letter 846. See also "Price of South Sea Stock from 2^d. August to 30th. November 1720," Cambridge U. L. Ch(H) Papers 88/27.
29. Cambridge U. L. Ch(H) Papers 88/99; Plumb, *Sir Robert Walpole*, 1:321–324; Dickson, *Financial Revolution*, 159, n. 7.
30. Tho. Houghton to Robert Walpole, Dec. 24, 1720, Cambridge U. L. Ch(H) Letters 868; Langford, *Excise Scheme*, 31–32.
31. [Robert Walpole], Notes on Ingraftment Scheme, [1720–1721], Cambridge U. L. Ch(H) Papers 88/34a. William Sloper, MP for Great Bedwyn, was an outspoken South Sea Company critic and a member of the Committee of Secrecy. "Mill^r" is ambiguous.
32. Archibald Hutcheson, "A State of the South Sea Stock," MS copy, with comments by [Robert Walpole], late 1720–early 1721, Cambridge U. L. Ch(H) Papers 88/118. Much of this material was printed in Hutcheson, *Some Computations Relating to the Proposed Transferring of Eighteen Millions of the Fund of the South Sea Company, to the Bank, and East-India Company* (London: *s.n.*, 1720/1), [3]^r, [4]^r.
33. Hutcheson, "State of the South Sea Stock," MS with Walpole comments, Cambridge U. L. Ch(H) Papers 88/118, f. [3]^r.

34. Hutcheson and Walpole were not the only contemporary figures to make use of multi-scenario calculation techniques. For another example: W[illiam] L[owndes], "Comparison Effects of a New Bank Contract," Lowndes Papers, TNA T 48/6.

35. "A Computation of the Present Value of South Sea Stock of the Old Capitall in Particular," enclosed in Samuel Hyde to Robert Walpole, May 2, 1721, Cambridge U. L. Ch(H) Letters 884.

36. Hyde, "A Computation," Cambridge U. L. Ch(H) Letters 884, f. [1]^r.

37. "A View of the Value of 100 Pounds Capitall in Sundry States of South Sea Stock," [1721], Cambridge U. L. Ch(H) Papers 88/94a.

38. BL Add. MS 74,066, item nos. 18/5, 18/10–13, 18/15, 22.

39. Emily Nacol, *An Age of Risk: Politics and Economy in Early Modern Britain* (Princeton, NJ: Princeton University Press, 2016), 2; Hacking, *Emergence of Probability;* Daston, *Classical Probability.*

40. Wm. Van Laitz, "A Proposal to Ease the Subject, and to Restore the Credit of the Nation; So That Within 6 Months Bank Bills and Tallies Will Be as Currant as Ever They Were" [on or before 1696], Charles Montagu Papers on the Bank of England, 1682–1696, NLS Adv. MS 31.1.7, f. 44. A slightly different version of Van Laitz's plan, with a different payment schedule, was printed in 1696: "W. V.," *A Proposal to Ease the Subject, and to Restore the Credit of the Nation* (London: *s.n.,* [1696]). On Van Laitz's plan and its context, see Desan, *Making Money,* 334–336.

41. Van Laitz, "A Proposal to Ease the Subject," NLS Adv. MS 31.1.7, f. 44^r.

42. On the fascinating figure of de Witt, and especially his contributions to finance and political economy, see Herbert H. Rowen, *John de Witt, Grand Pensionary of Holland, 1625–1672* (Princeton, NJ: Princeton University Press, 1978), chap. 9, esp. pp. 172–176; Rowen, *John de Witt: Statesman of the "True Freedom"* (Cambridge: Cambridge University Press, 1986), 60–63; Soll, "Accounting for Government," 226–228; Soll, *The Reckoning,* 85–86. For an overview of de Witt's mathematical career, see Joy B. Easton, "Witt, Jan De," in *DSB,* 14:465–467. See also N. A. Brisco, *The Economic Policy of Robert Walpole* (New York: Columbia University Press, 1907), 39–40; E. L. Hargreaves, *The National Debt* (London: Edward Arnold & Co., 1930), chap. 2.

43. *A Letter to a Member of the Late Parliament, Concerning the Debts of the Nation* (*s.l.: s.n.,* 1700), 1, 4. A second, revised edition was published in 1701 with the imprint "London: Printed for Edward Poole." The 1701 pamphlet is sometimes attributed to Paterson, but not in the English Short Title Catalogue.

44. [William Pulteney], *Some Considerations on the National Debts, The Sinking Fund, and the State of Publick Credit: In a Letter to a Friend in the Country* (London: Printed for R. Franklin, 1729), 67–68.

45. On Walpole's 1717 financial reforms and the introduction of the sinking fund, see Brisco, *Economic Policy of Robert Walpole,* 35–41; Plumb, *Sir Robert Walpole,* 1:247; Dickson, *Financial Revolution,* 82–87.

46. On interest-rate reduction, see [Robert Walpole], "M^r. Lowndes Savings, by Reductions of Redeemable Fonds to 5$^\epsilon$. p^r. C^t.," [c. 1716?], Cambridge U. L. Ch(H) Papers 49/1a/5; Untitled Notes in Robert Walpole's Hand, on Interest Savings,

Cambridge U. L. Ch(H) Papers 49/1a/9; Cambridge U. L. Ch(H) Papers 49/1a/16, f. [1]v; "A Computation of the Savings by Reducing such Debts as Carry 6ᵉ p[er] Cᵗ or upwards to 5ᵉ p[er] Cᵗ," Cambridge U. L. Ch(H) Papers 49/4. On the problem of irredeemable debts, see "State of the Present Value of the Several Annuitys Computed," [1716?], Cambridge U. L. Ch(H) Papers 49/1a/13; Cambridge U. L. Ch(H) Papers 49/1a/16.

47. Cambridge U. L. Ch(H) Papers 49/1a/10, recto.

48. [Walpole], Cambridge U. L. Ch(H) Papers 49/1a/16, f. [3]ᵛ. (Foliation is estimated.)

49. Plumb, *Sir Robert Walpole*, 1:247; Brisco, *Economic Policy of Robert Walpole*, 40.

50. Plumb, *Sir Robert Walpole*, vol. 2, chaps. 2–3.

51. On the political thought of the Walpolean Whigs, though with an emphasis on the latter half of Walpole's ministry, see Simon Targett, "Government and Ideology during the Age of Whig Supremacy: The Political Argument of Sir Robert Walpole's Newspaper Propagandists," *Historical Journal* 37, no. 2 (June 1994): 289–317, esp. 291, 302. Targett notes that there was actually considerable diversity among Walpolean Whig thinkers. See also Reed Browning, *Political and Constitutional Ideas of the Court Whigs* (Baton Rouge: Louisiana State University Press, 1982); Shelley Burtt, *Virtue Transformed: Political Argument in England, 1688–1740* (Cambridge: Cambridge University Press, 1992), chap. 6.

52. Note that Walpolean thinking was diverse, and not all Walpolean thinkers across the period were equally enamored of calculation. James Pitt, for example, wrote in 1730 that "Human Affairs are not to be reduced to Mathematical Calculations" (quoted in Targett, "Government and Ideology," 292).

53. For another example of this argumentative structure, see "A Member of the House of Commons," *A Letter to a Freeholder, on the Late Reduction of the Land Tax to One Shilling in the Pound* (London: Printed for J. Peele, 1732).

54. *London Journal*, no. 252 (May 23, 1724).

55. *The History and Proceedings of the House of Commons*, vol. 6, *1714–1727* (London: Printed for Richard Chandler, 1742), 318–319.

56. Plumb, *Sir Robert Walpole*, 2:123. On the Patriot opposition, see Christine Gerrard, *The Patriot Opposition to Walpole: Politics, Poetry, and National Myth, 1725–1742* (Oxford: Clarendon Press, 1994); Kramnick, *Bolingbroke*. On Pulteney and public accountability, see Roseveare, *Treasury: Foundations*, 59.

57. [Nathaniel Gould?], *An Essay on the Publick Debts of This Kingdom* (London: Printed for J. Peele, 1726); *London Journal*, no. 360 (May 18, 1726). Gould was cited as the author of the *Essay* by 1782, when a new edition was printed (London: Sold by B. White, 1782). Adam Anderson attributed the pamphlet to John Adlam: *Historical and Chronological Deduction*, 316–317.

58. Compare Cambridge U. L. Ch(H) Papers 49/13, 18/1–5, esp. 49/18/2 for Walpole's own notes and marginalia relating to the recent history of the national debt.

59. Gould, *Essay on the Publick Debts*, 6–7, 11, 22–23, 37–40.

60. Note that introductory mathematics textbooks of the era, like *Cocker's Arithmetic*, advertised that they were "suitable to the Meanest Capacity." See Wardhaugh, *Poor Robin's Prophecies*, chap. 3.

61. [Anon.], *Remarks on a Late Book, Intitled,* An Essay on the Publick Debts of This Kingdom (London: Printed for A. Moore, 1727), iii, v, 8–9.

62. Plumb, *Sir Robert Walpole,* 2:122.

63. Pulteney, *A State of the National Debt,* 14, 29.

64. *History and Proceedings of the House of Commons,* vol. 7, *1727–1733,* 23.

65. Pulteney, *State of the National Debt,* 5, 14–15, and Appendices I and II.

66. For Gould's argument, see *Essay on the Publick Debts,* 25–27. For Pulteney's retort, see *State of the National Debt,* 18–19.

67. [Anon.], Untitled comments on Pulteney's *State of the National Debt,* c. 1727, Cambridge U. L. Ch(H) Papers 49 / 14, recto; *History and Proceedings of the House of Commons,* 7:23; [Nathaniel Gould], *A Defence of* An Essay on the Publick Debts of This Kingdom, *&c. In Answer to a Pamphlet, Entitled,* A State of the National Debt, *&c. By the Author of the Essay* (London: Printed for J. Peele, 1727), 13, 51.

68. *History and Proceedings of the House of Commons,* 7:40–41, 48. Thanks to Steve Pincus for his observations on Pulteney's rambling tendencies.

69. [Robert Walpole], "An intire Paragraph" concerning the Sinking Fund, [1727 or later?], Cambridge U.L. Ch(H) Papers 49 / 1a / 14. For Pulteney's critique, see Pulteney, *Some Considerations on the National Debts,* 67–68.

70. *The Humble Representation of the House of Commons to the King, with His Majesty's Most Gracious Answer Thereunto* (London: Printed for R. Knaplock et al., 1728), 142–143; [Anon.], *Great Britain's Speediest Sinking Fund is a Powerful Maritime War, Rightly Manag'd, and Especially in the West Indies* (London: Printed for J. Roberts, 1727).

71. [Anon.], *Remarks on the R—p—n of the H—of C—ns to the K—g; and His M—y's A—s—r. Address'd to All True Britons* (London: Printed for A. Moore, [1728]), 14.

72. Ibid., 17, 19–20, 25, 31.

73. Pocock, *Machiavellian Moment,* 440; Pocock, *Virtue, Commerce, and History,* 98. On debt and political futurity in the French context, compare Michael Sonenscher, *Before the Deluge: Public Debt, Inequality, and the Intellectual Origins of the French Revolution* (Princeton, NJ: Princeton University Press, 2009).

74. Timothy Mitchell, "Economentality: How the Future Entered Government," *Critical Inquiry* 40, no. 4 (Summer 2014): 479–507, on 507.

75. William Pulteney, *An Enquiry into the Conduct of Our Domestick Affairs, from the Year 1721, to the Present Time* (London: Printed by H. Haines, 1734), 14. On Walpole's pillaging the sinking fund, see Soll, *The Reckoning,* 113–114.

76. William Pulteney, *Case of the Sinking Fund, and the Right of the Publick Creditors to It Considered at Large* (London: Printed for H. Haines, 1735).

77. Anderson, *Historical and Chronological Deduction,* 316–318.

78. Carl B. Cone, "Richard Price and Pitt's Sinking Fund of 1786," *Economic History Review* 4, no. 2 (1951): 243–251; Donald F. Swanson and Andrew P. Trout, "Alexander Hamilton, 'the Celebrated Mr. Neckar,' and Public Credit," *William and Mary Quarterly* 47, no. 3 (July 1990): 422–430; Max M. Edling, " 'So Immense a

Power in the Affairs of War': Alexander Hamilton and the Restoration of Public Credit," *William and Mary Quarterly* 64, no. 2 (Apr. 2007): 287–326, esp. 317–320.

79. Adam Smith, *Lectures on Justice, Police, Revenue and Arms, Delivered in the University of Glasgow,* ed. Edwin Canaan (Oxford: Clarendon Press, 1896), 210–211.

80. H. T. Dickinson, *Walpole and the Whig Supremacy* (London: The English Universities Press Ltd., 1973), 100.

CHAPTER SEVEN

Figures, Which They Thought Could Not Lie: The Problem with
Calculation in the Eighteenth Century

1. Walpole, *Some Considerations Concerning the Publick Funds,* 7–8.

2. Langford, *Excise Crisis;* Clyve Jones, "The House of Lords and the Excise Crisis: The Storm and the Aftermath, 1733–5," *Parliamentary History* 33, no. 1 (Feb. 2014): 160–200.

3. In *The Averaged American,* Sarah Igo has offered a powerful study of how new forms of quantitative inquiry were received by and influenced general audiences. Documentation illuminating similar trends in the eighteenth century is more elusive. On strategies for studying readership and the cultural reception of texts, notably through the study of marginalia, see two exemplary studies: John Brewer, *The Pleasures of the Imagination: English Culture in the Eighteenth Century* (London: HarperCollins, 1997), chap. 4; James A. Secord, *Victorian Sensation: The Extraordinary Publication, Reception, and Secret Authorship of* Vestiges of the Natural History of Creation (Chicago: University of Chicago Press, 2000).

4. *The Critical Review, or, Annals of Literature* 27 (Mar. 1769), 202–205, on 205.

5. Hume, *Political Discourses,* 81.

6. Bernard Peach, *Richard Price and the Ethical Foundations of the American Revolution* (Durham, NC: Duke University Press, 1979), 9.

7. Innes, *Inferior Politics,* 130.

8. William Allen, *Ways and Means to Raise the Value of Land: or, The Landlord's Companion* (London: Printed for J. Roberts, 1736), 18–20.

9. E. Hatton, *An Intire System of Arithmetic* (London: Printed by Henry Woodfall and Sold by Mr. Mount, 1721); Allen, *Ways and Means,* 19; Charles Davenant, *The Political and Commercial Works of That Celebrated Writer Charles D'Avenant, LL.D. Collected and Revised by Charles Whitworth, Member of Parliament,* vol. 1 (London: Printed for R. Horsfield, [1771]), [A3]; John Harris, *Lexicon Technicum: or, an Universal English Dictionary of Arts and Sciences; Explaining Not Only the Terms of Art, but the Arts Themselves,* vol. 2 (London: Printed for Dan. Brown, 1710), *s.v.* "POLITICAL Arithmetick"; Ephraim Chambers, *Cyclopaedia,* 4th ed., vol. 2 (London: Printed for D. Midwinter, 1744 [1728]), *s.v.* "POLITICAL Arithmetic"; Innes, *Inferior Politics,* 130.

10. Daniel Defoe, *A Journal of the Plague Year: Being Observations or Memorials, of the Most Remarkable Occurences, as Well Publick as Private, which Happened in London*

during the Last Great Visitation in 1665 (London: Printed for E. Nutt, 1722); Jonathan Swift, *A Modest Proposal for Preventing the Children of Poor People from Being a Burthen to Their Parents, or the Country, or for Making Them Beneficial to the Publick* (Dublin: Printed by S. Harding, 1729). On Defoe, see Campe, *Game of Probability,* 220–234; Charlotte Sussman, "Memory and Mobility: Fictions of Population in Defoe, Goldsmith, and Scott," in *A Companion to the Eighteenth-Century English Novel and Culture,* ed. Paula R. Backscheider and Catherine Ingrassia (Oxford: Blackwell, 2005), 191–213; Nicholas Seager, "Lies, Damned Lies, and Statistics: Epistemology and Fiction in Defoe's *Journal of the Plague Year,*" *Modern Language Review* 103, no. 3 (July 2008): 639–653. On Swift, see Sussman, "The Colonial Afterlife of Political Arithmetic: Swift, Demography, and Mobile Populations," *Cultural Critique* 56 (Winter 2004): 96–126; Sean Moore, "Devouring Posterity: *A Modest Proposal,* Empire, and Ireland's 'Debt of the Nation,'" *PMLA* 122, no. 3 (May 2007): 679–695.

11. Allen, *Ways and Means,* 19; Innes, *Inferior Politics,* 112.

12. D. V. Glass, *Numbering the People: The Eighteenth-Century Population Controversy and the Development of Census and Vital Statistics in Britain* (Farnborough, UK: D. C. Heath, 1973); Frederick G. Whelan, "Population and Ideology in the Enlightenment," *History of Political Thought* 12, no. 1 (Spring 1991): 35–72; Blum, *Strength in Numbers,* chap. 2; Rusnock, *Vital Accounts,* chap. 7.

13. Glass, *Numbering the People,* 49, 57–61.

14. Timothy Alborn, *Regulating Lives: Life Insurance and British Society, 1800–1914* (Toronto: University of Toronto Press, 2009), 105–110.

15. For representative writings by Young, see *A Six Weeks Tour Through the Southern Counties of England and Wales* (London: Printed for W. Nicoll, 1768); *Political Arithmetic: Containing Observations on the Present State of Great Britain; and the Principles of Her Policy in the Encouragement of Agriculture* (London: Printed for W. Nicoll, 1774). On Young, see Robert C. Allen and Cormac Ó Gráda, "On the Road Again with Arthur Young: English, Irish, and French Agriculture during the Industrial Revolution," *Journal of Economic History* 48, no. 1 (Mar. 1988): 93–116; Simon Schaffer, "The Earth's Fertility as a Social Fact in Early Modern Britain," in *Nature and Society in Historical Context,* ed. Mikuláš Teich, Roy Porter, and Bo Gustafsson (Cambridge: Cambridge University Press, 1997), 124–147; Liam Brunt, "Rehabilitating Arthur Young," *Economic History Review* 56, no. 2 (May 2003): 265–299; S. J. Thompson, "Parliamentary Enclosure, Property, Population, and the Decline of Classical Republicanism in Eighteenth-Century Britain," *Historical Journal* 51, no. 3 (Sept. 2008): 621–642, esp. 635–639; P. M. Jones, "Arthur Young (1741–1820): For and Against," *English Historical Review* 127, no. 528 (Oct. 2012): 1100–1120; Glass, *Numbering the People,* 55–57; Innes, *Inferior Politics,* 146–147, 154–155, 169–170.

16. Arthur Dobbs, *An Essay on the Trade and Improvement of Ireland* (Dublin: Printed by A. Rhames for J. Smith and W. Bruce, 1729); Innes, *Inferior Politics,* 132, 136–137; Stanley H. Palmer, *Economic Arithmetic: A Guide to the Statistical Sources of English Commerce, 1700–1850* (New York: Garland Publications, 1977), 6.

17. Council of Trade and Plantations, "State of Your Majesty's Plantations on the Continent of America," entry 656, Sept. 8, 1720, *CSP, Colonial America and West Indies,* vol. 32: *1720–1721,* ed. Cecil Headlam (London: HMSO, 1933), 408–449; Cohen, *A Calculating People,* 78; Hoppit, "Political Arithmetic," 526–528.

18. Bennet, *Two Letters and Several Calculations*; Joseph Massie, *A Computation of the Money That Hath Been Exorbitantly Raised upon the People of Great Britain by the Sugar-Planters, in One Year, from January 1759 to January 1760 (s.l.: s.n., 1760).*

19. Franklin's essay was first published as an appendix to William Clarke, *Observation on the Late and Present Conduct of the French with Regard to Their Encroachments upon the British Colonies in North America . . . To Which is Added, Wrote by Another Hand, Observations Concerning the Increase of Mankind* (Boston: Printed and Sold by S. Kneeland, 1755). On Franklin, see Alan Houston, *Benjamin Franklin and the Politics of Improvement* (New Haven, CT: Yale University Press, 2008), chap. 3; Joyce Chaplin, *Benjamin Franklin's Political Arithmetic: A Materialist View of Humanity* (Washington: Smithsonian Institution, 2009).

20. Otis, "By the Numbers," 235–250; Rusnock, *Vital Accounts,* 49–70.

21. Innes, *Inferior Politics,* 13; Jessica Warner, "Faith in Numbers: Quantifying Gin and Sin in Eighteenth-Century England," *Journal of British Studies* 50, no. 1 (Jan. 2011): 76–99.

22. Recent work by Ted McCormick has illuminated both the diverse sites and sources of quantitative practice in the early modern period, as well as the entanglements of quantification and early modern colonialism. See "Governing Model Populations: Queries, Quantification, and William Petty's 'Scale of Salubrity,'" *History of Science* 51, no. 2 (June 2013): 179–197; McCormick, "Political Arithmetic's Eighteenth-Century Histories"; McCormick, "Statistics in the Hands of an Angry God."

23. As an example, see John Watkinson, *An Examination of a Charge Brought against Inoculation, by De Haen, Rast, Dimsdale, and Other Writers* (London: Printed for J. Johnson and J. Sewell, 1777).

24. John Sinclair, *The History of the Public Revenue of the British Empire,* Part III (London: Printed by A. Strahan, 1790), 53; B. R. Mitchell, *Abstract of British Historical Statistics* (Cambridge: Cambridge University Press, 1962), 381; Palmer, *Economic Arithmetic,* 45–48; Roseveare, *Treasury: Foundations,* 58–60; Ian Hacking, *Taming of Chance,* chaps. 3–4; Soll, *The Reckoning,* 115.

25. George Thomas Keppel, Earl of Albemarle, *Memoirs of the Marquis of Rockingham, and His Contemporaries,* vol. 1 (London: R. Bentley, 1852), 67; P. J. Kulisheck, "Pelham, Henry (1694–1754)," *Oxford DNB.*

26. "Proceedings in Parliament relating to the Supply from 1660 to 1688," Clements Library, Townshend Papers 8 / 3B / 16; "Papers, Accounts, and Calculations made up by Townshend for his own use; chiefly on financial matters," Clements Library, Townshend 8 / 21 / 1–12; "The French Debt at 1762," Clements Library, Townshend 8 / 34 / 65; "Gauging beer," Clements Library, Townshend 298 / 7 / 6.

27. For commentaries on Thomas Whately's *Considerations,* see Clements Library, Townshend 8 / 20 / 19–26. The original publication is [Whately], *Considerations on*

the Trade and Finances of This Kingdom, and on the Measures of Administration (London: Printed for J. Wilkie, 1766).

28. On food, see "General Accounts of Corn Consumed," Clements Library, Townshend 8/14/10; "An Account of the true Market-price of Wheat, and Malt, at Windsor for 100 years," Clements Library, Townshend 8 / 14 / 18, also 8 / 14 / 19. On change-over-time, see "Customs: Account of the Gross Produce and Medium Gross Produce of the Customs from 1711 to 1765 both Inclusive," Clements Library, Townshend 8 / 21 / 8; Clement Library 8 / 21 / 60; Innes, *Inferior Politics,* 140. On the history of "time series" in general, see Judy L. Klein, *Statistical Visions in Time: A History of Time Series Analysis, 1662–1938* (Cambridge: Cambridge University Press, 1997).

29. Clements Library, Shelburne 102.

30. Both Alain Desrosières and Mary Poovey argue that "liberal" tendencies in British politics militated against the development of centralized data-gathering: Desrosières, *Politics of Large Numbers,* 24–29; Poovey, *History of the Modern Fact,* 144–157.

31. Glass, *Numbering the People*; Peter Buck, "People Who Counted: Political Arithmetic in the Eighteenth Century," *Isis* 73, no. 1 (Mar. 1982): 28–45, esp. 32–35; Desrosières, *Politics of Large Numbers,* 24–25; Innes, *Inferior Politics,* 137–138.

32. *Gentleman's Magazine* 23 (Nov. 1753), 500.

33. *Gentleman's Magazine* 23 (Dec. 1753), 549–552.

34. Richard Forster to Thomas Birch, BL Add. MSS 4,440, ff. 1760–1185, published in Glass, *Numbering the People,* 78–88, quotation on 85–86. See also Glass, *Numbering the People,* 47–51.

35. [Joseph Massie], *Calculations of Taxes for a Family of Each Rank, Degree or Class: For One Year* (London: Printed for Thomas Payne, 1756), 11–12, 15–16.

36. Peter Mathias, "The Social Structure in the Eighteenth Century: A Calculation by Joseph Massie," *Economic History Review* 10, no. 1 (1957): 30–45, on 33.

37. J[oseph] Massie, *A Representation Concerning the Knowledge of Commerce as a National Concern; Pointing Out the Proper Means of Promoting Such Knowledge in This Kingdom* (London: Printed for T. Payne, 1760), [i]–[ii], 14–15, 18–19, 24–25. See also Hoppit, "Political Arithmetic," 521. On early modern "information overload," compare Ann Blair, *Too Much to Know: Managing Scholarly Information before the Modern Age* (New Haven, CT: Yale University Press, 2010).

38. See, for example, *Lloyd's Evening Post,* no. 1043 (Mar. 16–19, 1764); *London Chronicle,* no. 1130 (Mar. 17–20, 1764); *St. James's Chronicle or British Evening Post,* no. 474 (Mar. 17–20, 1764); *Gazetteer and London Daily Advertiser,* no. 10,923 (Mar. 19, 1764). On Grenville's policies, see Justin DuRivage, "Taxing Empire: Political Economy and the Ideological Origins of the American Revolution, 1747–1776" (Ph.D. diss., Yale University, 2013), chap. 3.

39. *The Great Financier, or British OEconomy for the Years 1763, 1764, 1765* ([London]: "Publ^d. according to Law," [1765]).

40. David Hartley, *The Budget: Inscribed to the Man Who Thinks Himself Minister* (London: Printed for J. Almon, 1764), 10–11.

41. Ibid., [3], 5, 7.

42. DuRivage, "Taxing Empire," 182, quotation in n. 176.

43. [Anon.], *An Answer to* The Budget: *Inscribed to the Coterie* (London: Printed for E. Sumpter, 1764), [3], 4–5, 14; [Anon.], *The Wallet: A Supplementary Exposition of the Budget* (London: Printed for Williams and Vernor, 1764), 18; [Thomas Whately], *Remarks on* The Budget; *or, a Candid Examination of the Facts and Arguments Offered to the Public in That Pamphlet* (London: Printed for J. Wilkie, 1765), 4.

44. Margaret Schabas, "Hume on Economic Well-Being," in *The Bloomsbury Companion to Hume,* ed. Alan Bailey and Dan O'Brien, reprint ed. (London: Bloomsbury, 2015 [2012]), 332–348, esp. 332; Hume to William Mure, Aug. 4, 1744, in *The Letters of David Hume,* ed. J. Y. T. Greig, vol. 1 (Oxford: Oxford University Press, 2011 [1932]), 58.

45. On Hume's conservatism, see Sheldon S. Wolin, "Hume and Conservatism," *American Political Science Review* 48, no. 4 (Dec. 1954): 999–1016; James Conniff, "Hume on Political Parties: The Case for Hume as a Whig," *Eighteenth-Century Studies* 12, no. 2 (Winter 1978–1979): 150–173; J. Joseph Miller, "Neither Whig Nor Tory: A Philosophical Examination of Hume's Views on the Stuarts," *History of Philosophy Quarterly* 19, no. 3 (July 2002): 275–308; Neil McArthur, *David Hume's Political Theory: Law, Commerce, and the Constitution of Government* (Toronto: University of Toronto Press, 2007), 3–7, and chap. 6.

46. [David Hume], *Essays, Moral and Political* (Edinburgh: Printed by R. Fleming and A. Alison for A. Kincaid, 1741), 43–44; Hume, *Political Discourses,* 1–3; Nicholas Phillipson, *David Hume: The Philosopher as Historian,* revised ed. (New Haven, CT: Yale University Press, 2012 [1989]), 53–55; James Conniff, "Hume's Political Methodology: A Reconsideration of 'That Politics May Be Reduced to a Science,'" *Review of Politics* 38, no. 1 (1976): 88–108.

47. Hume, *Philosophical Essays Concerning Human Understanding* (London: Printed for A. Miller, 1748), 59–60, 64–65, 175. On Hume's epistemological ideas about probable knowledge and the challenges of achieving it, see Nacol, *Age of Risk,* 69–80, esp. 79. Mary Poovey and Emily Nacol both argue that Hume's use of the essay genre was a deliberate epistemological choice. See Poovey, *History of the Modern Fact,* 197–213; Nacol, *Age of Risk,* 85–97.

48. Hume, *Political Discourses,* 3; David Hume, *Essays, Moral and Political,* 2nd ed., vol. 2 (Edinburgh: Printed for A. Kincaid, 1742), 205.

49. Hume, *Political Discourses,* 29, 123.

50. Ibid., 159, 211–212; Whelan, "Population and Ideology," 50–52.

51. Hume, *Political Discourses,* 212–213, 216, 218; also, for example, 249–256, 260.

52. David Hume, *The History of Great Britain,* vol. 2, *Containing the Commonwealth, and the Reigns of Charles II and James II* (London: Printed for A. Millar, 1757), 197–198, 244.

53. Hume, *Political Discourses,* 126, 132, 134–136. On Hume's implicit reference to Walpole's sinking fund, see Hume, *Essays, Moral, Political, and Literary,* ed. Eugene F. Miller (Indianapolis: Liberty Fund, 1987), II.IX.6, n. 100, online at http://www.econlib.org/library/LFBooks/Hume/hmMPL32.html. The classic study of Hume's

ideas on debt is Istvan Hont, "The Rhapsody of Public Debt: David Hume and Voluntary State Bankruptcy," in *Political Discourse in Early Modern Britain,* ed. Nicholas Phillipson and Quentin Skinner (Cambridge: Cambridge University Press, 1993), 321–348.

54. Hume, *Political Discourses,* 80. On Hume's economics, see Margaret Schabas and Carl Wennerlind, "Hume on Money, Commerce, and the Science of Economics," *Journal of Economic Perspectives* 25, no. 3 (Summer 2011): 217–230; Schabas and Wennerlind, eds., *David Hume's Political Economy* (Abingdon, UK: Routledge, 2008).

55. David Hume to James Oswald, Nov. 1, 1750, in Greig, ed., *The Letters of David Hume,* 1:142–144, on 144; Hume, *Political Discourses,* 86; Nacol, *Age of Risk,* 89–90.

56. Hume, *Political Discourses,* 80–81.

57. Nacol, *Age of Risk,* 93.

58. For comprehensive studies of Price, see Carl B. Cone, *The Torchbearer of Freedom: The Influence of Richard Price on Eighteenth Century Thought* (Lexington: University Press of Kentucky, 1952); D. O. Thomas, *The Honest Mind: The Thought and Work of Richard Price* (Oxford: Clarendon Press, 1977); Henri Laboucheix, *Richard Price as Moral Philosopher and Political Theorist,* trans. Sylvia Raphael and David Raphael (Oxford: Voltaire Foundation, 1982 [1970]); Yiftah Elazar, "The Liberty Debate: Richard Price and His Critics on Civil Liberty, Free Government, and Democratic Participation" (Ph.D. diss., Princeton University, 2012). Laboucheix is the most attentive to Price's mathematical thinking.

59. Richard Price, *A Review of the Principal Questions and Difficulties in Morals* (London: Printed for A. Millar, 1758). On Price's ethics, see Cone, *Torchbearer,* chaps. 1–3, esp. p. 2; Peach, *Richard Price,* 14–17. On Price's Platonism, see Martha K. Zebrowski, "Richard Price: British Platonist of the Eighteenth Century," *Journal of the History of Ideas* 55, no. 1 (Jan. 1994): 17–35.

60. Richard Price, *Four Dissertations* (London: Printed for A. Millar and T. Cadell, 1767), 361–439, on 390.

61. The exact problem, as Bayes stated it, was: "*Given* the number of times in which an unknown event has happened and failed: *Required* the chance that the probability of its happening in a single trial lies somewhere between any two degrees of probability that can be named." See Richard Price, "An Essay towards Solving a Problem in the Doctrine of Chance. By the Late Rev. Mr. Bayes, F. R. S. Communicated by Mr. Price, in a Letter to John Canton, A. M. F. R. S.," *Philosophical Transactions of the Royal Society of London* 53 (1763): 370–418, on 376.

62. Ibid., 371, 373–374; Daston, *Classical Probability,* chap. 6.

63. Richard Price, "A Demonstration of the Second Rule in the Essay towards the Solution of a Problem in the Doctrine of Chances," *Philosophical Transactions* 54 (1764): 296–325, on 297.

64. Richard Price, *Observations on Reversionary Payments: on Schemes for Providing Annuities for Widows, and for Persons in Old Age; on the Method of Calculating the Values of Assurances on Lives; and on the National Debt* (London: Printed for T. Cadell, 1771), vii, 128, 131; Daston, *Classical Probability,* 168–181.

65. Price, *Observations on Reversionary Payments*, xi, xv; Thomas Reid to Richard Price, ca. 1772 / 3, in *The Correspondence of Richard Price: Volume I: July 1748–March 1778*, ed. W. Bernard Peach and D. O. Thomas (Durham, NC: Duke University Press, 1983), 153–154; Buck, "People Who Counted," 28–45; Laboucheix, *Richard Price*, 18–19.

66. Jonathan Levy, *Freaks of Fortune: The Emerging World of Capitalism and Risk in America* (Cambridge, MA: Harvard University Press, 2012), chap. 3.

67. For examples of Price's advice to Shelburne, see "Revenue Notes & Calcula[tio]ns by Dr. Price &c.," Clements Library, Shelburne 117. On Price's political affiliations in general, see Robbins, *Eighteenth-Century Commonwealthman*, 335–346; Verner W. Crane, "The Club of Honest Whigs: Friends of Science and Liberty," *William and Mary Quarterly* 23, no. 2 (Apr. 1966): 210–233; Elazar, "The Liberty Debate," 11–16, 37–38.

68. Price, *Observations on Reversionary Payments*, 135, 160–161.

69. Ibid., 140, 155, 163, 165.

70. Buck, "People Who Counted," 38–40, quotation on 40. See also Laboucheix, *Richard Price*, 19–21.

71. Richard Price, *Observations on the Nature of Civil Liberty, the Principles of Government, and the Justice and Policy of the War with America*, 7th ed. (London: Printed for T. Cadell, 1776), 2–6. On reconciling Price's ethics and politics, see Susan Rae Peterson, "The Compatibility of Richard Price's Politics and Ethics," *Journal of the History of Ideas* 45, no. 4 (Oct.–Dec. 1984): 537–547; Gregory I. Molivas, "Richard Price, the Debate on Free Will, and Natural Rights," *Journal of the History of Ideas* 58, no. 1 (Jan. 1997): 105–123; Ronald Hamowy, "Two Whig Views of the American Revolution: Adam Ferguson's Response to Richard Price," *Interpretation: A Journal of Political Philosophy* 31, no. 1 (Fall 2003): 3–36.

72. Price, *Observations on . . . Civil Liberty*, 4; Price, *Additional Observations on the Nature and Value of Civil Liberty, and the War with America* (London: Printed for T. Cadell, 1777), 31.

73. Price, *Observations on . . . Civil Liberty*, 60, 62, 67; Price, *Additional Observations*, 30, 44.

74. Price, *Observations on . . . Civil Liberty*, 32, 39, 44.

75. Peach, *Richard Price*, 125–175.

76. Price, *Additional Observations*, quotation on x–xii; also 89–170.

77. Price, *Observations on . . . Civil Liberty*, 106–108; Price, *Additional Observations*, 175.

78. Price, *Additional Observations*, 48; Price, *Two Tracts on Civil Liberty, the War with America, and the Debts and Finances of the Kingdom* (London: Printed for T. Cadell, 1778), xvi–xxiii.

79. Richard Price, *Observations on the Importance of the American Revolution, and the Means of Making It a Benefit to the World* (London: s.n., 1784), 57. On Price as social-scientific reformer, see Laboucheix, *Richard Price*, 39, 140, 146. On Price, democracy, and political participation, see Elazar, "The Liberty Debate," quotation on 18, and chap. 5.

80. Carl B. Cone, "Richard Price and the Constitution of the United States," *American Historical Review* 53, no. 4 (July 1948): 726–747; Cone, "Richard Price and Pitt's Sinking Fund"; Donald F. Swanson and Andrew P. Trout, "Alexander Hamilton's Hidden Sinking Fund," *William and Mary Quarterly* 49, no. 1 (Jan. 1992): 108–116.

81. Adam Smith, *An Inquiry into the Nature and Causes of the Wealth of Nations*, 2:124 (section IV.5.69). On Smith and numbers, see Poovey, *History of the Modern Fact*, 236–249; Robert W. Dimand, " 'I Have No Great Faith in Political Arithmetick': Adam Smith and Quantitative Political Economy," in *Measurement, Quantification and Economic Analysis: Numeracy in Economics*, ed. Ingrid H. Rima (London: Routledge, 1995), 22–30, esp. 24.

82. Edmund Burke, *Reflections on the Revolution in France, and on the Proceedings in Certain Societies in London Relative to that Event* (London: Printed for J. Dodsley, 1790), 15, 76, 113, 265, 272–273; Burke, *Observations on a Late State of the Nation* (London: Printed for J. Dodsley, 1769); Burke, "On Conciliation with the Colonies," in Peach, *Richard Price*, 229–234, esp. 230; "Articles of Charge of High Crimes and Misdemeanors against Warren Hastings," in *The Works of Edmund Burke*, vol. 9 (London: John C. Nimmo, 1887), 16–17; Frederick Dreyer, "The Genesis of Burke's Reflections," *Journal of Modern History* 50, no. 3 (Sept. 1978): 462–479; Laboucheix, 18, 139.

Conclusion

1. Daston and Galison, *Objectivity*, 27–35.

2. Theodore M. Porter, "Thin Description: Surface and Depth in Science and Science Studies," *Clio Meets Science: The Challenges of History*, ed. Robert E. Kohler and Kathryn M. Olesko, *Osiris* 27 (2012): 209–226; Sally Engle Merry, *The Seductions of Quantification: Measuring Human Rights, Gender Violence, and Sex Trafficking* (Chicago: University of Chicago Press, 2016), 3. On the history of arguments about the "superficiality" of quantification, see also Porter, *Trust in Numbers*, 84–86.

3. For Charles Dickens on "facts and figures," see *The Chimes: A Goblin Story of Some Bells That Rang an Old Year Out and a New Year In*, 4th ed. (London: Chapman and Hall, 1845), 41–42.

4. The classic study of statistics and the emerging concept of "society" in the nineteenth century is Hacking, *Taming of Chance*. On the professionalization of calculation, see Timothy Alborn, "A Calculating Profession: Victorian Actuaries among the Statisticians," *Science in Context* 7, no. 3 (Sept. 1994): 433–468.

5. Lorraine Daston, "Enlightenment Calculations," *Critical Inquiry* 21, no. 1 (Autumn 1994): 182–202, on 185. See also Simon Schaffer, "Babbage's Intelligence: Calculating Engines and the Factory System," *Critical Inquiry* 21, no. 1 (Autumn 1994): 203–227.

6. Kathrin Levitan, *A Cultural History of the British Census: Envisioning the Multitude in the Nineteenth Century* (New York: Palgrave Macmillan, 2011), 5; Oz Frankel,

States of Inquiry: Social Investigations and Print Culture in Nineteenth-Century Britain and the United States (Baltimore, MD: Johns Hopkins University Press, 2006), 8; Crook and O'Hara, eds., *Statistics and the Public Sphere.*

7. Appadurai, "Number in the Colonial Imagination," 320.

8. Helen Verran, *Science and an African Logic* (Chicago: University of Chicago Press, 2001), 115 (emphasis mine).

9. On the comparative status of quantification in British versus American civic epistemology, see, to begin, Porter, *Trust in Numbers,* chap. 5.

10. Olivier Zunz, *Why the American Century?* (Chicago: University of Chicago Press, 1998), 49; Porter, *Trust in Numbers,* chaps. 7–8; Carson, *The Measure of Merit,* chaps. 5–6; Josh Lauer, "From Rumor to Written Record: Credit Reporting and the Invention of Financial Identity in Nineteenth-Century America," *Technology & Culture* 49, no. 2 (Apr. 2008): 301–324; Martha Poon, "What Lenders See: A History of the Fair Isaac Scorecard" (Ph.D. diss., University of California, San Diego, 2012); Bouk, *How Our Days Became Numbered;* Igo, *Averaged American,* 6; Marks, "Trust and Mistrust"; Jasanoff, "Acceptable Evidence"; Jasanoff, *Designs on Nature,* 18–19, 259.

11. See Miller, "Civic Epistemologies," 1899.

12. Cohen, *A Calculating People,* 4.

13. On Franklin, see Chaplin, *Benjamin Franklin's Political Arithmetic;* Bruce H. Yenawine and Michele R. Costello, *Benjamin Franklin and the Invention of Microfinance* (London: Pickering & Chatto, 2010); Paul C. Pasles, *Benjamin Franklin's Numbers: An Unsung Mathematical Odyssey* (Princeton, NJ: Princeton University Press, 2008). On Hamilton, see Alexander Hamilton, *Report of the Secretary of the Treasury to the House of Representatives: Relative to a Provision for the Support of the Public Credit of the United States . . . Presented to the House on Thursday the 14th day of January, 1790* (New York: Printed for Francis Childs and John Swaine, 1790); Edling, "So Immense a Power"; Thomas K. McCraw, *The Founders and Finance: How Hamilton, Gallatin, and Other Immigrants Forged a New Economy* (Cambridge, MA: Harvard University Press, 2012), esp. 61–62; Richard Sylla, "Hamilton and the Federalist Financial Revolution, 1789–1795," *New-York Journal of American History* 65, no. 3 (Spring 2004): 32–39. On Jefferson, see Gary Wills, *Inventing America: Jefferson's Declaration of Independence,* reprint ed. (Boston: Mariner Books, 2002 [1978]), chap. 9, esp. p. 138. See also, for example, McCormick, "Statistics in the Hands of an Angry God" (on Cotton Mather).

14. Thomas Cooper, "Political Arithmetic, No. I," in *Political Essays, Originally Inserted in the* Northumberland Gazette, *with Additions* (Northumberland, PA: Printed by Andrew Kennedy, 1799), 39. On Cooper, see Eugene Volokh, "Thomas Cooper, Early American Public Intellectual," *New York University Journal of Law & Liberty* 4, no. 2 (2009): 372–381. Oliver Wolcott, Jr. (1760–1833) was Secretary of the Treasury from 1795 to 1800. James McHenry (1753–1816) was Adams's Secretary of War. "Pinckney" was probably an error for Timothy Pickering (1745–1829), Adams's Secretary of State.

15. Noah Webster, *Effects of Slavery on Morals and Industry* (Hartford, CT: Printed by Hudson and Goodwin, 1793), 42–43.

16. This literature is vast. To begin, see Bailyn, *Ideological Origins;* Gordon S. Wood, *The Creation of the American Republic 1776–1787,* reprint ed. (Chapel Hill: University of North Carolina Press for the Omohundro Institute, 1998 [1969]), chaps. 1–3; Pocock, *Machiavellian Moment,* chap. 15.

17. Cohen, *A Calculating People,* chap. 3.

18. See *The Correspondence of Richard Price,* ed. Peach and Thomas (3 vols.).

19. Porter and Jasanoff, for example, both tend to frame the American fashion for quantification as an effect of the particular tensions inherent in "democracy, American style." Both draw upon the characterization of American democracy offered by political theorist Yaron Ezrahi, who argues that American democratic culture places a high value on governmental *visibility* and on transparent, instrumental demonstrations of governmental efficacy and virtue. See Porter, *Trust in Numbers,* 86; Jasanoff, *Designs on Nature,* 77, 259; Yaron Ezrahi, *The Descent of Icarus: Science and the Transformation of Contemporary Democracy* (Cambridge, MA: Harvard University Press, 1990), chaps. 4, 8.

20. Porter, *Trust in Numbers,* 149.

21. Cohen, *A Calculating People,* chap. 6.

22. Arwen Mohun, "On the Frontier of the *Empire of Chance:* Statistics, Accidents, and Risk in Industrializing America," *Science in Context* 18, no. 3 (June 2005): 337–357, quotation on 343.

23. As was typical, these calculative debates were not one-sided. Southerners made their own numerical case for the superiority of Southern approaches, especially slavery. See, for example, Josiah Nott, "Statistics of Southern Slave Population. With Especial Reference to Life Insurance," *The Commercial Review of the South and West,* ed. J. D. B. De Bow, IV, no. 3 (Nov. 1847): 275–290.

24. Desrosières, *Politics of Large Numbers,* 188–194, quotation on 192.

25. Vincanne Adams, "Introduction," in *Metrics: What Counts in Global Health,* ed. Adams (Durham, NC: Duke University Press, 2016), 9.

26. Shore and Wright, "Audit Culture Revisited," 421.

27. Zeynep Tufekci, "Engineering the Public: Big Data, Surveillance and Computational Politics," *First Monday* 19, no. 7 (July 2014); William Davies, "How Statistics Lost Their Power—and Why We Should Fear What Comes Next," *Guardian* (Jan. 19, 2017).

28. Teresa Bejan, *Mere Civility: Disagreement and the Limits of Toleration* (Cambridge, MA: Harvard University Press, 2017).

29. Diane M. Nelson, *Who Counts? The Mathematics of Death and Life after Genocide* (Durham, NC: Duke University Press, 2015), 10.

30. Bruno, Didier, and Vitale, "Statactivism," and accompanying articles in the special issue of *Partecipazione & Conflitto* 7, no. 2 (2014) on "Statistics and Activism," esp. Boris Samuel, "Statistics and Political Violence: Reflections on Social Conflict in 2009 in Guadeloupe" (237–257) and Alain Desrosières, "Statistics and Social

Critique" (348–359). For further discussions of how quantitative practices might be used as modes of resistance, and the potential challenges entailed, see Marlee Tichenor, "The Power of Data: Global Malaria Governance and the Senegalese Data Retention Strike," in Adams, ed., *Metrics,* 105–124; Molly Hales, "Native Sovereignty by the Numbers: The Metrics of Yup'ik Behavioral Health Programs," ibid., 125–146.

31. Davies, "How Statistics Lost Their Power."
32. On the importance of humility in policy-making and governance around technoscientific questions, see Sheila Jasanoff, "Technologies of Humility," *Nature* 450, no. 7166 (Nov. 1, 2007): 33.

Acknowledgments

One lesson of *Calculated Values* is that it can be immensely complicated to account accurately for debts. This is no different when it comes to all the debts that I have accrued in the course of researching and writing this book. It seems impossible to calculate the number of different people who have helped me on this project, but I will try to offer a thorough and accurate reckoning. (As always, this model has certain assumptions and conventions built in. For example, for reasons of space I only list each name once, though there are many people who ought to have multiple entries in the ledger.)

I have had the great fortune to develop this project as a member of three remarkable scholarly communities. The first was the Department of History, and specifically the Program in History of Science, at Princeton University, where this project began. Michael Gordin graciously agreed to advise a project that was chronologically and geographically distant from his primary areas of interest. He has been a consummate mentor, both during and after my time at Princeton. Most of all, he taught me that historical scholarship should be both careful and adventurous. Linda Colley guided me through the (often combative) worlds of British history and historiography, and taught me that precise and elegant writing is central to the historian's craft. Graham Burnett, Jonathan Levy, and Steve Pincus offered generous guidance and incisive thoughts. Angela Creager has been a gracious teacher and mentor, and I have learned much from her insights on the challenges and rewards of an academic life.

I chose to begin my scholarly career in Princeton's History of Science program because of its unmatched commitment to communal conversation as a scholarly practice. My wager proved a winning one. Thanks to all of the participants in the weekly History of Science Program Seminar, especially Michael Barany, Dan Bouk, Hannah-Louise Clark,

Yulia Frumer, Lindy Baldwin Fulford, Tony Grafton, Ben Gross, Nathan Ha, Vera Keller, Bob Macgregor, Chris McDonald, Erika Milam, Sue Naquin, Eileen Reeves, Aviva Rothman, Margaret Schotte, Ksenia Tatarchenko, Emily Thompson, Keith Wailoo, Iain Watts, and Adrian Young. Thanks also to the many teachers and scholars at Princeton whose lessons and suggestions helped shape this project, including David Bell, Emmanuel Kreike, Peter Lake, Yair Mintzker, and Dan Rodgers. I also wish to recognize Elizabeth Bennett, Judy Hanson, Lauren Kane, Debbie Macy, Reagan Maraghy, and Kristy Novak for their generous help. Most special about Princeton were the many great friends and fellow historians I met there, all of whom have left their mark on my thinking. In addition to the historians of science already mentioned, I would like to thank Nimisha Barton, Alex Bevilacqua, Alex Bick, Sarah Brooks, Frederic Clark, Paul Davis, Rohit De, Catherine Evans, Chris Florio, Matt Growhoski, Natalie Holstein, Zack Kagan-Guthrie, Jamie Kreiner, Dael Norwood, Andrei Pesic, Ronny Regev, Padraic Scanlan, Ben Schmidt, Chris Shannon, and Annie Twitty. Among my Princeton colleagues, three—Henry Cowles, Kyrill Kunakhovich, and Sarah Milov—merit special recognition for their insight, humor, and lasting friendship.

The second community in which this book was developed was the Society of Fellows in the Humanities at Columbia University. I could not have imagined a more exciting scholarly environment in which to refine my thinking and expand my vision. The first acknowledgment is due, of course, to Eileen Gillooly, for all she does to make the Society the extraordinary place that it is. This book has been shaped in innumerable ways by conversations with my brilliant co-Fellows: Vanessa Agard-Jones, Teresa Bejan, Benjamin Breen, Maggie Cao, Dana Fields, Brian Goldstone, David Gutkin, Hidetaka Hirota, Murad Idris, Ian McCready-Flora, Emily Ogden, Dan-el Padilla Peralta, Carmel Raz, David Russell, Edgardo Salinas, Yanfei Sun, Rebecca Woods, and Grant Wythoff. Being at the Society also allowed me to learn from the tremendous scholars at Columbia and around the New York area. Special recognition is due to Chris Brown, Matt Jones, and Carl Wennerlind, all of whom offered crucial thoughts on this book and generous guidance on postdoctoral life. Thanks also to Tim Alborn, Arunabh Ghosh, David Johnston, Abram Kaplan, Eugenia Lean, Mark Mazower, Michele Moody-Adams, Tim Shenk, Pamela Smith, Adam Tooze, and Lee Vinsel. This book also benefited greatly from the feedback I received from participants in the Society of Fellows Thursday Lecture Series, the Columbia University Seminar on British History, the New York History of Science Lecture Series, and the Columbia History Department's "Money, Numbers, and Power" series. Thanks finally to Jonah Cardillo, Clarence Coaxum, Christina Dawkins, Conley Lowrance, and Sarah Monks for helping make the Heyman Center such an inspiring place.

Third, I have been fortunate enough to complete this book as a member of the Program in Science, Technology, and Society at MIT. Thanks to my new colleagues for all they have done to welcome me to MIT and to push my thinking in new directions—in particular to Dwai Banerjee, Jon Durant, Deborah Fitzgerald, Malick Ghachem, Stefan Helmreich, Caley Horan, Dave Kaiser, Jen Light, Clapperton Mavhunga, Anne McCants, Heather Paxson, Jeff Ravel, Harriet Ritvo, Roe Smith, Chris Walley, and Roz Williams. Special thanks to Robin Scheffler and Hanna Rose Shell for making my transition to MIT such a convivial one. As I worked through the final stages of this book, I had the great fortune to think with an exceptional collection of graduate students in MIT's HASTS program and neighboring departments, specifically Marc Aidinoff, Marion Boulicault, Rijul Kochhar, Crystal Lee, Xavier Nueno, Carlos Sandoval Olascoaga, Michelle Spektor, Erik Stayton, and Jamie Wong. Thanks also to Carolyn Carlson, Karen Gardner, Paree Pinkney, Judy Spitzer, and Gus Zahariadis for their support and cordiality.

Beyond these three communities, I have benefited from the generosity of many scholars, whose keen thoughts have improved this book immeasurably. First and foremost, let me thank two anonymous reviewers for Harvard University Press, who offered exceptionally thoughtful and incisive comments on the manuscript; both of them, in different ways, steered me away from dangerous conceptual pitfalls and helped to make this book tremendously better. Jacob Soll has been a steadfast supporter and inspiring interlocutor. Ted Porter, whose scholarship has long been a model for me, kindly invited me to UCLA to present in the History of Science colloquium and to discuss a substantial section of my book with his graduate seminar. Sophus Reinert has graciously invited me to participate in his working group on the history of political economy at Harvard Business School; thanks to him for his feedback and encouragement, and to the other members of that most entertaining group (not already mentioned): John Brewer, Hannah Farber, Martin Giraudeau, John Shovlin, Andre Wakefield, and Christine Zabel. A conversation with Emily Nacol got me started thinking about David Hume. Chris Phillips offered a discerning critique of an earlier version of the Introduction. At various steps along the way, conversations with David Armitage, Thomas Broman, Emmanuel Didier, Justin DuRivage, Fredrik Albritton Jonsson, Mordechai Levy-Eichel, Mary Morgan, Steve Shapin, and Rachel Weil helped inspire the ideas in this book.

Thanks to those scholars who have kindly invited me to present my work in venues they have organized—Richard Bourke, Yiftah Elazar, James Livesay, Nicholas Mulder, Anne O'Donnell, Jamie Pietruska, Emma Rothschild, and Cornel Zwierlein—and to those who have offered thoughtful commentary on my work in various settings— Matthias Dörries, Daniel Hulsebosch, Ted McCormick, Jeffrey Sklansky, Phil Stern, and Chris Tomlins. This book has benefited greatly from the astute questions and comments I have received from audiences at the German Historical Institute in Paris, Harvard, the Institute of Historical Research in London, the London School of Economics, Rutgers, UCLA, the University of Dundee, USC, and Yale, as well as at meetings of the American Historical Association, the American Society for Eighteenth Century Studies, the Consortium for the Revolutionary Era, the History of Science Society, the North American Conference on British Studies, and the Omohundro Institute for Early American History and Culture.

Harvard University Press has proven to be the perfect home for a book that falls in between a few different academic fields. Michael Aronson first saw the promise in the book and encouraged me to sign on with HUP. Special acknowledgment is owed to my editor Ian Malcolm for his kind encouragement, thoughtful feedback, and skillful management of the editorial process. Thanks also to Joy Deng and Olivia Woods of HUP for their assistance, and to Annamarie Why for her excellent design for the cover.

Mikala Guyton and the team at Westchester Publishing Services guided the book through the production process deftly and swiftly. Robert Koelzer copyedited the manuscript with care and precision. Thanks also to Elliot Linzer for excellent work on the index.

This project has drawn me to several different libraries and archives, where I have relied on the kind and knowledgeable assistance of their professional teams. I would like to thank the professionals at the British Library Manuscripts Reading Room, Butler Library of Columbia University, the Cambridge University Library Manuscripts Department, Edinburgh University Library's Centre for Research Collections, Firestone Library at Princeton University, the Huntington Library, MIT Libraries, the National Archives (UK), the National Library of Scotland, the National Records of Scotland, the Royal Bank of Scotland Archives, and the William L. Clements Library at the University of Michigan. I also wish to recognize His Grace the Duke of Montrose; Sir Robert Clerk, Bt.; Sir Hew Hamilton-Dalrymple, Bt.; the Syndics of Cambridge University Library; and the Royal Bank of Scotland for their gracious permission to quote from their manuscript archives. Thanks to the following organizations for their kind assistance and permission in reproducing images from their collections: the Beinecke Rare Book and Manuscript Collection, Yale University; Edinburgh University Library Special Collections; the Kress Collection of Business and Economics, Baker Library, Harvard Business School; the National Archives (UK); the Rare Book and Manuscript Library, Columbia University; Senate House Library, University of London; the Syndics of Cambridge University Library; and the Trustees of the British Museum. Chapter 1 explores in greater detail key concepts initially examined in "Finding the Money: Public Accounting, Political Arithmetic, and Probability in the 1690s," *Journal of British Studies* 52, no. 3 (July 2013): 638–668, and I am grateful to Cambridge University Press for affording a primary venue for developing the larger discussion.

The research and writing of this book has been made possible by generous financial support from several organizations. I wish to thank the Graduate School, the Department of History, the Program in International and Regional Studies, and the Alumni Council, all of Princeton University; the Society of Fellows and the Heyman Center for the Humanities, as well as the Andrew W. Mellon Foundation, for support during my time at Columbia University; the School of Humanities, Arts, and Social Sciences and the Leo Marx Career Development Fund at MIT; the Huntington Library; and the William L. Clements Library.

Before I close this account, a few people merit exceptional recognition for all they have contributed to this book—and to my life, scholarly and otherwise—over the past ten years. One of the greatest pieces of good fortune in my academic career is that I have been able to share it with my two closest friends, Jonathan Gienapp and Evan Hepler-Smith, both of whom I met long before I officially became a historian, and both of whom, against whatever odds and whatever good judgment, also chose to become historians themselves. Thanks to you both; I doubt there is a single (good) idea in this book that has not been enlivened by a conversation with one or both of you. Additional thanks to Evan for diligently reading and commenting on the entire manuscript; my debt in butler-dollars only continues to accrue. Thanks to my sisters Emily, Kate, and

Molly for all of your support, and for teaching me at an early age that I probably ought to have evidence for the statements I make. Thanks to my mom and dad for their consistent love and encouragement, and for teaching me early on about the value of ideas.

My final acknowledgment goes to Susanna Brock, herself a brilliant educator and scholar of mathematical thinking. When we had our first conversation—about mathematical problem solving, no less—I could not possibly have projected how much joy you would bring to my life. The most valuable things are, after all, incalculable. This book is for you (and G).

Index